Wallace R Beardsley

Feb 24, 1976

29^{50}

Image Science
*principles, analysis and evaluation
of photographic-type imaging processes*

Image Science

principles, analysis and evaluation of photographic-type imaging processes

J. C. DAINTY
Department of Physics,
Queen Elizabeth College,
University of London,
London, England

and

R. SHAW
Xerox Corporation,
Webster Research Center,
N.Y. 14580, U.S.A.

Academic Press
London · New York · San Francisco
A Subsidiary of Harcourt Brace Jovanovich, Publishers

1974

ACADEMIC PRESS INC. (LONDON) LTD.
24/28 Oval Road
London NW1

United States Edition published by
ACADEMIC PRESS INC.
111 Fifth Avenue
New York, New York 10003

Copyright © 1974 by
ACADEMIC PRESS INC. (LONDON) LTD.

All Rights Reserved

No part of this book may be reproduced in any form by photostat, microfilm or any other means without written permission from the publishers

Library of Congress Catalog Card Number: 74 5667
ISBN : 0 12 200850 2

Text set in 10/11 pt. Monotype Times New Roman, printed by letterpress, and bound in Great Britain at the Pitman Press, Bath.

Preface

The origins of this book can be traced back to 1967. Although long-interested in such topics from a research viewpoint, it was only then that I found it necessary to compile a formal series of lectures on photographic image analysis. These lectures were for final-year students reading for a B.Sc. Honours Degree in Photographic Technology at the Polytechnic of Central London. Chris Dainty was one of the earliest recipients of these original lectures and, as fate will have it, responsibility for the lecture course eventually fell on his own shoulders. The two authors having delivered the course over six academic years, and sharing similar views as well as common experience, it seemed a natural conclusion to co-operate in the preparation of a formal text. The lack of any such standard text was one of the disadvantages in those early days.

In spite of these origins, and although many features of the lecture course are included in the text, we have consciously attempted to make the presentation also suitable for those who are first introduced to these topics at a postgraduate level. The latter represent perhaps the majority of those concerned with image analysis and evaluation, either from within the photographic manufacturing industry where the interest is in photographic processes and their image properties *per se*, or in the various fields of applied and scientific photography where the emphasis is on the imaging process purely as a recording medium. However, whether for students or pure or applied researchers, our intent has been to provide a fundamental treatment of the principles and analysis of imaging processes and the evaluation of their images.

A deliberate attempt has been made to keep the treatment as general as possible, and hence applicable to any process which provides a spatially-recorded image. This is not solely on account of the increasing number of unconventional processes which are already here or just around the corner, but also to emphasize that all such processes are subject to much the same fundamental principles and limitations. Optical, chemical, and electronic imaging processes may involve quite different technologies, but the appropriate input/output relationships may be much the same, and in any case can be treated within the same scientific framework. Some aspects of image analysis which grew up with conventional photography were immersed in silver-halide technology to such an extent that the various coefficients and criteria which still lurk in the literature may need patience to be fully appreciated by those not already initiated.

The science of photographic image analysis dates back to the end of the nineteenth century and the classical studies of Hurter and Driffield, leading to the concept of the characteristic curve on the macro-scale. However, the substantial advances in the analysis of image properties on the micro-scale which have taken place since the 1950s had their origins outside photographic science. Ideas due to Duffieux, Wiener, Shannon, Rose and Schade are among the most prominent of these, and were mainly concerned with electrical communication systems. Following their adoption in optics, these ideas were eventually applied to the analysis of photographic images due to the lead of Fellgett, Linfoot, Jones, Zweig and others. Since the bases of these new ideas are now well-documented we have attempted to avoid mere repetition, and to emphasize their application to image analysis. For example, the implications of Fourier theory and information theory as evaluation techniques are stressed, while the formal theories are summarized as briefly as possible.

It is scarcely possible, let alone fruitful, to include all aspects of image theory and practice in a single text. Subjective attributes of images and the mechanisms of the visual process have been largely ignored. Nor is colour reproduction and evaluation discussed in any detail since there are standard texts in this field, at least concerning the macro-scale. The photographic process is treated as a *de facto* system of image-forming sub-systems, and technological details of parts of the system such as the development process are neglected. We are aware of these and other omissions, but hope that on balance this is beneficial towards stressing the underlying scientific principles of image recording.

We have kept the ten chapters fairly self-contained for those who may be interested in only a limited area. Thus each chapter has its own independent set of literature references and students exercises. Since the treatment is essentially quantitative, mathematical details are unavoidable. However, we have attempted to stress the implications of the theory, and it is hoped that those not having the necessary mathematical background may still find the text beneficial towards understanding the nature of image properties and their inter-relationships.

Symbols, units and nomenclature proved a major problem during preparation of the manuscript, since there is little in the way of accepted standards in this field. We have aimed at uniformity and have fallen in with common usage where possible. Perhaps those who find inconsistencies or confusions will be spurred on to new suggestions for future standardization.

March 11th, 1974

RODNEY SHAW
Webster, New York

Acknowledgements

We wish to thank Dr. R. H. Ericson, Dr. T. H. James, and Dr. A. Rose for providing copies of the photographs shown in Figs 5 and 8, Chapter 2, and Fig. 12, Chapter 4 respectively. We are grateful to Mr. G. Horsnell for the computer programming for Figs. 6, 7, and 8, Chapter 6. We are also grateful to the following for permission to copy: Dr. L. O. Hendeberg (Fig. 10, Chapter 7), American Institute of Physics (Fig. 1, Chapter 9), Prof. E. Wolf (Fig. 5, Chapter 9), Optical Society of America (Figs 7 and 11, Chapter 9), and the Society of Photographic Scientists and Engineers (Figs 12 and 13, Chapter 9).

Contents

Preface   v

Chapter One
Spatially-Recorded Images: Some Fundamental Statistical Limitations

		Page
1.1	Ideal arrays of photon counters	1
	Introduction	1
	Input/Output Relationships	1
	The Comparative Noise Level	6
	Threshold Limitation	9
1.2	Image density characteristics	12
	Density and Gamma	12
	Density Fluctuations and Comparative Noise Level	14
1.3	Random array of receptors	17
	Three-Dimensional Layers	17
	Image Density and Gamma	20
	Density Fluctuations and the Comparative Noise Level	22
1.4	Image resolution	24
1.5	Detective quantum efficiency	28
	References	30
	Exercises	31

Chapter Two
Input/Output Relationships for Conventional Photographic Processes: Experimental Observables

2.1	Properties of silver halide grains	33
	Latent Image Formation	33
	Quantum Yield and Quantum Sensitivity	35
	The Grain Size Distribution	38
2.2	Sensitometric properties of photographic layers	39
	Image Density and the Nutting Formula	39
	The Exposure Scale	46
	The Characteristic Curve	47

	Page
2.3 Micro-image properties	53
Adjacency Effects	53
Image Resolution and the Spread Function	55
Image Density Fluctuations	58
Detective Quantum Efficiency	61
References	64
Exercises	65

Chapter Three
Input/Output Relationships for Conventional Photographic Processes: Analytical Models

3.1 Models for single-level grains	68
Introduction	68
DQE for Uniform Grain Array	69
DQE for Random Grain Array	70
Influence of Sensitivity Distribution	72
Model including Grain Size Distribution	74
3.2 The characteristic curve	76
Basic Theory	76
Constant Grain Size and Sensitivity	79
Variable Grain Sensitivity	81
Variable Grain Size	83
Number and Size of Image Grains	88
Grain-Volume Absorption of Quanta	89
Grain-Size Amplifications	92
Exposure Distribution in the Layer	94
Influence of Fog Grains	94
3.3 Image noise characteristics	96
Basic Theory	96
The Siedentopf Relationship	98
Influence of Layer Parameters	99
Size Amplification Factors	103
3.4 Detective quantum efficiency	106
Formulation	106
Influence of Layer Parameters	107
3.5 Modelling unconventional processes	111
References	111
Exercises	113

Chapter Four
Quantum Sensitivity and Ultimate Photographic Sensitivity

	Page
4.1 Measurement of quantum sensitivity	116
Experimental Techniques	116
Analysis of Results	118
Sensitivity Distributions	123
4.2 Interpretation of quantum sensitivity	125
Problems of Interpretation	125
Simple Model of Sensitivity Distribution	126
Quantum Sensitivity and the Characteristic Curve	129
Quantum Sensitivity and DQE	132
4.3 Ultimate photographic sensitivity	135
Some Estimates	135
Speed and Resolution	137
Ultimate Sensitivity and DQE	141
Exposure Addition and Multiplication	144
4.4 Sensitivity and DQE for two-stage processes	145
Two-Stage Imaging	145
DQE Transfer	146
Implications	147
References	148
Exercises	150

Chapter Five
Detective Quantum Efficiency, Signal-to-Noise Ratio, and the Noise-Equivalent Number of Quanta

5.1 Detective quantum efficiency and signal-to-noise ratio	152
Introduction	152
S/N Ratio Analysis	152
Noise-Equivalent Number of Quanta	156
S/N Ratio Optimization	158
5.2 Detection problems	162
Contrast and Energy Detectivity	162
Pre-Exposure Advantages	165
Practical Detection Techniques	168
5.3 Noise-equivalent quanta and quantum noise	171
X-Rays and Unconventional Exposures	171
Role of Photon Noise	176
Photon Noise Amplification	178
Photographic Amplifying Factors	182
References	185
Exercises	188

Chapter Six
Fourier Transforms, and the Analysis of Image Resolution and Noise

	Page
6.1 Fourier transforms	190
Introduction	190
Definitions	191
Two Dimensional Transforms	194
Properties of Transforms	196
The Sampling Theorem	197
Computational Methods	198
Computational Examples	201
6.2 Input-output relationships for linear stationary systems	204
Basic Theory	204
The Line Spread Function	209
The Modulation Transfer Function	211
Symmetry Properties	213
6.3 Noise analysis	215
Random Processes	215
First Order Statistics	217
The Autocorrelation Function	220
The Wiener Spectrum	222
Symmetry Properties	223
Computational Examples	224
References	227
Recommended Reading	229
Exercises	229

Chapter Seven
The Modulation Transfer Function

7.1 Linearity of the photographic process	232
Input and Output Parameters	232
A One-Stage Model	233
A Two-Stage Model	236
Adjacency Effects	237
A Three-Stage Model	246
7.2 Measurement of the MTF	241
Sine Wave Methods	241
Spread Function Methods	244
Coherent Light Methods	247
Practical Results	249
Influence of Image Noise	255
7.3 Analysis and application of the MTF	258
Empirical MTF Models	258
Monte Carlo MTF Models	260

CONTENTS

	Page
Practical Resolution Criteria	263
Systems Cascade of MTFs	267
References	269
Exercises	273

Chapter Eight
Image Noise Analysis and the Wiener Spectrum

8.1 Photographic Wiener spectrum relationships	276
Introduction	276
One-Dimensional Scans	277
The Wiener Spectrum and the Noise Parameter, G	280
Transmittance and Density Fluctuations	283
8.2 Measurement of the Wiener spectrum	284
The Scale Value	284
Analogue Methods	288
Digital Methods	292
Practical Results	298
8.3 Analysis and application of the Wiener spectrum	303
Wiener Spectrum Models	303
Wiener Spectrum Transfer	307
The Wiener Spectrum and DQE	311
References	314
Exercises	317

Chapter Nine
Microdensitometry

9.1 Design of conventional microdensitometers	320
Single and Double Beam Instruments	320
Digital Systems	324
9.2 Theory of imaging in microdensitometers	325
Effects of Partial Coherence	325
The Microdensitometer Transfer Function	330
Correction for Microdensitometer Degradation	334
9.3 Special features and instruments	336
Automatic Focussing and Alignment	336
Other Special Features	339
References	340
Exercises	342

Chapter Ten
Image Assessment by Information Theory

	Page
10.1 The information theory approach	344
Introduction	344
Relevance to Photography	345
10.2 Shannon's theorems	346
Measure of Information	346
The Zero-Memory Information Source	348
Markov Information Sources	350
Communication Channels	352
The Influence of Noise	353
Power-Limited Signals, Gaussian Noise	356
10.3 Photographic applications: discrete signals. . . .	357
Information Storage Cells	357
Binary and Multilevel Recording	360
Problem of Coding	363
10.4 Photographic applications: continuous signals . . .	364
Spatial Frequency Analysis	364
Information and DQE	366
Small Signals	368
Maximum Information Capacity	372
Practical Applications	374
References	376
Exercises	387
Appendix	381
Author Index	387
Subject Index	395

1. Spatially-Recorded Images: Some Fundamental Statistical Limitations

1.1 Ideal Arrays of Photon Counters

Introduction

The basic principles involved in recording a two-dimensional spatial image can be understood without reference to the technology of any specific imaging device, whatever the chemical, physical or electronic mechanisms which are involved in the imaging process. Similarly, it is possible to establish fundamental criteria by which we may measure the imaging efficiency of any device. These criteria are based largely on the concept of the ideal image, and ways in which we may assess departures from this ideal for practical imaging devices.

This is not to say that any precise definition of an ideal image is straightforward, and it would be possible to engage in considerable speculation as to whether or not we can attach physical meaning to any general statement that an image is ideal. Even within the limited framework whereby the imaging device is considered merely as a recorder of exposure quanta, there are still difficulties—and sometimes contradictions—depending on the practical application which is involved. For example, an imaging device may be required for such divergent tasks as facsimile pictorial reproduction, or storing information in the form of binary digits. Tasks such as these may involve quite different balances between image criteria which—as we shall see later—may be reciprocal in nature, such as input/output amplification and resolution, or image contrast and dynamic exposure latitude. These and other important factors are closely related, and their relative importance cannot be detached from practical requirements.

Many of these practical implications of imaging properties will be explored in later chapters. Here the properties of images and the concept of the ideal image are analysed as though the imaging device were simply a spatial array of individual photon receptors and counters. In this way we shall see that it is possible to arrive at expressions for the input/output relationships which go a long way towards providing a basis for evaluating practical imaging devices and understanding the nature of their evaluated image properties.

Input/Output Relationships

A reasonable basic specification for an ideal imaging device might be that there should be a one-to-one relationship between incident quanta and some measurable output state of the image. If the device consisted of a spatial array of individual photon counters, each having identical properties, then the spatial resolution of the image would be determined by the spatial dimensions of the receptors. Figure 1 illustrates the spatial structure of such an idealized array.

Each receptor is assumed to record incident quanta in a manner independent of its neighbours and to assume a unique and distinguishable image state dependent only on the number of incident exposure quanta. In order to keep this assumption reasonable, each receptor is restricted to a finite number of

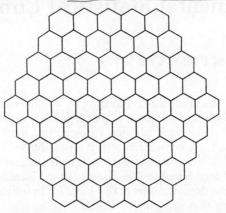

FIG. 1. Ideal spatial array of identical photon counters.

distinguishable image states, so that after the final state has been attained, all further incident quanta go unrecorded. Up to this fixed level, L, each receptor acts with 100% counting efficiency, and above this level with zero efficiency. At this stage there is no need to specify the spatial dimensions of the receptors; the implications of these will be investigated later in this chapter.

Any initial concept of an ideal process has already been compromised by limiting the receptors to a finite number of output states. To investigate the effect of this we shall analyse the input/output relationships according to various statistical criteria, but as a preliminary it is necessary to consider the statistics of the incident quanta themselves. Suppose that from a classical description, the exposure is uniform over the entire spatial extent of the imaging device. From a quantum viewpoint the number of incident quanta will be governed by Poisson statistics with a sufficient degree of validity for the present purposes. If the exposure level is such that the average number of quanta per receptor is q, then the distribution of quanta among the receptors will be as follows:

Number of quanta	0	1	2	...	r	...
Proportion of counters receiving this number	e^{-q}	qe^{-q}	$q^2 \dfrac{e^{-q}}{2!}$	...	$q^r \dfrac{e^{-q}}{r!}$	...
if $q=3$	e^{-3}	$3e^{-3}$	$\dfrac{9e^{-3}}{2}$	$\dfrac{27e^{-3}}{6}$	$\dfrac{81e^{-3}}{24}$	
	$\dfrac{1}{e^3}$	$\dfrac{3}{e^3}$	$\dfrac{9}{2e^3}$	$\dfrac{9}{2e^3}$	$\dfrac{27}{8e^3}$	

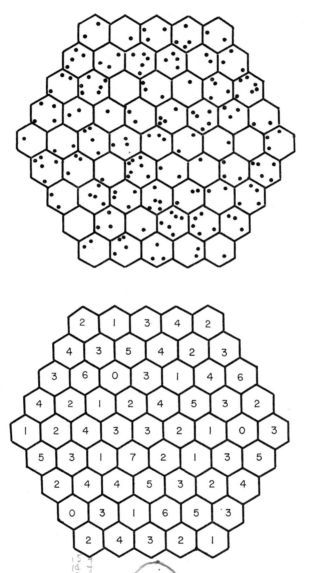

FIG. 2. Lower: typical distribution of numbers for Poisson statistics with an average value of 3. Upper: these same numbers shown as random dots representing incident quanta.

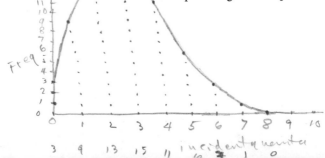

Figure 2 illustrates the implications of the above probabilities. The numbers in the lower part show a typical Poisson distribution corresponding to an average number of three. In the upper part these numbers are represented as a series of random dots, where each dot corresponds to an incident quantum. These statistics will be seen to have an important influence on image characteristics, and it is interesting to note that in this simple example, whereas some

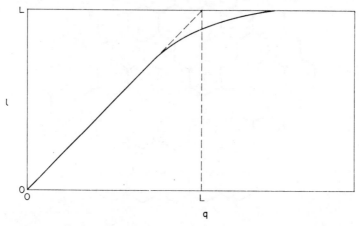

FIG. 3. Relationship between average recorded counts, l, and average quantum exposure, q, in terms of the counter saturation level, L.

receptors have still not received any quanta some have been exposed to as many as six or seven.

From the probability distribution it is simple to calculate the proportion of receptors which have received saturation level, since this will be given by the sum from L to infinity of the individual probabilities; i.e.,

$$\sum_{r=L}^{\infty} q^r \frac{e^{-q}}{r!}$$

This summation will provide an overall measure of the inefficiency of the process. However various other forms of specifying, the input/output relationships prove to be of greater significance, and we shall investigate some of these in detail.

First we consider the relationship between the average count level, l, and the average quantum exposure level, q. Since by definition the average value is given by the sum of the products of each value with its probability of occurrence, (i.e., the first moment of the distribution) we may write

$$l = \sum_{r=1}^{L-1} r\, q^r \frac{e^{-q}}{r!} + \sum_{r=L}^{\infty} L\, q^r \frac{e^{-q}}{r!}. \tag{1}$$

Figure 3 shows a plot of the average count level in relation to the saturation level, as a function of the average exposure level. As might be anticipated there will be a linear relationship between primary quanta and the secondary number of image counts at exposure levels where q is much less than L. This linearity breaks down as the exposure level approaches the saturation level, and at exposures high enough for all receptors to have received at least L incident quanta, each receptor will record exactly L counts.

Inspection of the summation in equation (1) reveals that this equation may also be written as

$$l = \sum_{r=1}^{\infty} q^r \frac{e^{-q}}{r!} + \sum_{r=2}^{\infty} q^r \frac{e^{-q}}{r!} + \ldots + \sum_{r=L}^{\infty} q^r \frac{e^{-q}}{r!}. \qquad (2)$$

There are L summation terms on the RHS of equation (2) and since these are all summed to infinity each term will be equal to unity less the missing first terms. Hence we can define

$$l = L(1 - f_1 e^{-q}), \qquad (3)$$

where

$$f_1 = \frac{1}{L}\left(1 + \sum_{0}^{1} \frac{q^r}{r!} + \sum_{0}^{2} \frac{q^r}{r!} + \ldots + \sum_{0}^{L-1} \frac{q^r}{r!}\right). \qquad (4)$$

Equation (3) is in a more convenient form than equation (1) for evaluating the next characteristic of interest, which is the rate of change of count level with exposure, or gradient. The gradient of the curve shown in Fig. 3 provides another useful way of expressing the same input/output relationships, and as will be seen shortly, is closely related to a parameter used as a practical measure of the so-called contrast in a photographic image. Denoting the gradient by g, differentiation of equation (3) gives

$$g = \frac{dl}{dq} = Le^{-q}\left(f_1 - \frac{df_1}{dq}\right). \qquad (5)$$

The function f_1 contains L summation terms which are the first terms in the natural expansion of e^q up to successive limits ranging from 0 to $L - 1$. Now the derivative of $q^r/r!$ is $q^{r-1}/(r-1)!$, so differentiation of f_1 with respect to q will merely truncate each summation by removing the last term. Subtraction of the derivative of the function from the function itself will leave only the final term of each of the summations, so that equation (5) reduces to

$$g = L f_2 e^{-q} \qquad (6)$$

where

$$f_2 = \frac{1}{L} \sum_{r=0}^{L-1} \frac{q^r}{r!}. \qquad (7)$$

In Fig. 4 the gradient as defined in equation (6) is shown in relation to the saturation level L, as a function of the quantum exposure. The gradient will be unity at low exposure levels, indicating a one-to-one input/relationship, or unit gain: as q approaches L the gradient decreases and eventually falls to zero at some exposure level above L where all the receptors have reached saturation. A plot such as this shows clearly the exposure region over which the image is ideal according to the criterion of unit gain. There are however other criteria which concern the spatial fluctuations in the image or noise, and

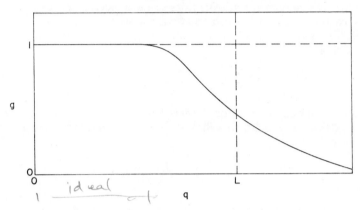

FIG. 4. The gradient, g, as a function of the average quantum exposure, q, in terms of the counter saturation level, L.

which have important practical consequences in defining the quality of the image.

The Comparative Noise Level

There will be a spatial fluctuation in the image which arises in the present ideal case from the statistical distribution of the exposure quanta. Such image noise can be evaluated in terms of the mean-square fluctuation in the number of counts, denoted by $\overline{\Delta l^2}$. This can be calculated using a well-known statistical result for the second moment of a distribution, which in the present notation is

$$\overline{\Delta l^2} = m_2 = m_2' - l^2, \tag{8}$$

i.e., the second moment m_2 of the distribution about the mean level is equal to the second moment m_2' about the origin, minus the square of the first moment.

The second moment of the image counts about the origin can be expressed as

$$m_2' = \sum_{r=0}^{L-1} r^2 q^r \frac{e^{-q}}{r!} + \sum_{r=L}^{\infty} L^2 q^r \frac{e^{-q}}{r!}. \tag{9}$$

Following some manipulation of these summations[1], this can also be written in the form

$$m_2' = L^2(1 - f_3 e^{-q}), \tag{10}$$

where

$$f_3 = \frac{1}{L^2}\left(1 + 3\sum_0^1 \frac{q^r}{r!} + 5\sum_0^2 \frac{q^r}{r!} + \ldots + (2L-1)\sum_0^{L-1} \frac{q^r}{r!}\right). \tag{11}$$

1. SPATIALLY-RECORDED IMAGES

It is interesting to note the close correspondence between equations (1) and (9), (3) and (10), and (4) and (11).

Combining equations (3), (8), (10) and (11):

$$\overline{\Delta I^2} = L^2 \left(1 - f_3 e^{-q}\right) - (1 - f_1 e^{-q})^2), \qquad (12)$$

expresses the noise level as a function of the exposure level, q, and the saturation level, L. Figure 5 shows the mean-square fluctuation as defined by equation

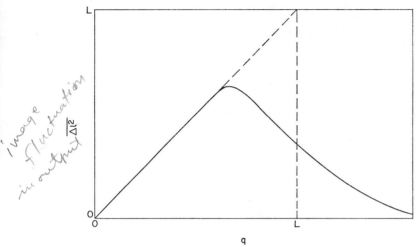

FIG. 5. The mean-square fluctuation in the number of image counts, $\overline{\Delta I^2}$, as a function of the average quantum exposure, q, in relation to the saturation level, L.

(12) and plotted as a function of the exposure level. At low exposure levels the noise follows the straight-line defined by $\overline{\Delta I^2} = q$ before decreasing as the exposure level approaches the saturation level, and eventually falling to zero. When all the receptors record exactly L counts, there will obviously be no image fluctuation.

The question arises as to whether the image noise can be measured on some absolute scale. To answer this we consider the noise associated with the input due to the quantum fluctuations. At the exposure level q the second-moment equation for the mean-square fluctuation in the number of quanta will be:

$$\overline{\Delta q^2} = \sum_{r=0}^{\infty} r^2 q^r \frac{e^{-q}}{r!} - q^2. \qquad (13)$$

The summation on the RHS can be expressed as

$$\sum_{r=0}^{\infty} r^2 q^r \frac{e^{-q}}{r!} = q e^{-q} \sum_{r=0}^{\infty} r \frac{q^{r-1}}{(r-1)!} = q e^{-q} (e^q (1+q)),$$

and hence equation (13) reduces to

$$\overline{\Delta q^2} = q, \qquad (14)$$

which is the well-known result for the mean-square fluctuation about the mean of a Poisson distribution. This last equation explains the observed straight-line relationship at low exposures, $\overline{\Delta l^2} = q$, already referred to. By equation (14) it is now seen that this relationship is in fact $\overline{\Delta l^2} = \overline{\Delta q^2}$, i.e., when the exposure level is low enough to ensure that no individual receptor has received more than L quanta, the output noise merely mirrors the input noise.

This noise level in the input provides the absolute noise scale we are seeking since it is unreasonable to expect that the output fluctuation referred back to the equivalent fluctuation in the input can be less than was actually there in the input. This would imply destruction of statistical entropy, which is forbidden according to the second law of thermodynamics. Fuller implications of

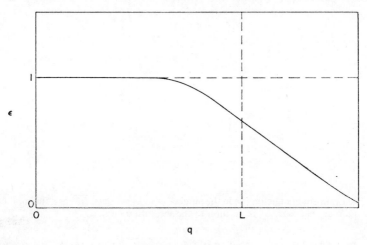

FIG. 6. The comparative noise level, ε, as a function of the average quantum exposure, q, in terms of the saturation level, L.

the relationships between the transfer of signal-to-noise ratio during image recording and information and entropy are left until Chapter 10.

The noise in the image referred back in terms of the exposure will be:

$$\overline{\Delta l^2} \left(\frac{dq}{dl}\right)^2 = \frac{\overline{\Delta l^2}}{g^2}. \tag{15}$$

This will have a lower limit defined by equation (14), and so the ratio

$$\varepsilon = \frac{\overline{\Delta q^2}}{\overline{\Delta l^2}/g^2} = q \frac{g^2}{\overline{\Delta l^2}}, \tag{16}$$

will have a upper limit of unity.

Substitution of equations (6) and (12) in equation (16) leads to:

$$\varepsilon = \frac{q(f_2 e^{-q})^2}{(1 - f_3 e^{-q}) - (1 - f_1 e^{-q})^2}. \tag{17}$$

1. SPATIALLY-RECORDED IMAGES

Through the course of the following chapters we shall see in detail the important role played in general image theory by this concept of the comparative noise level, which is a dimensionless ratio by the nature of its definition. Here we see in Fig. 6 the form of its variation with the exposure level in this ideal case. This variation is seen to be similar though not identical in form to that of the gradient. However in many practical cases to be studied later the gradient and the comparative noise level have quite different forms.

This concept of the comparative noise level is particularly useful in giving intuitive understanding to the level of image noise. In general a plot such as that of Fig. 5 conveys little about the scale limitations of noise, whereas a plot such as that of Fig. 6 is greatly instructive in this respect. From the viewpoint of image noise an image may defined as ideal when the comparative noise level is close to unity.

Threshold Limitation

Already the idealized array of photon receptors has been restricted by an upper limit to the count level. The influence of a further count limit is now investigated, this being the restriction to a lower limit in addition to the upper limit. We shall see in due course that practical photographic processes are usually limited by such a threshold effect.

Adjustment to the previous analysis may be made as follows. All receptors are still assumed to have identical properties, but now the first $(T-1)$ incident quanta are not recorded by a receptor, their influence being only to activate some threshold mechanism. The Tth incident quantum then records as a single count, as do subsequent incident quanta up to a saturation level now denoted by S. The symbol L is retained to denote the total number of counting levels, and so

$$L = S - T + 1. \tag{18}$$

The proportion of receptors now recording a particular number of counts will be as shown below:

Number of incident quanta	0	1	..	T	..	r	..	S	$S+1$	..
Proportion of receptors receiving this number	e^{-q}	qe^{-q}	..	$\dfrac{q^T e^{-q}}{T!}$	..	$\dfrac{q^r e^{-q}}{r!}$	..	$\dfrac{q^S e^{-q}}{S!}$	$\dfrac{q^{S+1} e^{-q}}{(S+1)!}$	..
Number of recorded counts	0	0	..	1	..	$r-T+1$	.	L	L	..

Equation (1) for the average count level will now be replaced by

$$l = \sum_{r=T}^{S-1}(r-T+1)\frac{q^r e^{-q}}{r!} + \sum_{r=S}^{\infty} L \frac{q^r e^{-q}}{r!}. \tag{19}$$

However it is straightforward to show[1] that equation (3) for the average count level remains valid, provided that the function f_1 is now redefined as

$$f_1 = \frac{1}{L}\left(\sum_0^{T-1}\frac{q^r}{r!} + \sum_0^{T}\frac{q^r}{r!} + \sum_0^{T+1}\frac{q^r}{r!} + \ldots + \sum_0^{S-1}\frac{q^r}{r!}\right). \tag{20}$$

Similarly it can be shown[1] that equations (6) and (12) for the gradient and mean-square count fluctuation also remain valid if the functions f_2 and f_3 are now redefined according to:

$$f_2 = \frac{1}{L} \sum_{T-1}^{S-1} \frac{q^r}{r!}; \qquad (21)$$

$$f_3 = \frac{1}{L^2} \left(\sum_0^{T-1} \frac{q^r}{r!} + 3 \sum_0^{T} \frac{q^r}{r!} + 5 \sum_0^{T+1} \frac{q^r}{r!} + \ldots + (2L-1) \sum_0^{S-1} \frac{q^r}{r!} \right). \qquad (22)$$

With functions f_1, f_2 and f_3 redefined in this way to include the threshold, the comparative noise level also then remains as defined previously by equation (17).

It is instructive to look again at the various input/output relationships in terms now of both the threshold T and saturation S. For the sake of illustration we consider two examples, having respective values of $T = 4$, $S = 8$ and $T = 2$, $S = 16$, thus having $L = 5$ and $L = 15$ count levels respectively. Curves appropriate to these examples are shown in Fig. 7. The top plot shows the curves for the average count level. Due to the influence of the threshold these curves no longer have a linear portion at low exposures. The maximum count levels to which the curves tend at high exposure levels correspond to the respective L-values. The gradient curves reveal that in the example $T = 2$, $S = 16$ there is still a portion of the curve where there is a one-to-one relationship or unit gain, but that there is no such portion for the example $T = 4$, $S = 8$. The reasons for this can be understood qualitatively as follows. When the threshold and saturation levels are sufficiently far apart there will be an exposure region wherein the probability of any receptor having received less than T or more than S incident quanta is negligibly small. In this case any additional quanta will be recorded in a one-to-one manner. The gradient curve for $T = 4$, $S = 8$ always falls well short of unity since there is no average exposure level for which all the receptors have attained the threshold without some having reached saturation, so there can never be such a one-to-one relationship.

The mean-square fluctuation in the number of image counts is shown in Fig. 7 (c), and again due to the influence of the threshold there is no longer an exposure region over which $\overline{\Delta I^2} = \overline{\Delta q^2}$. Finally, the comparative noise level is shown in Fig. 7(d). The form of these curves is again similar though not identical to the gradient curves, and for $T = 2$, $S = 16$ there is an exposure region over which the image is ideal in the comparative noise sense, although no such region exists for $T = 4$, $S = 8$.

To summarize, the effect of introducing an operating threshold below which a receptor produces no recorded count need only limit the exposure region over which the image can be defined as ideal from either a unit gain criterion or a comparative noise criterion. If the threshold and saturation levels are sufficiently separated then it is still possible for the image to be ideal over a wide exposure range.

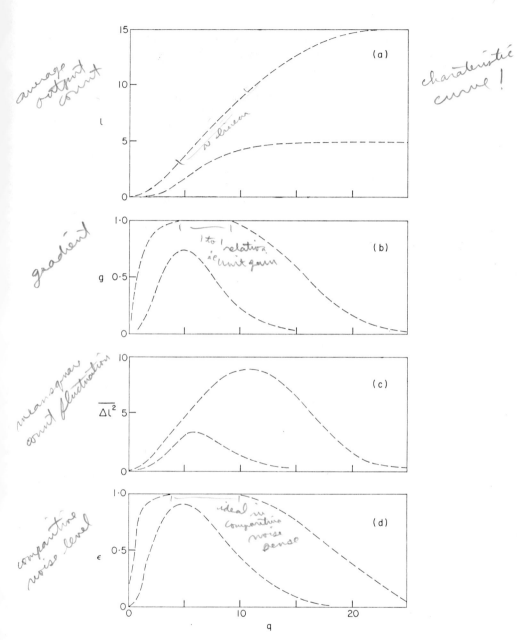

Fig. 7. Image characteristics for fixed counting threshold and saturation levels: (a) average output count; (b) the gradient; (c) mean-square count fluctuation; (d) comparative noise level. These examples are for $T = 2$, $S = 16$ (upper curves in each case), and $T = 4$, $S = 8$ (lower curves).

It should be noted that this analysis was only concerned with identical receptors each having fixed values of threshold and saturation, and the examples for the input/output characteristics are only appropriate for these assumptions. The effects of T and S being random variables among the receptors, or the output counts not being sequential to the exposure quanta due to various quantum inefficiencies, would be to modify these characteristics. Although a more general analysis including factors such as these would introduce unnecessary complexities at this stage, it is possible to use qualitative arguments to understand the general form of the modifications to the image characteristics which would be involved. Some of these are suggested as an exercise at the end of the chapter. Also, for practical reasons connected with the behaviour of conventional silver halide grains, the influence of a variation in T for single-level receptors will be analysed in some detail in Chapter 3.

1.2 Image Density Characteristics

Density and Gamma

Whereas considering the input and output in terms of individual receptor counts reduces the specification of image characteristics to the most basic forms, for practical imaging systems other forms of expressing the input/output

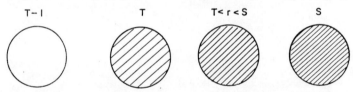

FIG. 8. Illustration of the change in opacity of an individual receptor with increase in quantum exposure level. From T to S incident quanta there is an increase in opacity.

relationships are usual. In practice, due to the very large numbers which are involved (a typical 5 × 5 cm photograph might contain around 10^{10} separate image grains arising from 10^{12} exposure quanta), the image relationships are usually specified in terms of larger-scale measurements.

Suppose that the degree of "blackness" or opacity of an individual receptor provides the image measure of the receptor count level after exposure, as illustrated in Fig. 8. The image level might then be deduced by shining light through the receptor and measuring the amount of light transmitted, I_T, in terms of that incident, I_O, as illustrated in Fig. 9, and defining opacity by the ratio

$$O = \frac{I_O}{I_T}.$$

A more useful practical measure is often found to be that of the density, D, defined on a logarithmic scale according to

$$D = \log_{10} O = \log_{10} \frac{I_O}{I_T}. \tag{23}$$

Rather than measuring the opacity or density of individual receptors, a large number of them may be measured simultaneously, as illustrated in Fig. 10. Here it is supposed that an image-measuring aperture of area A is used to

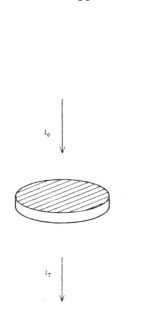

FIG. 9. Incident and transmitted light intensities during measurement of the recorded image level of a receptor.

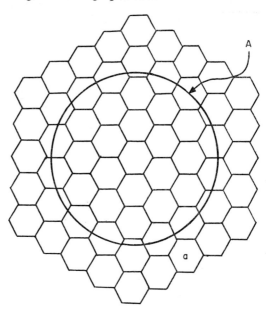

FIG. 10. Simultaneous sampling of receptors during image measurement. The scanning aperture A measures N receptors of area a at any one aperture position.

determine the density. The aperture area is then related to the receptor area, a by the equation

$$A = Na, \qquad (24)$$

where N is the number of receptors covered by the aperture.

It is now necessary to specify the way in which the count level determines the transmitted intensity during image read-out. Suppose that it is assumed that the amount of light subtracted from the incident beam during transmission through a receptor is proportional to the count level. Below threshold the amount subtracted will be zero, i.e., $I_T = I_O$; at threshold $I_T = I_O - bI_O$, where b denotes some fixed proportion of the incident intensity; at saturation $I_T = I_O - LbI_O$. The average amount subtracted by many receptors, as measured for example with image scanning-aperture A, will be lbI_O, where l is the mean count level. Thus the measured image opacity will be

$$O = \frac{I_O}{I_T} = \frac{I_O}{I_O - lbI_O} = \frac{1}{1 - lb}. \qquad (25)$$

and hence
$$D = -\log_{10}(1 - lb)$$

By substitution of the mean count level from equation (4),

$$D = -\log_{10}(1 - bL(1 - f_1 e^{-q})). \tag{26}$$

From equations (25) and (26) the limits of opacity are seen to range from 1 to $1/(1 - bL)$, while those of density range from 0 to $\log_{10}(1 - bL)$. A convenient scale emerges by setting $b = 9/10L$, since this gives an opacity range from 1 to 10, and a density range from 0 to 1.

Since the output is now in terms of density rather than count level, the gradient might now be defined as dD/dq. However we shall see that in practice the image density characteristics are usually expressed as a function of the logarithm of the exposure level, it being more appropriate to work in terms of so-called gamma, where

$$\gamma = \frac{dD}{d(\log_{10} q)}. \tag{27}$$

Since $d(\log_{10} q) = \log_{10} e \, d(\log_e q) = \log_{10} e \, \frac{dq}{q}$,

it follows that

$$\gamma = \frac{q}{\log_{10} e} \frac{dD}{dq}. \tag{28}$$

Differentiation of equation (26) leads to

$$\gamma = \frac{q f_2 e^{-q}}{f_1 e^{-q} + \left(\frac{1 - bL}{bL}\right)}. \tag{29}$$

An example of the image characteristics in terms of density and gamma is given in Fig. 11 for the case $T = 2$, $S = 16$, which was one of the cases considered in Fig. 7 for the output count-level characteristics. The density scale has been normalized according to the assumption already discussed, so that $D_{max} = 1$. Gamma is seen to reach a peak before returning to zero at high exposure levels. It is simple to reason that in general the greater the number of recording levels, the greater will be the exposure range over which gamma is non-zero, but the less will be the maximum value of gamma. Thus there is reciprocity between image contrast and the recording latitude.

Density Fluctuations and the Comparative Noise Level

In spite of practical utility, expressing the input/output characteristics in terms of density further obscures the extent to which the image can be compared with the ideal case. However the comparative noise level is still a direct and useful criterion for this purpose. If we denote the mean-square fluctuation

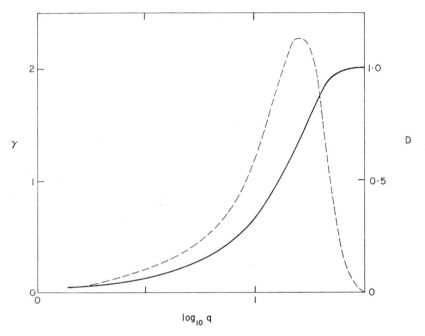

FIG. 11. Image density (solid curve) and gamma (broken curve) shown as functions of the logarithm of the exposure level in the case $T = 2$, $S = 16$.

in density as measured with scanning-aperture A by $\overline{\Delta D_A^2}$, then when referred back to the equivalent exposure fluctuation by use of equation (28):

$$\overline{\Delta D_A^2}\left(\frac{dq}{dD}\right)^2 = \overline{\Delta D_A^2}\left(\frac{q_A}{\gamma \log_{10}e}\right)^2, \qquad (30)$$

where q_A denotes the average number of exposure quanta per image area A. The fluctuations associated with the exposure quanta will be defined by

$$\overline{\Delta q_A^2} = q_A,$$

leading to a comparative noise level of

$$\varepsilon = \frac{(\log_{10}e)^2 \gamma^2}{q_A \, \overline{\Delta D_A^2}}. \qquad (31)$$

The problem now remains to calculate $\overline{\Delta D_A^2}$, and we can do this in terms of the count fluctuations already evaluated. Since

$$D = -\log_{10}(1 - bl),$$

$$\frac{dD}{dl} = \frac{b \log_{10}e}{1 - bl} = \frac{\Delta D_A}{\Delta l_A}, \qquad (32)$$

where this final step assumes that the finite fluctuations are small enough to be represented by the differential equation. This condition will be satisfied if the aperture A averages over a large enough number N of individual receptors. The relationship between mean-square fluctuations will thus be:

$$\overline{\Delta D_A^2} = \left(\frac{b \log_{10} e}{1 - bl}\right)^2 \overline{\Delta l_A^2}. \tag{33}$$

Since the aperture averages over N receptors each having independent image counts, the mean-square fluctuation measured by A will be related to the

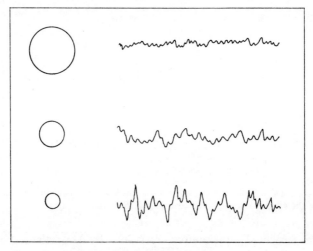

FIG. 12. Relationship between aperture size and density fluctuation during image scanning.

mean-square fluctuation when measured for individual receptors by the equation:

$$\overline{\Delta l_A^2} = \frac{1}{N} \overline{\Delta l^2}, \tag{34}$$

where $\overline{\Delta l^2}$ is already defined by equation (12). Combining equations (24), (33) and (34) leads to

$$A \overline{\Delta D_A^2} = a \left(\frac{b \log_{10} e}{1 - bl}\right)^2 \overline{\Delta l^2}. \tag{35}$$

Equation (35) demonstrates that although the measured density fluctuations depend on the size of the measuring aperture, the product $A \overline{\Delta D_A^2}$ is independent of the aperture. We shall see in later chapters that this is a result of great practical significance. The relationship between A and $\overline{\Delta D_A^2}$ is illustrated in Fig. 12. The density fluctuations recorded by the aperture during image scanning are shown for three aperture sizes: the largest aperture will record the least fluctuation and vice versa.

If the product $A\overline{\Delta D_A{}^2}$ is denoted by G, then the comparative noise level can be redefined in terms of G, since in equation (31) $q_A = Nq = Aq/a$, and hence

$$\varepsilon = \frac{(\log_{10}e)^2 \gamma^2}{\frac{q}{a} \cdot G}. \qquad (36)$$

Since q is the average number of exposure quanta per receptor, q/a is the average number per unit image area.

Equation (36) provides the relationship commonly used for evaluating practical imaging processes. For the present analysis of the ideal process, substitution of the expressions already obtained for gamma and G leads to

$$\varepsilon = \frac{q(f_2 e^{-q})^2}{(1 - f_3 e^{-q}) - (1 - f_1 e^{-q})^2}. \qquad (37)$$

Equation (37) is seen to be identical with equation (17) which was deduced for the comparative noise level in terms of input and output counts rather than density relationships. This is another result of significance. For when the input/output relationship is expressed in a form such as the density-log exposure curve, it becomes difficult to interpret the extent to which the image departs from ideal from inspection of either the curve or its slope, gamma. However, the above result demonstrates in a striking manner that the concept of the comparative noise level is free from this drawback.

We shall in fact see later that the comparative noise level is not always independent of the way in which the output is defined, since some definitions effectively introduce additional fluctuations which themselves contribute to the noise. But the important fact is that the comparative noise level always has an absolute scale with an upper limit of unity, no matter how the output is defined and measured.

Figure 13 shows the comparative noise level as defined by equation (37) for a range of receptor threshold and saturation levels. It is seen that in spite of the threshold and saturation restrictions there may still be an exposure range over which the image is ideal in this sense, and the greater the separation between threshold and saturation the greater will be this range.

1.3 Random Array of Receptors

Three-dimensional Layers

So far the spatial configuration of the individual receptors has been considered to be in the form of a close-packed uniform mesh. We now investigate another situation which might be regarded as the other extreme, whereby the receptors are distributed entirely at random. It is then assumed that, like the exposure quanta, the receptors themselves obey a Poisson distribution law in space.

The physical structure of a spatial imaging device which has receptors obeying this new assumption is illustrated in Fig. 14. For a two-dimensional Poisson distribution there will always be a finite probability of any two receptors occupying the same space, this probability increasing as the average

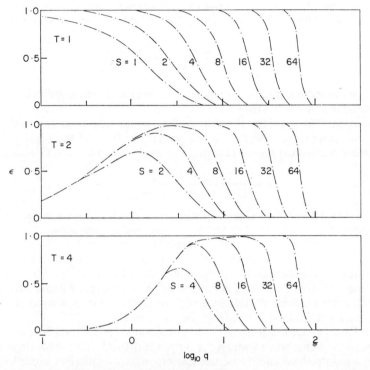

FIG. 13. The comparative noise level plotted on a log exposure scale for an ideal mesh of receptors, for a range of threshold and saturation levels.

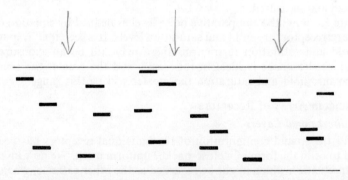

FIG. 14. Cross-section of an imaging device having identical receptors in the form of flat tablets which are randomly-distributed in three-dimensional space.

number of receptors per unit area increases. Since practical types of receptor have a physical structure which forbids such overlapping, it is assumed that the imaging device is now three-dimensional, with cross-section as indicated in Fig. 14. In this way the three-dimensional distribution will be such that the coordinates of the receptors are approximately Poisson-distributed when viewed as a two-dimensional array from the surface, so long as the number of receptors per unit volume is not too great. In other words, there will be two-dimensional overlapping in effect and a high maximum density will be possible without contravening the assumption of randomness. Questions such as the

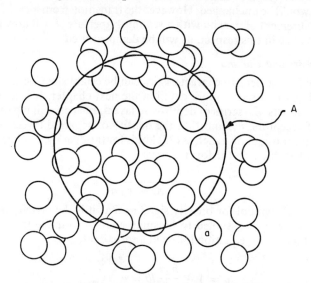

FIG. 15. The overlapping distribution of receptors of area a, viewed from the surface of the three-demensional layer. The number of receptors covered by the aperture A varies with the position of A.

physical manner in which receptors might be suspended in such a three-dimensional way are left until practical consideration of actual devices.

It is assumed that the individual receptors still have identical properties, and are in the form of flat discs of area a, parallel to the surface as indicated in Fig. 14. When the image obtained by such a device is measured with a scanning aperture of area A, as illustrated in Fig. 15, the number of receptors included by the aperture at any one position will no longer be constant, but will vary according to Poisson statistics, with average number denoted by N_A.

The new image characteristics can be evaluated if we make a few more simple assumptions. First we must specify the number of exposure quanta reaching each receptor as a function of position within the three-dimensional layer. To do this out of practical context, (e.g., knowledge of the light scattering and absorption properties of the receptors and support medium etc., as will be discussed in later chapters for conventional photographic processes) is unrealistic, but since the aim at this stage is to define the influence of the change to a random structure from an ideal mesh, it will be assumed that all receptors still have the same statistical access to exposure quanta. In other words, the only exposure fluctuation from receptor to receptor will be due to the statistical fluctuation of the exposure quanta, independent of receptor position.

If the image were still to be evaluated by counting the output levels of each individual receptor, the relationships between mean and mean-square fluctuation in counts, l and $\overline{\Delta l^2}$, and the exposure level q, would remain unchanged since equations (19) to (22) would still be valid, and hence the comparative noise level would be unchanged. However the transition from individual counts to the measurement of density with a scanning aperture A will now involve an essential change in this respect, as will be demonstrated.

Image Density and Gamma

A model for the density characteristics can be established using an analysis which will be discussed in more detail in the next chapter. It is assumed that the image consists of a series of depth-wise layers, each containing only a small number of receptors, n_A, and that the density of separate layers is additive. For each layer an equation such as (25) holds for the opacity, but it is now more convenient to write it in the form

$$o = \frac{A}{A - n_A lb}, \tag{38}$$

where the small symbol for opacity signifies that it is the opacity due only to one layer. Since n_A is assumed to be small enough to make the ratio $n_A a/A$ much less than unity, equation (38) will approximate to:

$$o = 1 + \frac{n_A a}{A} lb = e^{\frac{n_A a}{A} lb}, \tag{39}$$

The corresponding density, d, will thus be

$$d = \left[\log_{10} e\right] \frac{n_A a}{A} lb, \tag{40}$$

and hence the total density of the image layer will be

$$D = \sum d = \left[\log_{10} e\right] \frac{N_A a}{A} lb. \tag{41}$$

Substituting the expression for the average level, l, from equation (3)

$$D = \left[\log_{10} e\right] \frac{N_A a}{A} bL(1 - f_1 e^{-q}), \tag{42}$$

where f_1 is defined by equation (20).

The image density can again be given a convenient scale, this time by assuming that $b = 1/L$ and that the population of receptors is such that

$$\frac{N_A a}{A} = \frac{1}{\log_{10} e} = 2 \cdot 30, \tag{43}$$

i.e., all receptors packed closely side-by-side would occupy an area 2·30 times that of the surface area of the image. Equation (42) then reduces to:

$$D = 1 - f_1 e^{-q}. \tag{44}$$

Since q still denotes the average number of exposure quanta per receptor, the exposure per unit image area will now be $\dfrac{N_A q}{A}$, and if $\dfrac{N_A a}{A} = 2\cdot 30$, this exposure will be equal to $2\cdot 30\, q/a$. In other words, if an average of $2\cdot 30\, q/a$ quanta are incident per unit image area, each receptor will absorb an average of q quanta if it is assumed that all incident quanta are absorbed by receptors.

Figure 16 shows a set of density characteristics for a threshold of $T = 2$ and a range of saturation levels, compared with those for the uniform mesh as

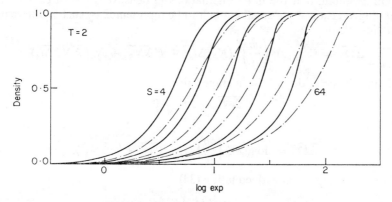

FIG. 16. Density—log exposure characteristics for a uniform mesh (solid curves) and random array (broken curves) of receptors, for threshold $T = 2$ with saturation levels $S = 4, 8, 16, 32$ and 64.

analysed previously. For comparative purposes the logarithms referred to on the exposure scale are those of the ratio q/a for the uniform mesh, and $2\cdot 30\, q/a$ for the random array, with an arbitrary value of $a = 1$ in each case. The uniform mesh is seen to give a higher slope, or gamma, in all cases. As the number of recording levels increases the curves are moved towards higher exposures. This is because with a larger number of recording levels the change in output associated with a single level becomes smaller (i.e., b is less).

The analytical expression for gamma follows by differentiation of equation (42):

$$\gamma = \frac{q}{\log_{10} e}\frac{dD}{dq} = \frac{N_A a}{A} bL q \frac{d}{dq}(1 - f_1 e^{-q})$$

$$= \frac{N_A a}{A} bL f_2\, q e^{-q}. \tag{45}$$

Thus after normalization of the density scale as described,

$$\gamma = \frac{1}{\log_{10}e} f_2 q e^{-q}, \tag{46}$$

where f_2 is defined by equation (21).

Density Fluctuations and the Comparative Noise Level

If it is assumed that the average number, l, of recording levels, and the average number, N_A, of receptors are independent variables, then the differential of equation (41) may be expressed as

$$d\mathrm{D}_A = \log_{10}e \, \frac{ab}{A}(N_A + l \, dN_A).$$

Again assuming that the finite fluctuations in density as measured with the scanning aperture A are small enough to be represented by such a differential equation, then

$$\overline{\Delta \mathrm{D}_A^2} = (\log_{10}e)^2 \left(\frac{ab}{A}\right)^2 (N_A^2 \overline{\Delta l_A^2} + l^2 \overline{\Delta N_A^2} + N_A l \, \overline{\Delta N_A} \, \overline{\Delta l_A}),$$

and since by definition of the average values,

$$\overline{\Delta N_A} = \overline{\Delta l_A} = 0,$$

it follows that

$$\overline{\Delta \mathrm{D}_A^2} = (\log_{10}e)^2 \left(\frac{ab}{A}\right)^2 (N_A^2 \overline{\Delta l_A^2} + l^2 \overline{\Delta N_A^2}). \tag{47}$$

Now by comparison with equation (34),

$$\overline{\Delta l_A^2} = \frac{1}{N_A} \overline{\Delta l^2}, \tag{48}$$

and since l and $\overline{\Delta l^2}$ are defined by equation (3) and (12) and $\overline{\Delta N_A^2} = N_A$, equation (47) leads to

$$\mathrm{G} = A \overline{\Delta \mathrm{D}_A^2} = (\log_{10}e)^2 (abL)^2 \frac{N_A}{A}(1 - f_3 e^{-q}), \tag{49}$$

where f_3 is defined by equation (22). Thus with normalization of the density scale as described,

$$\mathrm{G} = a \log_{10}e \, (1 - f_3 e^{-q}). \tag{50}$$

Since the ratio N_A/A is independent of A, then G is again independent of the area of the scanning aperture, whether defined by equation (49) or equation (50).

The relationship between the image noise level as defined by G and the image density level is one which proves to be of practical interest for conventional photographic processes. With $b = 1/L$, combining equations (42) and (49) leads to:

$$\mathrm{G} = a \log_{10}e \, \mathrm{D} \, \frac{(1 - f_3 e^{-q})}{(1 - f_1 e^{-q})}. \tag{51}$$

Figure 17 shows a set of curves for the image noise as measured by G as a function of density, and for a threshold of $T = 2$ and a range of saturation levels as indicated. As can be seen from the density scale, these curves are for the normalized case where D_{max} equals unity. The comparative set of curves for the image noise of the uniform mesh is also shown. An arbitrary receptor area of $a = 1$ has again been used in calculating each set of curves. In this case equation (52) predicts that for the random array the value of G at maximum density will be equal to $\log_{10}e = 0.434$, as is seen from the plotted curves.

The main difference between the noise for the uniform and random arrays of receptors is seen to be that for the uniform mesh the noise returns to zero

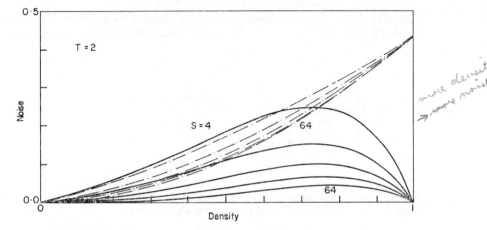

FIG. 17. Noise—density characteristics for uniform mesh (solid curves) and random array (broken curves) of receptors, for threshold $T = 2$ with saturation levels $S = 4, 8, 16, 32$ and 64.

at maximum density when every individual receptor contributes to the image, rather than approaching some upper value such as 0.434 as calculated in this example. The reason for this difference at maximum density is quite obvious, representing the difference between a uniformly dense image and one made up of random units. For both cases the noise at a given density level decreases as the number of recording levels per receptor increases. But as we saw from the density-exposure characteristics a higher exposure level is necessary to obtain the same density level as the number of recording levels increases. In what way might it be considered worth increasing the required exposure to obtain this reduction in noise? To answer this question we again turn to the comparative noise level.

The comparative noise level may now be calculated using the definition of equation (36), remembering that the average number of exposure quanta per unit area is now $N_A q/A$. Substitution of the appropriate values for gamma and the noise, G, leads to the result:

$$\varepsilon = \frac{q(f_2 e^{-q})^2}{(1 - f_3 e^{-q})}. \tag{52}$$

It is interesting to note that this result may be arrived at from the general equations (45) and (49), or from those corresponding to a normalized density scale, equations (46) and (50). This demonstrates that the comparative noise level is unaffected by mere scale changes in the measurement of the image output.

It follows from equations (37) and (52) that the comparative noise level for the random array, ε_r, is less than that for the uniform array, ε_u, at all exposure levels, since

$$\varepsilon_r = \varepsilon_u \left(1 - \frac{(1 - f_1 e^{-q})^2}{(1 - f_3 e^{-q})}\right). \tag{53}$$

Figure 18 shows the comparative noise level defined by equation (52) for a threshold value of $T = 2$ with a series of saturation levels, and these curves

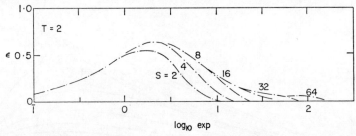

Fig. 18. Comparative noise level as a function of log exposure for a random array of receptors with threshold $T = 2$ and saturation levels as shown.

may be compared with the equivalent $T = 2$ set of Fig. 13 for the uniform mesh of receptors. There is no longer any exposure region for which the image is ideal in the comparative noise sense, no matter how far apart are the threshold and saturation levels. This highest value is now in the region of 0·6 in this example.

This conclusion is significant in illustrating the influence of the form of the output measurement on the comparative noise level. For the uniform mesh there was no difference in the comparative noise level whether the output was expressed in terms of counting individual receptor levels or larger-scale density measurements, but this is no longer true for the random array of receptors. The difference arises because for the uniform array exactly N receptors are always measured for one position of the aperture, while for the random array N_A receptors are measured at a time, where N_A is a random variable. Measuring N_A receptors at a time as in density measurements thus introduces an additional source of fluctuation.

1.4 Image Resolution

We now consider the way in which the image resolution is related to the input/output characteristics. So far the size of an individual area of a receptor has been specified by its area, a, and in the examples shown an arbitrary value of $a = 1$ has been assumed. It is now supposed that the receptors may vary in area, as indicated in Fig. 19. The image resolution of the uniform mesh may be defined by the linear dimensions of the receptors as indicated. The resolution

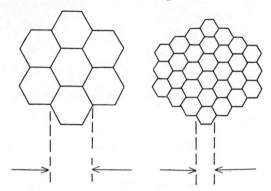

FIG. 19. Influence of receptor area on linear image resolution.

for the random three-dimensional distribution of receptors, where scattering occurs in the layer prior to image formation, can only be specified within the practical context, as will be investigated in some detail in Chapter 7.

The general relationship between resolution and receptor area for the uniform mesh will be as indicated in Fig. 20. The resolution can be related to

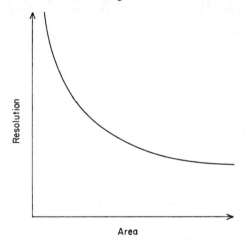

FIG. 20. Image resolution as defined by the reciprocal of the square-root of the receptor area.

the density and noise characteristics in a simple manner. The exposure scale will change for both as a changes, since the same average quantum exposure per receptor will now correspond to different quantum exposures per unit image area. Also, the noise will change in absolute value as the receptor area varies.

Suppose, for example, that the receptor area is increased from a to $2a$. The linear resolution will be in the ratio of $1 : 1/\sqrt{2} = 1\cdot41 : 1$. If both have identical density scales the density characteristics for the receptors of area $2a$ will be

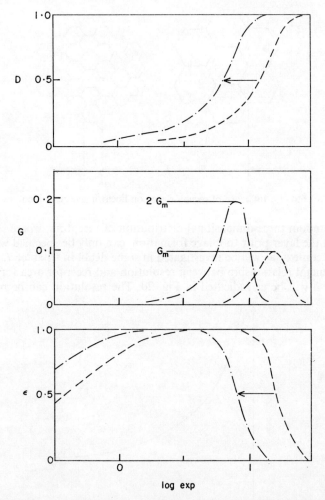

FIG. 21. Illustration of the changes in characteristics which accompany a change in image resolution due to an increase in receptor area from a to $2a$: Top: the density curve is moved as shown by a constant amount of $\log_{10}2$; Middle: the noise is scaled up by a factor of 2, as well as moved by the same amount along the log exposure scale; Bottom: the comparative noise level is unchanged apart from the shift of $\log_{10}2$ as shown.

shifted by $\log_{10}2 = 0\cdot30$ log exposure units towards lower exposure levels. This follows from equation (26), since in effect the only change is in q, and by a factor of 2. This difference in the density characteristics associated with receptors of area a and $2a$ is illustrated in the top diagram of Fig. 21.

The change in the noise characteristics as measured by G follows directly from equation (50). As well as the change in q, there will be a scaling-up of the noise by a factor of 2 when the receptor area increases from a to $2a$ since G is directly proportional to the receptor area. This change in G is shown in the middle diagram of Fig. 21. However from the definition according to equation (36) we can deduce that the comparative noise level will be unchanged in magnitude, although moved on the log exposure scale due to the effective change in q. This is because although the value of G is twice as high when the area changes from a to $2a$, it occurs at half the absolute exposure level, and the product of the two in the denominator of equation (36) remains unchanged.

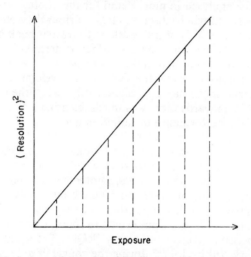

FIG. 22. Relationship between the square of the linear resolution and the exposure needed to produce a given output density, for a quantum-limited imaging system. The operating region for practical systems must fall in the shaded region to the right.

This mere exposure scale change of the comparative noise level is illustrated in the bottom diagram of Fig. 21.

It is concluded that an increase in resolution made possible by a decrease in receptor area is accompanied by less image density and less image noise at a given level of exposure. However, the comparative noise level is unchanged apart from the change in exposure scale, and this invariance in magnitude again stresses its suitability as a general measure of the image noise.

The absolute exposure level required to achieve some fixed output response is often used as a means of specifying the sensitivity of an imaging device. For the ideal mesh required exposure will be proportional to the reciprocal of the receptor area, as is the square of the linear resolution. A plot of the square of linear resolution against the required exposure will result in a straight-line, as shown in Fig. 22. If the constant of proportionality is that for the uniform mesh of receptors with a threshold of $T = 1$ and a high saturation level, (i.e., each exposure quantum is recorded as a separate output event), then all other cases will fall to the right of this quantum-limited boundary in the shaded region. In the practical case the working region will be defined by both quantum inefficiencies and resolution inefficiencies, the latter when the resolution is defined by dimensions greater than the areas of individual receptors. A more detailed practical study of these sensitivity, resolution and quantum efficiency relationships will be given in Chapter 4.

1.5 Detective Quantum Efficiency

During the course of this chapter the comparative noise level has emerged as an image evaluation parameter of fundamental significance. In subsequent chapters it will be analysed in more detail for the photographic process, and its general implications will be discussed from various theoretical and practical aspects. In view of this, it is of interest here to trace it back to its conceptual origins, and also to establish a more satisfactory terminology. Although we have used the term comparative noise level since it arises by comparison of input and output noise levels on the same scale of reference, this is rather an unsatisfactory term since it takes its highest value when the noise is lowest.

From the 1940's onwards there was much scientific activity concerning the classification of the performance of radiation detectors, and especially in the comparison of different types of radiation detector. Amongst the leading work in this respect was that of Rose[2,3], Jones[4,6] and Fellgett[5], who were concerned with the performance of such widely different detector types as represented by television camera tubes, photoemissive and photoconductive devices, the photographic process, and even the human eye. Up to that time the most universal way of specifying performance was by the quantum efficiency as measured by the ratio of output events to input events, this usually being termed the responsive quantum efficiency (RQE). The unsatisfactory aspects of RQE were surveyed by Jones[7] during the course of an extensive review of the previous work on the evaluation of radiation detectors. One main drawback of RQE is that it need not have an upper limit of unity: for example, photomultiplier tubes may have values much greater than unity, and we might summarize this and other drawbacks by saying that RQE links the input/output numbers in quantity but not in quality. As we have seen, the comparative noise level relates the input/output *fluctuations* rather than the input/output *numbers*.

It was Rose[2] in 1946 who proposed the concept that we have termed the comparative noise level as a measure of the "useful" quantum efficiency. To distinguish it from RQE it is usually termed the detective quantum efficiency (DQE), and this term will be adopted in this book from now on. Other terminology to be found in the literature includes detective efficiency, noise-equivalent quantum efficiency and equivalent quantum efficiency (EQE). It is also useful to note at this stage that DQE is usually expressed as a percentage, an ideal detector thus having a DQE of 100%. The relationship between RQE and DQE for the photographic process will be explored in more detail in Chapter 4. General reviews of the significance and utility of DQE have been given by Zweig[8] and Jones[9].

At first it might seem remarkable that the same concept can be applied to such diverse detector types, for which the output may vary from a two-dimensional spatial picture to a one-dimensional time-varying electrical current. However the principle of referring back the output fluctuations through the input/output operating characteristics and expressing them in terms of the equivalent input fluctuations can be applied to all of these, no matter what the units and physical measurements which are involved. The ultimate basis of DQE is the quantum nature of radiation.

Our analysis so far has been in terms of exposure quanta, but the input/output relationships of practical interest may for example be in terms of energy.

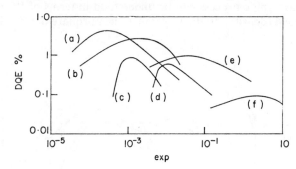

FIG. 23. DQE as a function of exposure level[7] for: (a) Image Orthicon 6849; (b) Image Orthicon 5820; (c) Royal-X film; (d) Plus-X film; (e) human vision; (f) 6326 Vidicon. The exposure scale is in ergs cm^{-2}.

However the conversion from numbers of quanta to energy enables the two to be equated. For example the energy of one quantum associated with radiation of wavelength λ is equal to $(1 \cdot 986/\lambda) \times 10^{-12}$ ergs, if λ is expressed in terms of microns. Some further useful conversions between different forms of input/output quantities will be given in the following chapter.

Results from one of the earliest comparisons of detector types[7] are shown in Fig. 23. DQE is shown as a function of exposure level for three types of television camera tube, two types of photographic film and human vision. These values were representative of the state-of-the-art for these detector types when the results were first published in the 1950s, and were measured under fairly optimum operating conditions, including wavelength of radiation. Figure 24 shows the DQE values for the same detector types, but now the DQE values are plotted against the wavelength of the radiation, at optimum exposure levels in each case.

A striking conclusion from these sets of curves is that in the photographic case the maximum values of DQE are in the region of 1%, compared with values up to 100% we have calculated in this chapter for hypothetical and idealized devices. The reasons for this discrepancy will become clearer during the course of the next two chapters, but can be summarized here into three classes: first, in the practical case there usually exists a substantial fraction of exposure quanta which are not absorbed by any individual receptor; second, for conventional photographic processes the number of receptor recording levels is normally only one; third, there are other sources of statistical fluctuations, such as the size of the individual receptors, which show up in the output. It will be seen how these can reduce the DQE values to the order of a few percent.

In spite of the theoretical nature of this initial study of ideal spatial-imaging devices, various relationships have already emerged which will be of benefit in understanding the properties of practical imaging processes during the remaining chapters. The concept of DQE is in many ways the most important one, but we shall also see that many other observed and empirical relationships for the photographic process can be understood in terms of the concepts we have developed (density, gradient, gamma, mean-square density fluctuations,

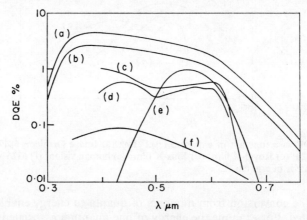

FIG. 24. DQE as a function of wavelength of radiation[7] for the same detector types as Fig. 23.

noise G, etc) and the relationships between these concepts. In fact most of the well-known empiricisms associated with the photographic image will be seen to follow naturally from these relationships.

It will be seen that the other significant class of image properties not covered by the analysis so far is that associated with the microscopic spatial detail in the image. For although the spatial variations of the noise have been analysed, these have only been considered on a relatively large scale, for example as averaged with a scanning aperture which includes a large number of individual receptors at a time.

The latter might be considered as macro-fluctuations, whereas the micro-fluctuations need an additional analytical framework in which they may be studied, as do the resolution inefficiencies. This framework, based largely on Fourier analysis, will be developed in Chapter 6.

References

1. Shaw, R. (1972). Multilevel grains and the ideal photographic detector. *Photogr. Sci. Eng.*, **16**, 192.

2. Rose, A. (1946). A unified approach to the performance of photographic film, television pick-up tubes, and the human eye. *J. Soc. Motion Picture Engrs.*, **47**, 273.
3. Rose, A. (1948). Sensitivity performance of the human eye on an absolute scale. *J. Opt. Soc. Amer.*, **38**, 196.
4. Jones, R. C. (1949). A new classification system for radiation detectors. *J. Opt. Soc. Amer.*, **39**, 327.
5. Fellgett, P. B. (1949). On the ultimate sensitivity and practical performance of radiation detectors. *J. Opt. Soc. Amer.*, **39**, 970.
6. Jones, R. C. (1952). "Detectivity": the reciprocal of the noise equivalent input of radiation. *Nature*, **170**, 937.
7. Jones, R. C. (1959). Quantum efficiency of detectors for visible and infrared radiation. *Advances in Electronics and Electron Physics*, **11**, 87, Academic Press, New York and London.
8. Zweig, H. J. (1964). Performance criteria for photo-detectors—concepts in evolution. *Photogr. Sci. Eng.*, **8**, 305.
9. Jones, R. C. (1968). How images are detected. *Scientific American*, **219**, Number 3, 110.

Exercises

(1) The DQE of ideal arrays of identical grains has been analysed for identical receptors with a fixed threshold (T) and saturation (S). Use empirical arguments to indicate how the DQE might be modified according to the following alternative assumptions:

(a) T is still fixed but S is a random variable;
(b) S is still fixed but T is a random variable;
(c) between fixed values of T and S there is a change in recording level only for every n^{th} exposure quantum (where n is constant), rather than for every single quantum;
(d) as for (c), but with n a random variable.

(2) Use equation (46) to show that for single-level receptors ($T = S$),

$$\gamma = \frac{1}{\log_{10} e} \frac{q^T e^{-q}}{(T-1)!},$$

and hence that the maximum value is given by

$$\gamma_{\max} = \frac{1}{\log_{10} e} \frac{T^T e^{-T}}{(T-1)!}.$$

Plot $\gamma_{\max}$ as a function of T.

(3) Show that for randomly-distributed single-level receptors there is a straight-line relationship between image noise, G, and image density, and determine the influence on this relationship of the area of an individual receptor.

How would you expect the relationship to be modified for receptors with a distribution of areas?

32 IMAGE SCIENCE

1.2 (4) An imaging device consists of a uniform array of photoreceptors, each having only one recording level for which one exposure quantum produces a density of unity. Plot the transmitted intensity and opacity of the image as a function of the average quantum exposure level, when measured with an aperture which samples a large number of receptors simultaneously.

Use equations (26), (29) and (35) to show that at low exposure levels the image density, gamma and noise are all directly proportional to the average number of exposure quanta per receptor, and that the DQE approximates to 100%.

1.5 (5) Discuss in general terms the difference between the concepts of DQE and RQE.

Suppose that two different imaging devices operate in such a way that for the first each input quantum yields exactly n countable output events, while the second yields an average of n. Discuss the difference in RQE and DQE of the two devices, and derive an expression for the DQE of the second in terms of n. Assume that each quantum produces at least one output count, while the fluctuation in counts approximates to a Poisson distribution ($n \gg 1$), and that the output measurement is in the form of simultaneous counting of output events from a large number of input quanta.

1.1 (6) Derive equation (10) from equation (9), where f_3 is defined by equation (11).

1.3 (7) Starting from equation (46) show that for a random distribution of receptors having only two recording levels (T and $T + 1$), the maximum value of gamma occurs at an average exposure level midway between these two levels if T is large.

1.4 (8) An imaging process consists of an ideal mesh of identical receptors having threshold and saturation levels of $T = 2$, $S = 64$, with the highest level corresponding to a density of 4. If the area of the receptors must be such that a critical average exposure level, $E_c = 10^8$ quanta/cm², produces an image density of at least 0·5, estimate the maximum linear image resolution which is possible (assume the density-exposure characteristics are based on the appropriate curve of Fig. 16, but note that the maximum density is now 4).

Sketch the form of the relationship between E_c and image resolution in the ideal case.

1.3 (9) Adapt equation (53) for single-level receptors ($T = S$), and show that in this case

$$\frac{\varepsilon_r}{\varepsilon_u} = \sum_{r=0}^{T-1} \frac{q^r e^{-q}}{r!}.$$

Comment on the physical significance of the value of this ratio when q is small.

1.5 (10) For a random distribution of imaging receptors it was seen that D, γ, and G were all directly dependent on the ratio $N_A a/A$. Discuss in general terms the reasons for this dependence, and why in spite of this the DQE is independent of the ratio. Are there reasons why the analysis for DQE on which such a conclusion is based might prove invalid, especially when the ratio is small?

2. Input/Output Relationships for Conventional Photographic Processes: Experimental Observables

2.1 Properties of Silver Halide Grains

Latent Image Formation

Models of conventional photographic processes are usually based on the assumption that silver halide grains are simply photon receptors and recorders of specified light absorption, size, shape, and photon threshold. When the threshold number of photons has been absorbed, the grains become activated in such a way that after development they have a new absorption, size and shape. A grain in this new state is a single image unit, and further absorption of photons in excess of the threshold makes no additional contribution to this image.

The above simplified picture can provide the basis for models which go a long way towards an understanding of the properties and limitations of conventional photographic processes. Here and in the remainder of this section we shall survey the reasons why grains in silver halide emulsion layers may be assumed to behave in this way, and also consider the type of quantitative details which are found experimentally.

A silver halide crystal (or "grain") consists of a cubic lattice of silver ions and halide ions. For a medium-speed negative emulsion the halide is typically 95% bromide by weight, and 5% iodide. The crystal structure has imperfections and impurities which act as trapping sites for the photoelectrons which are produced following absorption of photons, and which assist in the efficiency of the formation of a developable grain. The energy levels of a silver bromide crystal are shown schematically in Fig. 1. An electron may be raised from the valence to the conduction band by absorption of a photon of sufficient energy ($hv > 2\cdot5\ eV$, or $\lambda < 495\ nm$, approximately), or by thermal fluctuations. Electrons in the conduction band may then be trapped in an energy level corresponding to some crystal defect or impurity.

Following absorption and utilization of a sufficient number of light quanta, the silver halide grain has the property such that it may be preferentially reduced to silver when brought into contact with a suitable reducing agent (the developer). Although all grains will eventually be reduced to silver if developed for sufficient time, the rate of reduction is very much greater for those grains which have absorbed the threshold number of quanta during exposure. This increased rate is due to the presence of the "latent image". A latent image can only be detected by development, and consequently its existence is defined by the particular chemical reducing agent which is used. The theory of the development process is therefore closely interwoven with that of the nature of the latent image.

In 1938 Gurney and Mott[1] suggested the mechanism of latent image formation in general terms, and although there has been much discussion[2] of the details of their theory, there is strong experimental evidence in favour of their basic principle[3]. This is that the latent image is formed as a result of alternate arrival of photoelectrons and interstitial silver ions at specific sites in the grain. A photoelectron released into the conduction band is temporarily trapped at the site. A silver ion is attracted to the site and combines with the electron to

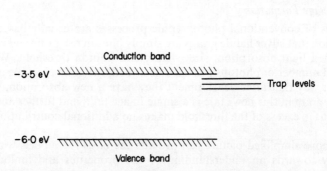

FIG. 1. Schematic representation of the energy levels in silver bromide crystals.

form a silver atom. This process is repeated and the number of silver atoms at the site increases until a stable latent image is formed. At least three or four silver atoms are required to achieve this stable state, and this implies that a minimum of three or four quanta are required to form a developable grain. In general more than this number will be required since not all the photoelectrons which are produced make a contribution to the latent image.

Once a latent image has been formed in a silver halide grain, ensuring a high probability of development, any further quanta which are absorbed by the grain are wasted in so far as they make no further contribution to image formation. Thus silver halide grains are essentially on/off photoreceptors with the transition occurring at the quantum threshold, and with a single image output level. This behaviour is quite different to that of the ideal photoreceptors analysed in Chapter 1, for which there was a gradual transition of the grain image state between the threshold T and the saturation S, and in effect there is now a restriction to the special case $T = S$.

Latent image formation in silver halide grains may be extended to longer exposure wavelengths by the presence of adsorbed spectral sensitizers[4]. However as the sensitivity extends to longer wavelengths there is an increase in the number of grains that become developable as a result of naturally-occurring thermal fluctuations, and so spectral sensitization is generally only practical up to wavelengths of around 1·2 μm.

Other important variables in latent image formation include those of exposure time and intensity. For very long or very short exposure times the process of latent image formation is less efficient, and this is known as reciprocity law failure. This is because a fixed product of intensity and time, (i.e., a given number of exposure quanta) will be associated with different efficiencies of latent image formation as intensity and time are varied in a reciprocal manner. Latent image formation is also less efficient if an exposure is subdivided into many shorter exposures, and this is termed the intermittency effect. These, and other well-known latent image effects have been summarized by Berg[5]

Quantum Yield and Quantum Sensitivity

It is often useful to regard the sensitivity of silver halide grains to exposure light as a two-stage process. There is the primary process which results in the formation of the latent image, and there is the secondary process of development with the very large amplification which is necessary to give the final image. The secondary process of development is such that the order of 10^9 silver atoms are produced for each stable silver atom in the latent image.

The term quantum yield is sometimes used to describe the efficiency of the primary process, and may then be defined as the average number of contributions towards latent image formation per absorbed quantum. If each absorbed quantum produces a photoelectron, then this yield would be unity if every photoelectron always led to a stable silver atom in the latent image. In general however there are processes whereby some of the photoelectrons are wasted and make no contribution, and so in practice the quantum yield is always less than unity.

Sometimes the same term of quantum yield is used to describe the overall effect of both primary and secondary stages, and can then be very much greater than unity. To avoid this confusion of terminology we may call the secondary process the amplification, the overall process then being described by the primary quantum yield and secondary amplification.

It is the high amplification associated with the development process which accounts for the fundamental advantage in sensitivity over other photographic processes based on other chemical and physical mechanisms of image formation. Table I shows some comparative values in this respect. However, since the amplification factor does not include the influence of the primary process, and since the unit of image output (the developed grain for conventional processes) may make a different image contribution for different processes having the same amplification, these factors will not necessarily be proportional to the useful sensitivity, no matter how this is defined. In vesicular photography for example, which is a form of diazo process, the image unit is the light-scattering vesicle and is entirely different to the light-absorbing developed silver grain.

TABLE I. Amplification factors for different types of imaging process.

Process	Amplification factor
Silver halide	10^9
Electrophotography	10^7
Photopolymerization	10^5
Diazo	1

The quantum yield of the primary process will be defined by the reciprocal of the number of absorbed quanta (Q) which are necessary to make a grain developable. It is found experimentally that silver halide grains of nominally the same size and composition cannot be described by a single value of Q for exposures to visible light, but will in fact have a wide spread of Q-values. This is at least partly a consequence of the statistical processes involved in latent image formation, and in Chapter 4 a simple probabilistic model of latent image formation will be described which can account for a wide spread of observed Q-values for nominally identical grains. Since it is likely that there are also basic differences from grain to grain in the efficiency of latent image formation, the natural statistical spread of Q-values will be further broadened.

Due to the wide spread of Q-values it is usual to describe the grains of a given emulsion layer by their sensitivity distribution. This corresponds to the proportion, α_Q, of grains needing to absorb Q quanta to be made developable, where $\Sigma_Q \alpha_Q = 1$, and a plot of α_Q against Q for all observed Q values will provide a statistical description of grain sensitivity. If the average value of Q is $\bar{Q}$, as defined by $\bar{Q} = \Sigma_Q \alpha_Q Q$, then the average quantum yield of the primary process will be $1/\bar{Q}$. Since confusions exist in the literature, it should be stressed here that Q is the number of quanta absorbed by a grain for potential development, and not necessarily the number utilized in latent image formation. This point will be discussed in more detail in Chapter 4.

Absolute measurements of the distribution of Q-values[6,7] indicate that the distribution is not strongly dependent on the size of the grains, or the conventional speed-rating of the process. Two distributions deduced from experimental measurements[7] are shown in Fig. 2, and these are similar, although the speed-ratings of the two films concerned are quite different. The larger grains of the film with higher sensitometric speed will have a larger probability of absorbing quanta, and will also make a bigger unit image contribution. These relationships will be studied in more detail in Chapter 3.

The conclusion that the quantum sensitivities of grains are largely independent of grain sizes is confirmed by measurements of speed as a function of grain size. Figure 3 shows experimental results for a silver bromide emulsion exposed to light[8]. The dashed line shows the relationship which would be anticipated if Q were independent of grain projection area. For mean grain projection area greater than $1 \cdot 0$ μm^2 the experimental curve departs from the theoretical prediction. Experimental evidence indicates that for larger

2. EXPERIMENTAL OBSERVABLES

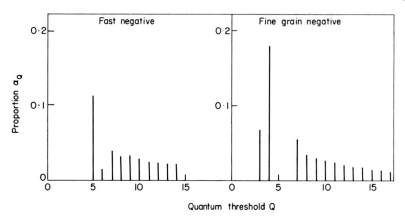

FIG. 2. Quantum sensitivity distributions for two silver halide emulsions[7].

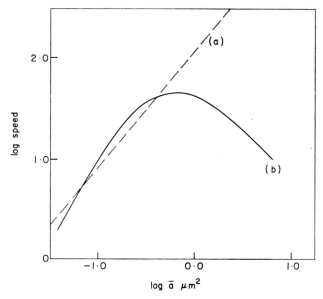

FIG. 3. The variation of speed with grain size for a silver bromide emulsion[8] (a) the theoretical relationship; (b) experimental results.

grains the aggregation of silver to developable latent image sites may be less efficient.

The Grain Size Distribution

Commercially available silver halide films usually contain a wide range of geometrical shapes and sizes. One convenient way of expressing the variation of these geometrical properties is by the distribution of the projection grain area. As we shall discuss shortly, the projection area of the undeveloped grain and the developed image grain are by no means identical, and usually the

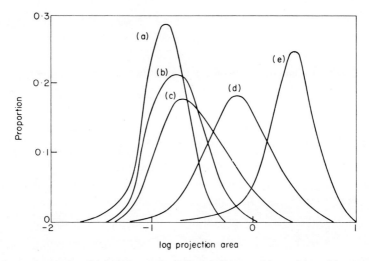

Fig. 4. Grain size distributions for typical silver halide emulsions: (a) positive-type film; (b) fine-grain roll film; (c) portrait film; (d) medium-speed orthochromatic film; (e) X-ray film. The area a is in terms of μm^2.

grain size distribution refers to the grains in the layer prior to development. Practical examples of the size distributions defined in this way are shown in Fig. 4. The proportion of grains having a given projection area is plotted against the logarithm of the area.

The nature of the curves in Fig. 4 is typical, and indicates that the size distribution is approximately log-normal. A survey of the parameters associated with theoretical log-normal distributions and their implications to grain size distributions can be found in reference 9. Examples of the mean and root mean-square projection areas are shown in Table II. Also shown are the average numbers of grains contained in an emulsion volume defined by a cube of side 10 μm. Although the average areas and numbers of grains vary by factors of 100 or so over the range of film types, the product of number times area only varies by a factor of about 2 or 3.

The size and shape of grains may be quite closely regulated by control of the stages involved in the technology of emulsion manufacture. Emulsions

2. EXPERIMENTAL OBSERVABLES

TABLE II. Grain size constants for various film types.[9]

Film type	Mean area μm^2	RMS area μm^2	Number
High-resolution	0·0019	0·00081	—
Motion-picture-pos.	0·07	0·045	576
Positive-type	0·31	0·25	118
Fine-grain roll	0·49	0·58	52
Portrait	0·61	0·75	26
High-speed roll	0·93	0·81	23
X-ray	2·30	1·03	6

containing grains of almost identical size and shape may be produced, as shown in the example of Fig. 5. We shall see later in this chapter and in the next chapter why grains of constant size are desirable to minimize the fluctuations in image density and maximize the detective quantum efficiency.

2.2 Sensitometric Properties of Photographic Layers

Image Density and the Nutting Formula

The most common quantitative measure of photographic output is the optical density of the developed layer. Light of intensity I_O is shone through the image and compared with the transmitted intensity I_T. As in Chapter 1, the density is then defined by

$$D = \log_{10} \frac{I_O}{I_T}. \tag{1}$$

This equation may also be expressed in the forms

$$D = -\log_{10} T = \log_{10} O \tag{2}$$

where T and O are the transmittance and opacity respectively. In practice the density of a given image depends on the geometry of the measuring system and the wavelength over which the density is measured.

The geometry of the measuring system is important because photographic image layers exhibit an appreciable degree of scattering of light. Density is referred to as specular if the illumination and collection angles are small, and as diffuse if one of these angles is large. Since small variations in the measurement system can lead to different measured density values, density is usually measured according to standard specifications[10]. Specular density is always greater than diffuse density, and the ratio of specular to diffuse density, which is thus always greater than unity, is known as the Callier coefficient. Figure 6 shows the variation of the Callier coefficient with density level for a range of image contrasts for a typical photographic layer. The Callier coefficient provides a useful indication of the scattering properties of the developed layer.

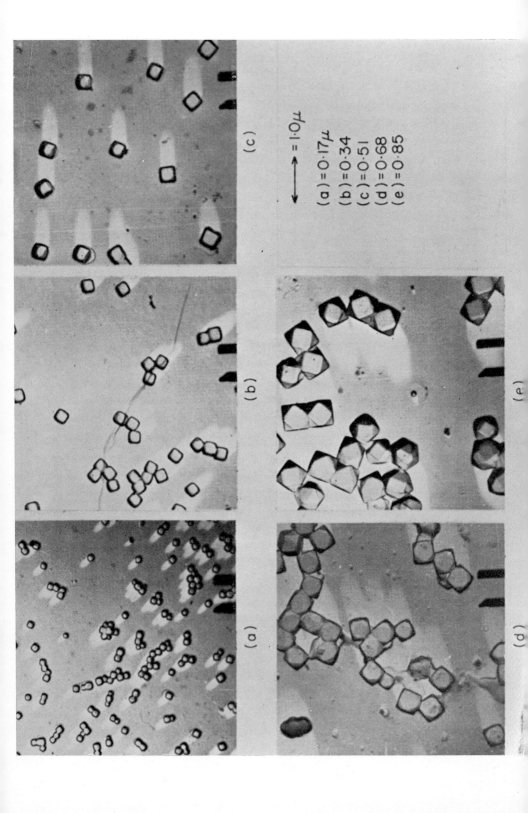

= 1·0 μ

(a) = 0·17 μ
(b) = 0·34
(c) = 0·51
(d) = 0·68
(e) = 0·85

The measured density depends on the wavelength range of the light with which the measurement is made since the developed layer is spectrally selective. If a photocell used for the comparison of intensities has a spectral sensitivity distribution S_λ, and the radiation incident on the image has an energy distribution P_λ, then the measured density will be given by

$$D = - \log_{10}\left(\frac{\int P_\lambda S_\lambda T_\lambda \, d\lambda}{\int P_\lambda S_\lambda \, d\lambda}\right). \tag{3}$$

The detailed design features of practical densitometers are discussed extensively in the literature (see for example reference 11). Densities of very small

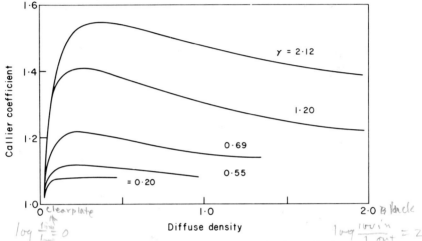

FIG. 6. The Callier coefficient, as a function of diffuse density for various gammas of a fine-grain film (reference 3, page 64).

image areas are measured using a microdensitometer. This is essentially a microscope plus photomultiplier and some means of scanning the image. The detailed features of practical microdensitometer systems will be discussed in Chapter 9.

In 1913 Nutting gave a simple model of the developed photographic layer, expressing the image density in terms of the number and area of the developed grains[12]. This model has proved of lasting benefit, and the Nutting formula and variations of the basic equation are still widely used in photographic science.

As in the simple outline of the Nutting model of Chapter 1, it is supposed that the developed emulsion layer may be divided into a number, k, of elementary layers, each about one grain thick. A random distribution of grains is then assumed for each layer, with the Poisson statistics predicting a negligible degree of overlapping of grains in any one layer. The fact that grains may fall partially in two successive layers is of no importance, and for analytical purposes they are considered to be in the layer that contains most of their volume. The situation is represented schematically in Fig. 7.

Suppose the aperture used to measure the image density has an effective area A at the surface of the image. The number of developed grains in the ith layer included in the area A is denoted by $n_{A,i}$, and the mean projection area

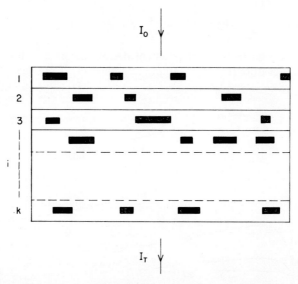

FIG. 7. Representation of the developed photographic layer according to the Nutting model. The layer is arbitrarily divided into k layers each about one grain thick.

of these image grains is denoted by $\bar{a}_i$. The part of area A not covered by grains in the ith layer will be

$$A_i = A - n_{A,i}\,\bar{a}_i,$$

and if the ith layer were isolated from the rest of the layers, its transmittance would be measured as

$$t_i = \frac{A_i}{A} = 1 - \frac{n_{A,i}\,\bar{a}_i}{A}.$$

If it is assumed that in the composite layer each separate layer acts independently in the transmittance sense, then the total transmission will be given by the product

$$T = \prod_{i=1}^{k} t_i = \prod_{i=1}^{k}\left(1 - \frac{n_{A,i}\bar{a}_i}{A}\right),$$

and since density is related to transmission in a logarithmic manner, the total density will be

$$D = -\log_{10} T = -\sum_{i=1}^{k} \log_{10}\left(1 - \frac{n_{A,i}\bar{a}_i}{A}\right). \tag{4}$$

2. EXPERIMENTAL OBSERVABLES

Equation (4) signifies that the total density is equal to the sum of the densities of each elementary layer.

If the product of the number of developed grains and their area is such that

$$\frac{n_{A,i}\bar{a}_i}{A} \ll 1,$$

then equation (4) may be written as

$$\begin{aligned} D &= -\sum_{i=1}^{k} \log_{10}\left(\exp\left(-\frac{n_{A,i}\bar{a}_i}{A}\right)\right) \\ &= \frac{\log_{10}e}{A} \sum_{j=1}^{k} n_{A,i}\bar{a}_i. \end{aligned} \quad (5)$$

We now assume that the mean area $\bar{a}$, is the same for the developed grains in all layers, and hence

$$D = \frac{\log_{10}e}{A} \bar{a} \sum_{i=1}^{k} n_{A,i} = \log_{10}e \frac{n_A \bar{a}}{A}, \quad (6)$$

where n_A now denotes the *total* number of image grains included in the image area A. Equation (6) is in the form usually referred to as the Nutting formula.

The main assumptions included in the Nutting formula are that the transmittances of the separate layers act independently on the light which is shone through the image layer to measure the total density, and that each image grain is opaque and has an absorption area equal to its projection area. The assumption concerning transmittance can be interpreted in two ways. It is as though the discrete grains in each layer were "diluted and blended" into a uniform filter of constant transparency, a succession of such filters giving an exponential decay of the transmitted intensity. The total density would then be given exactly by the additive densities of this series of filters. But such a picture is physically unrealistic, and alternatively, if the scattering of each layer were such that the light emerging from the layer could be considered as completely diffused, then this light would be incident on the next layer without the effects of any correlation of discrete grain opacities (or overlapping) in the successive layers. Since the Nutting formula often works quite well in practice for quite high image densities where the projection of the grains of all layers onto a single plane would involve a high degree of overlapping, it seems that this latter picture may be fairly realistic.

The assumption concerning the opacity and area of a single grain is found to hold quite well if a factor z is included [13-15], such that

$$\bar{a}_{\text{dev}} = z\,\bar{a}, \quad (7)$$

and $\bar{a}_{\text{dev}}$ is the value used in the Nutting formula. This value refers to the effective opaque absorption cross-section, and $\bar{a}$ to the projection area of the undeveloped grains, as taken from normal grain size distributions. The evidence is that z may be in the region of 2–5 for silver halide grains with normal development procedures.

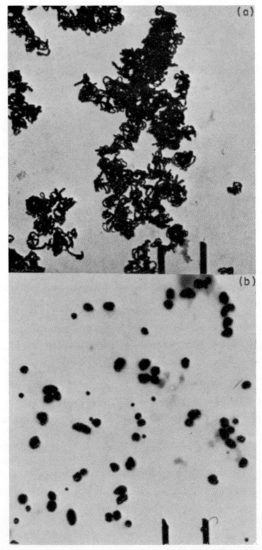

FIG. 8. The structure of developed silver for: (a) predominantly direct development; (b) predominantly physical development.

Examples of the structure of developed silver are shown in Fig. 8, for the two well-known types of development. So-called physical development tends to produce developed grains with compact structure resembling the undeveloped grain. On the other hand so-called direct development produces image grains with a filamentary structure which is quite different to that of the undeveloped grain. Developers in common use combine both physical and direct development, resulting in features of each in the image grain.

The developed grains scatter light as well as acting as absorbers. If z is to be incorporated in the Nutting formula as a constant factor it should be independent of grain area. Recent measurements due to Berry and Skillman[15] indicate that this may not be the case. Using a series of monolayer emulsions they measured the absorption and scattering cross-sections of grains as a function of their size, and some of their results are shown in Fig. 9. The absorption and scattering cross-sections are shown in proportion to the geometrical cross-section of the undeveloped grain. The absorption cross-section stays proportional to the projection area, but the scattering cross-section increases disproportionately with grain size.

A detailed investigation of the optical density of diffusing has been given by Salib et al.[40].

In spite of all the assumptions concerning the properties of grains and layers of grains, the Nutting formula works quite well in practice, and although the

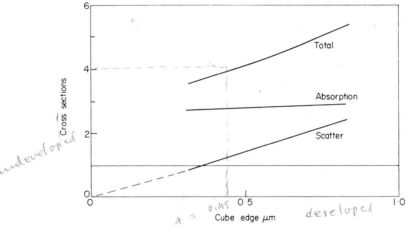

FIG. 9. The ratio of optical cross-sections for scattering and absorption to the geometrical cross-section, as a function of edge-length for developed cubic silver grains[15]. The geometrical cross-section is represented by the horizontal straight-line at unit cross-section.

exact physical significance of z may be dubious, it can at least be interpreted as a correction factor. The essential point of the Nutting formula is that the image density is roughly proportional to the product of number and area of image grains. We see now the significance of the first-order independence of this product for the film-types of Table II. The numbers there related to the total number of grains, indicating a fairly constant value of the maximum image density for conventional film types.

The covering power of a photographic layer is often used as a convenient measure of the efficiency of producing image density from a given total grain volume. It is defined as the ratio D/M, where M is the mass of silver per unit area, which, for an average of N_A grains per area A, will be given by

$$M = \frac{N_A}{A} \bar{v} \rho_{Ag},$$

where $\bar{v}$ is the mean grain volume and ρ_{Ag} is the mass per unit volume of silver. Using the Nutting formula modified by the inclusion of the factor z, and in terms of N_A to give the maximum density,

$$\frac{D}{M} = \frac{z\log_{10}e}{\rho_{Ag}} \frac{\bar{a}}{\bar{v}} \qquad (8)$$

The ratio of projection or surface area to volume increases with decrease of grain dimensions, and consequently the covering power is greater for smaller grains.

The Exposure Scale

The image density is usually referred to the amount of exposure which was necessary to produce it. The exposure E may be defined as the product of the intensity I of the exposure radiation and the duration t of the exposure:

$$E = It. \qquad (9)$$

We have already mentioned the latent image effects, whereby for a given photographic layer, a constant product of intensity and time does always lead to the same output density. However for a wide range of exposure times, typically from 10^{-3}–10^1 seconds, the density depends only on E, and not on I and t separately.

The exposure is usually specified in terms of one of three types of units:

(a) photometric units of metre-candela-seconds or lumens. s/m², etc;
(b) radiometric units such as ergs/cm²;
(c) total number of quanta, as in quanta/μm².

Photometric units are commonly used when referring the image density to the visual impression, since they make allowance for the visual properties of a standard observer. When the photographic process is used as a detector, radiometric units are often appropriate, as for example, in the measurement in terms of the number of exposure quanta. The latter is also useful to the photographic scientist wishing to relate the image density characteristics to the quantum mechanisms of latent image formation, or concerned with the practical evaluation of detective quantum efficiency.

If the spectral energy distribution, E_λ, of the exposure source is known, then it is possible to convert from any one measure of exposure to any other, although of course photometric exposures are only defined within the visible region of the spectrum. Useful working relationships may be established between the different units[16,17,18]. For example,

$$E_r = E_p \frac{1}{10^4 K_m} \frac{\int E_\lambda \, d\lambda}{\int E_\lambda V_\lambda \, d\lambda}, \qquad (10)$$

where

E_r denotes the radiometric exposure in units of ergs/cm²;
E_p denotes the photometric exposure in m.cd.s;
K_m is the maximum value of the luminous efficiency of radiation (approximately $6{\cdot}8 \times 10^{-5}$ cd.s/erg);
V_λ is the relative luminous efficiency of the eye of the standard observer (the maximum value is unity at $\lambda = 0{\cdot}55\ \mu$m).

The number of exposure quanta may be calculated from the radiometric exposure by

$$E_q = E_r \frac{10^{-8}}{hc} \frac{\int E_\lambda \lambda\, d\lambda}{\int E_\lambda\, d\lambda}, \qquad (11)$$

where E_q is the exposure in quanta/μm²;
E_r is again in units of ergs/cm²;
$h = 6{\cdot}62 \times 10^{-27}$ erg.s denotes Planck's constant;
$c = 3 \times 10^{10}$ cm/s denotes the velocity of light;
λ is in microns.

If we substitute the values of the constants in equations (10) and (11) we arrive at the approximate equations:

$$E_r = E_p\, 1{\cdot}47\, \frac{\int E_\lambda\, d\lambda}{\int E_\lambda V_\lambda\, d\lambda}; \qquad (12)$$

$$E_q = E_r\, 5 \times 10^3\, \frac{\int E_\lambda \lambda\, d\lambda}{\int E_\lambda\, d\lambda}; \qquad (13)$$

where the units of E_r, E_p, E_q and λ are as defined above.

When the absolute sensitivity of the photographic process is to be expressed in a fundamental way it is usual to relate the output to that for monochromatic exposures, and in such cases equation (13) reduces to

$$E_q = 5 \times 10^3\, E_r \lambda.$$

This allows for easy conversion between exposure energy and exposure quanta.

The Characteristic Curve

The most common form of expression for the operating characteristics of a photographic process is the curve relating (output) density to the logarithm of (input) exposure, and this is often referred to simply as the characteristic curve. A typical D-log$E$ curve for a medium speed film is shown in Fig. 10.

For a given photographic layer the exact form of the D-log$E$ curve depends on many variables, examples of which are:

(a) the developer composition;
(b) the development time, temperature and degree of agitation;
(c) the wavelength distribution of the exposure;
(d) the nature of variations in exposure over the image area (e.g., whether due to changes in exposure time or exposure intensity);
(e) the geometry and spectral characteristics of the densitometer which is used.

Due to these and other sources of variation, any given characteristic curve is that for the total system under the exact conditions of use, and all input/output relationships and deductions from it likewise relate to the overall system and not just to the properties of the photographic layer.

It is often convenient to extract various parameters from the characteristic curve and to specify the overall response by these selected parameters. A developed layer which has not been exposed will always have a small residual density above that of the base on which the layer is coated. This is known as the

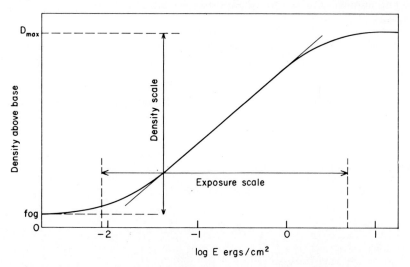

FIG. 10. A typical characteristic curve for a medium speed film.

fog level, and the difference between the maximum density, D_{max}, and fog density is known as the density scale. The exposure scale, or dynamic recording latitude, is usually defined as the $\log E$ range over which the process shows a change in density scale, as indicated in Fig. 10.

The slope of the D-log$E$ curve, $\gamma = \dfrac{d\mathrm{D}}{d(\log_{10}E)}$, is used as a measure of image contrast, and often is fairly constant over a large exposure range. Over such a range we may write

$$D = \gamma\,(\log_{10}E - \log_{10}E_0),$$

where $\log_{10}E_0$ is the intercept of the straight-line region of the D-log$E$ curve with the log$E$ axis. The contrast associated with photographic processes used for normal pictorial purposes is usually given by an average value of the slope of the characteristic curve over some specified exposure range[19].

The sensitometric speed of the photographic process is usually expressed by some measure of the position of the D-log$E$ curve on the log$E$ axis. For pictorial photography various standards exist[20], and naturally these are defined in terms of photometric units. For scientific purposes it is convenient to define

2. EXPERIMENTAL OBSERVABLES

TABLE III. Typical speeds for various types of imaging process.

Process	Speed (ergs/cm^2)$^{-1}$
Silver halide	10^2
Electrophotography	1
Photopolymerization	10^{-6}
Diazo	10^{-8}

speed as the reciprocal of the exposure needed to yield some specific output density, hence with units such as (ergs/cm^2)$^{-1}$ or (quanta/μm^2)$^{-1}$. Altman et al.[39] have discussed both the theoretical and practical relationships between radiometric and photometric speed ratings. Some typical comparative speed values are shown in Table III for the same types of process whose amplification factors were shown in Table I, and for reasons already explained it is seen that there is only weak correlation between amplifications and speeds.

Whilst parameters such as fog, gamma and speed provide a certain amount of useful information about the characteristic curve, they do not define the

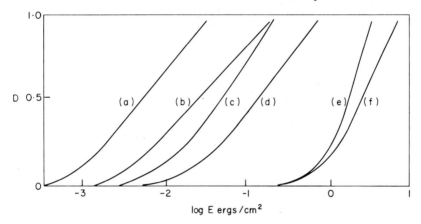

FIG. 11. Characteristic curves for a range of film types from a fast panchromatic film (a) to a microfilm (f), when exposed to radiation of wavelength 420 nm[28].

curve exactly. Using a technique known as eigenvector analysis, Simonds[21] has examined the variability of D-log$E$ curves of a large number of emulsion/development combinations, and has concluded that the shape of most curves may be completely specified by only four parameters, two of which correspond closely to fog and contrast. The curve for any particular emulsion/development combination may then be reconstructed with satisfactory accuracy given the four relevant parameters. The disadvantage of this and other curve-fitting methods is that a large amount of computational work may be required to extract the parameters.

The D-log$E$ curves for a range of conventional silver halide processes are shown in Fig. 11, having much the same shape but with an overall displacement of almost three log$E$ units.

The characteristic curve is particularly useful for displaying the input/output response of a photographic process over the entire exposure scale. The logarithmic nature of the definition of image density makes it a convenient measure of the output for pictorial photography, due to the approximate logarithmic response of the eye to variations in scene brightness. Hence when density is plotted against the logarithm of exposure the slope of the curve is closely related to the visual impression of image contrast. This is why the slope, gamma, is often referred to simply as the contrast.

Another form of input/output expression that is sometimes useful is the plot of density against exposure, rather than the logarithm of exposure. From

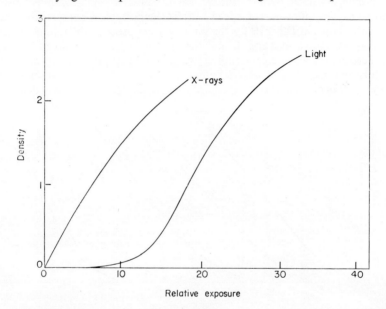

Fig. 12. Density—exposure curves for a typical film as exposed to X-rays and visible light.

equation (28) of Chapter 1, the slope of this curve, g, will be related to that of the D-log$E$ curve by

$$g = \gamma \frac{\log_{10} e}{E}.$$

Figure 12 shows typical D-E curves for normal exposures to light quanta, and also for X-ray quanta. For X-rays the slope is greatest at $E = O$, while that for light increases from zero at $E = O$ to some maximum value before decreasing to zero again. This difference is due to the fact that the quantum sensitivity distributions for photons such as shown in Fig. 2 are not applicable to X-ray recording. Due to the high energy of X-ray quanta, a single quantum can render one or more grains developable, and for a single-quantum process the D-E curve has maximum slope at $E = O$. Further consideration of the photographic characteristics for X-rays in comparison to photons will be given in Chapter 3.

When the photographic process is used in incoherent optical systems the transmittance —exposure $(T - E)$ curve may be of most relevance. This is because these systems are linear in intensity. On the other hand, coherent optical systems are linear in complex amplitude, and the amplitude transmittance is the most relevant quantity. In holography the curve of amplitude transmittance against log exposure $(T_A - \log E)$ is often the most useful

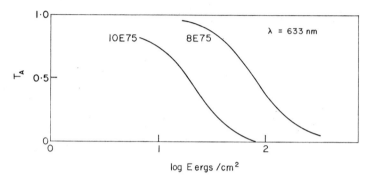

FIG. 13. Amplitude transmittance as a function of log exposure for two films used in holography (Agfa-Gevaert Technical Information, July 1971).

expression. Fig. 13 shows the $T_A - \log E$ curves for two emulsions used in holography.

The depth variation of the layer response plays an important role in determining the characteristic curve of the composite photographic layer. To examine this role we again consider the composite layer as made up of a series of elementary layers. It might be expected that the intensity of radiation during exposure would decrease exponentially with the depth of the layer. In this case the D-$\log E$ curves for each elementary layer would be identical in shape but displaced by equal increments along the $\log E$ axis, as indicated in Fig. 14.

It has been demonstrated that in practice[22] the high degree of exposure scattering within the emulsion layer may lead to an effective exposure such as that indicated in Fig. 15. The most favourably-placed layer of grains for absorbing exposure quanta is not the top layer, but a layer a short way below the surface. For practical depth-distributions such that of Fig. 15, the characteristic curves of the separate elementary layers will not be equally spaced along the $\log E$ axis.

The total displacement of the elementary characteristic curves will be equal to the density, Δ, of the unexposed and undeveloped emulsion for the wavelength of the exposure[23]. If this density is small, the separate curves of Fig. 14 will show little spread, and the shape of the composite curve will approximate to that of an elementary curve. As Δ increases the composite curve will become dominated by the spread of the elementary curves rather than by their

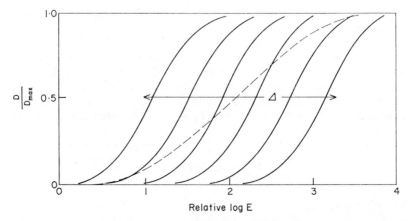

FIG. 14. Characteristic curves for elementary layers (solid lines) and for the composite layer (dashed). The density of the undeveloped layer is denoted by Δ.

shape. The unexposed emulsion density is typically 0·5–1·0, and both the shape and spread of the elementary curves have an influence on the composite curve.

In input/output terms the influence of the variation of the wavelength of the exposure on the quantum sensitivity of the grains is usually expressed in

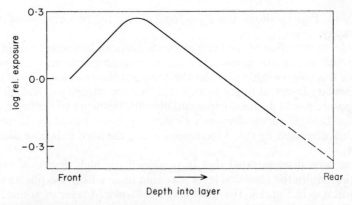

FIG. 15. Measured example of the effective exposure distribution in the emulsion layer, on a logarithmic scale (for details see reference 22).

terms of the variation of the slope or speed of the characteristic curve. Figure 16 shows spectral sensitivity curves for a typical silver halide process. This is based on the reciprocal of the energy needed to produce five different image density levels, as a function of exposure wavelength.

The absorption of quanta by silver halide grains in the near ultra-violet and blue regions of the spectrum will be approximately proportional to the

2. EXPERIMENTAL OBSERVABLES

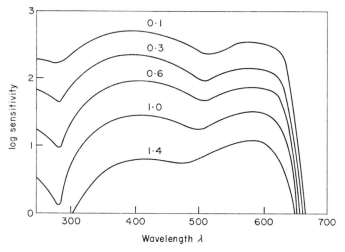

FIG. 16. Spectral sensitivity curves based on five different density levels of a medium-speed panchromatic emulsion (reference 3, page 432).

grain volume. In the region of spectral sensitization the absorption of light is more nearly proportional to the projection area of the grains. The nature of the dependence of gamma on wavelength has been studied by Farnell[24,25].

2.3 Micro-image Properties

Adjacency Effects

It is found that in practice the density at a given region in the image is not only a function of the exposure for that region, but also depends on the exposure and resulting density of adjacent areas. This shows up in various ways in the images of lines, edges, small dots, large dots, etc., and various names have been given to these different manifestations. Two well-known phenomena, the edge effect and the Eberhard effect, are illustrated in Fig. 17. We shall denote all these under the general classification of adjacency effects. A general review of adjacency effects and their image implications has been given by Barrows and Wolfe[26].

Adjacency effects are largely determined by the development conditions, and in particular by local changes in the rate of development and the formation of inhibiting development by-products. A simplified explanation of the edge effect shown in Fig. 17 is that the area just inside the high density side of the boundary will have closer access to fresh developer, while the area just inside the low density side of the boundary will be more affected by the flow of stale developer from the high density side. Both these causes will give image density kinks of positive and negative size on the respective sides of the boundary, and due to their different nature these kinks will generally be asymmetrical.

The implications of adjacency effects on the linearity of the photographic

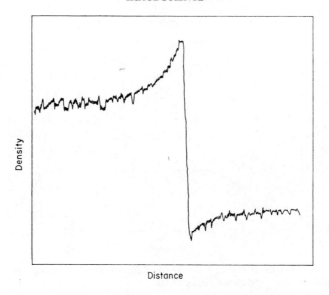
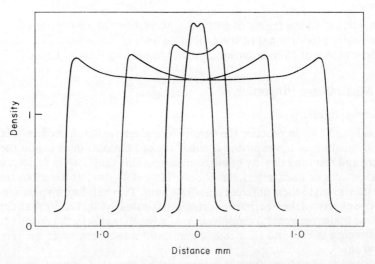

FIG. 17. Illustrations of the edge effect (top) and the Eberhard effect (bottom)[26].

process will be considered in detail in Chapter 7. Here we note that the characteristic curve will be influenced by the geometry of the image areas for which the density is determined[27], as shown in Fig. 18. For areas of less than around 1 mm² down to areas approaching the image resolution area, the density increases as the area of exposure decreases. In general adjacency effects may be minimized by use of a suitable developer type and employing a high degree of agitation during development to assist diffusion in the layer.

Image Resolution and the Spread Function

The resolution area of a photographic image is found to be greater than the area of a single output unit, or grain, by a factor of 10–100 or so. This is due to the scattering of light in the photographic layer prior to being absorbed by grains during exposure. A nominal "point" of exposure light will give rise to an image density distribution which is called the point spread function.

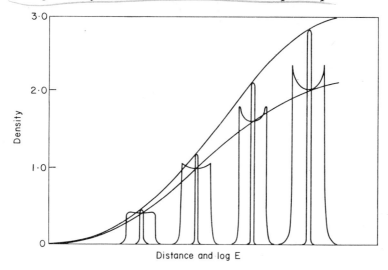

FIG. 18. Characteristic curves for isolated image areas of different size[27].

This may be thought of as the limiting statistical average for the image of many such points, and is usually given as a smooth function.

One of the factors having an important influence on the spread function is the wavelength of the exposure radiation. A model for studying the influence of the various scattering parameters will be studied in Chapter 7. The point spread functions for a range of film types are shown in Fig. 19. The diameters of these point spread functions range from around 3 μm to 20 μm, where the effective diameter is defined at the 25% response level.

An arbitrary spatial exposure distribution on the photographic layer may be considered as being made up of an assembly of small point-like exposures. Ignoring non-linearities the resulting image will be made up of the sum of the respective spread functions, each of the same shape but with scale determined by the intensity at the respective exposure points. Viewed in this manner the point spread function is the "building-block" out of which all images of extended objects are constructed. For one-dimensional objects (lines and edges) this building-block will be the equivalent line spread function. Figure 20 shows the image of an edge as built up from a series of overlapping line spread functions, each one considered as arising from a fine exposure line in the object. The sum of these line spread functions at any image point gives the magnitude of the image at that point.

56 IMAGE SCIENCE

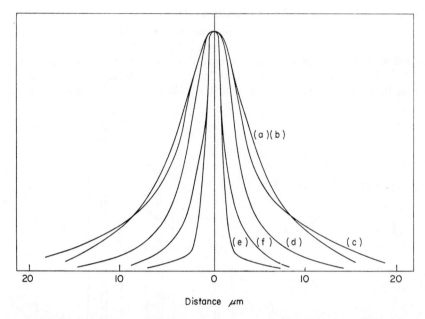

Fig. 19. Point spread functions[28] for the same film-types and exposure conditions as the D-log$E$ curves of Figure 11.

The point spread function sets the upper limit to the number of possible picture points per unit area, and provides the analogy with the resolution area discussed in Chapter 1. For the film types whose spread functions are

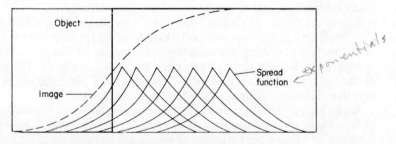

Fig. 20. The image of an edge considered as the summation of an array of line spread functions.

shown in Fig. 19 the number of resolution areas lies in the region of 10^5 to $10^7/cm^2$. Only the shapes of the spread functions are shown in Fig. 19, but scales can be given to them if they all relate to point exposures of equal intensity. In some applications of photography that involve point and line images, for example in the photography of bubble chamber tracks, the speed of a process may be most usefully expressed in terms of the central density of the spread function for a given input exposure. Figure 21 shows two line spread functions on such an absolute scale. The photographic layer with the lower large-scale sensitometric speed-rating proves to have a higher speed-rating so far as line imaging is concerned.

Due to non-linearities the shape of the density profile resulting from a point exposure will depend on both the background exposure level and the height

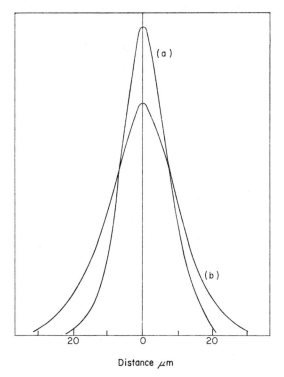

FIG. 21. The line spread functions for two films for line exposures of the same magnitude: (a) has less large-scale sensitometric speed than (b), but has a greater speed (as measured by the central density) for line and point objects.

of the point exposure above this background. Methods of analysis of the spread function and image resolution properties of the photographic process will be considered in detail in Chapter 7. Whereas the spread function defines the image for a given exposure distribution, calculating the image from the individual spread functions will involve a substantial summation problem for an exposure divided into a typical large number of exposure points. The inverse and more usual task of determining the spatial distribution in the exposure from the measured spatial image distribution and with knowledge of the relevant spread function, is even more arduous. This generally calls for the methods of Fourier analysis to be studied in Chapter 6.

Image Density Fluctuations

The utility of the parameter **G** as a measure of image noise was demonstrated in Chapter 1, with **G** defined by the product of the scanning aperture and the measured mean-square density fluctuation:

$$G = A \, \overline{\Delta D_A^2}.$$

Here, for abbreviation and following more common notation we shall write this as

$$G = A\sigma_A^2, \tag{14}$$

where σ_A^2 and σ_A are the mean-square and root-mean-square fluctuations in density.

It is found experimentally, that for conventional silver halide images, G is independent of A for a wide range of aperture sizes. Selwyn[29] established that this would be expected for a random distribution of image grains, as in the practical case, and defined the granularity coefficient

$$S = \sigma_A \sqrt{2A}, \tag{15}$$

which will thus be related to G according to

$$S^2 = 2G. \tag{16}$$

S is often referred to as the Selwyn granularity coefficient.

The ideal grain models studied in Chapter 1 were shown from first principles to have values of G independent of A, provided the aperture is big enough to sample a large number of image grains simultaneously.

According to the Nutting formula the density measured over some area A is proportional to the mean area of the image grains within that area. For a uniformly exposed and developed area of film, both these quantities will fluctuate from region to region as A scans the image, and these fluctuations will give rise to the measured value of σ_A.

Figure 22 shows the Selwyn coefficient plotted as a function of the diameter of the scanning aperture for three different film types, confirming the constancy of S in the practical case. It is also found experimentally that G is directly proportional to the mean developed grain area, and to the image density level. Figures 23 and 24 show typical experimental results. In Fig. 23 σ_A is plotted as a function of cube edge-length for an experimental emulsion with cubic grains of constant size[31], showing a straight-line relationship. In Fig. 24 the Selwyn coefficient is plotted as a function of density level for six film types, both on a logarithmic scale. If G is proportional to D, the slope of the log$S$ – logD curve should be $\tfrac{1}{2}$, which is usually found in practice to a first approximation.

In 1937 Siedentopf[32] derived the relationship

$$G = \log_{10}e \, \bar{a}_D \, D \left(1 + \frac{\overline{\Delta a_D^2}}{(\bar{a}_D)^2}\right). \tag{17}$$

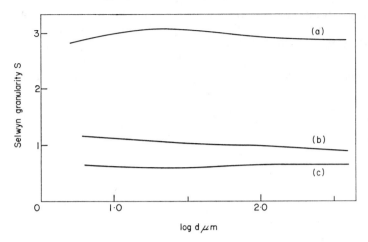

FIG. 22. The variation of the Selwyn granularity coefficient, S, with the diameter of the scanning aperture, d, for[30]: (a) a high-speed film; (b) a portrait film; (c) a fine-grain film.

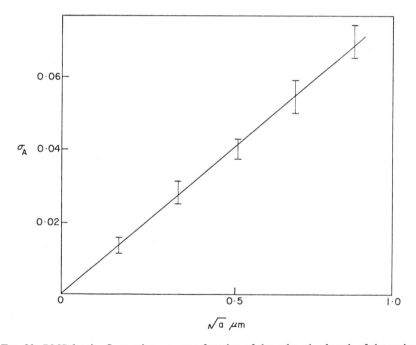

FIG. 23. *RMS* density fluctuation, σ_A, as a function of the cube edge-length of the grains, for a range of experimental emulsions with monodispersed cubic grains[31]. The values are all at an average density of unity and for a scanning aperture A defined by a diameter of 24 μm.

The new subscript ($_D$) has been used to denote the mean and mean-square grains areas for the proportion of grains which contribute to the image at density D. Both $\bar{a}_D$ and $\overline{\Delta a_D^2}$ will depend on the image density level, and the relationship with the undeveloped grain areas will be governed by the factor z. The average value $\bar{a}_D$ will be given by equation (7), where $\bar{a}_{\text{dev}}$ is the value as

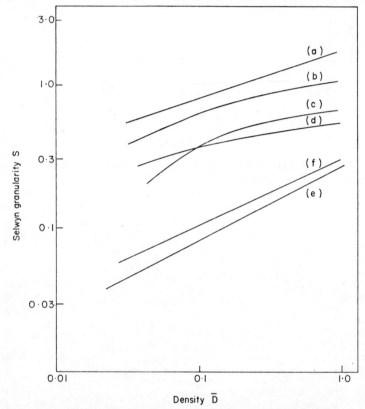

FIG. 24. The Selwyn granularity coefficient, S, as a function of density level[28], for the same film-types and exposure conditions as the D-log$E$ curves and point spread functions of Figs 11 and 19.

averaged for the size distribution of grains which contribute to the image at the density D. In Chapter 3 we shall analyse the dependence of the size distribution of the image grains on density level in more detail. At maximum density where all grains contribute to the image, we would of course expect the size distribution of the image grains to be identical with that for the original unexposed and undeveloped grains, except for the constant factor, z.

Equation (17) explains the nature of experimental results such as those of Fig. 23 and 24. The term in brackets provides a correction factor to the simple formula:

$$G = \log_{10} e \; a_D \, D, \tag{18}$$

2. EXPERIMENTAL OBSERVABLES

which would apply for mono-sized image grains. At a given density level the correction factor will be bigger as the mean-square fluctuation in grain area increases for emulsions of the same average grain size. For mono-sized grains we would expect a linear relationship between G and D, but for grains with a size distribution the relationship is more complex, and will depend on the variation with density of the product of the mean area of the image grains and the correction factor. Experimental results show that for grains with typical

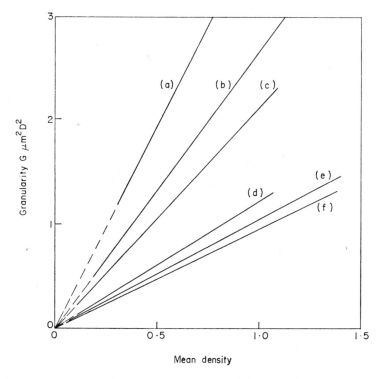

FIG. 25. Noise, as measured by G, as a function of density level for six types of aerial film[35], ranging from (a) a very fast film, to (f) a high-resolution film.

grain size distributions the linearity of G and D holds quite well, especially at low density levels. Figure 25 shows measured values for a range of six aerial films[35]. The slopes of these straight lines are in approximate agreement with the relative average grain sizes of the six different film types.

Detective Quantum Efficiency

As described in detail in Chapter 1, the detective quantum efficiency (DQE) has many advantages as an absolute expression of the image noise level. The practical DQE of photographic processes may be measured using a definition such as that of equation (36) of Chapter 1, which can be written as

$$\text{DQE} = \frac{(\log_{10} e)^2 \gamma^2}{E_q G}. \tag{19}$$

The parameters G and γ in equation (19) must refer to the same exposure level, E_q, which denotes the absolute quantum exposure level per unit image

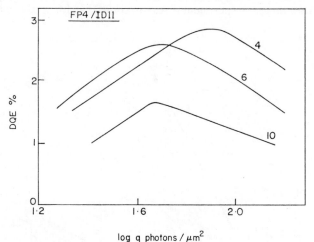

Fig. 26. DQE-log$E$ curves[36] for a medium-speed film (FP4) with development in ID-11 for 4, 6 and 10 minutes.

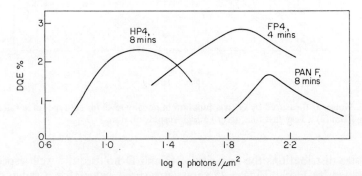

Fig. 27. DQE-log$E$ curves[36] for three films developed for optimum time in ID-11 developer.

area. The experimental measurement of DQE over the complete dynamic exposure latitude thus involves the measurement of G and γ over this range. The units of image area will be the same as those in which $G = A\sigma_A^2$ is defined, and if A is in μm^2, E_q will be in quanta/μm^2.

The first estimations of DQE for practical film types were due to Jones[33] and Fellgett[34] in 1958. These results indicated that the maximum DQE values of a typical range of black-and-white negative films fell within the range of 0·1%–1%. Subsequently Shaw[35] gave measured DQE curves for six aerial-film types, with DQE reaching to just over 1% for two of these films. More recent results of Shaw and Shipman[36] and Vendrovsky et al.[37] indicate that selected film/development combinations can produce a DQE in excess of 2%. Burton et al.[38]

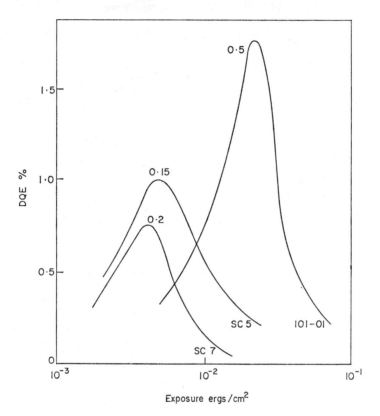

FIG. 28. Variation of DQE with exposure for three Kodak films, SC5, SC7, and 101–01, at $\lambda = 254$ nm[38]. The densities corresponding to the peak DQE values are indicated on each curve.

have published DQE—exposure curves for several Kodak films exposed in the ultra-violet at wavelengths of 120 nm and 254 nm.

Figure 26 shows DQE as a function of log$E$ for a medium speed film with a typical developer, for a range of development times. Figure 27 compares the DQEs of three film types at the respective optimum development times. Figure 28 shows the variation of DQE with exposure for three films exposed in the ultra-violet. The scale and shape of practical curves such as these, the governing factors, and the implications when the photographic process is used as a detector, will be discussed in Chapters 3, 4 and 5.

References

1. Gurney, R. W. and Mott, N. F. (1938). The theory of the photolysis of silver bromide and the photographic latent image. *Proc. Roy. Soc.*, **A164**, 151
2. Mitchell, J. W. (1957). Photographic sensitivity. *Rep. Prog. Phys.*, **20**, 433.
3. Mees, C. E. K. and James, T. H. (1966). 3rd Ed. "The Theory of the Photographic Process", Chapter 5, Macmillan, New York.
4. Reference 3; Chapter 12.
5. Berg, W. F. (1963). Latent image theories. *Photographic Theory*, Section III. Focal Press, London.
6. Farnell, G. C. and Chanter, J. B. (1961). The quantum sensitivity of photographic emulsion grains. *J. Photogr. Sci.*, **9**, 73.
7. Marriage, A. (1961). How many quanta? *J. Photogr. Sci.*, **9**, 93.
8. Farnell, G. C. (1969). The relationship between speed and grain size. *J. Photogr. Sci.*, **17**, 116.
9. Reference 3; page 39.
10. See for example *American National Standard PH. 2.19*: American National Standard diffuse transmission density.
11. Gorokhovskii, Y. N. and Levenberg, T. M. (1965). "General Sensitometry". Focal Press, London.
12. Nutting, P. G. (1913). On the absorption of light in heterogeneous media. *Phil. Mag.*, **26**, 423.
13. Farnell, G. C. and Solman, L. R. (1963). The relationship between covering power at saturation density and undeveloped grain size. *J. Photogr. Sci.* **11**, 347.
14. Romer, W. and Morawiski, T. (1966). The absorption cross sections for light of the silver grains in photographic deposits; II further experimental results. *J. Photogr. Sci.*, **14**, 278.
15. Berry, C. R. and Skillman, D. C. (1971). Cross sections for absorption and scattering of clumps of filamentary silver. *Photogr. Sci. Eng.*, **15**, 236.
16. Keitz, H. A. E. (1971). "Light Calculations and Measurements". Philips Technical Library; (2nd Revised Ed.), Macmillan, London.
17. Derr, A. J. (1963). Energy-sensitivity relationships in the photographic recording of photographic displays. *Photogr. Sci. Eng.*, **7**, 310.
18. Krochman, J. (1970). On the maximum value of the photometric equivalent of radiation. *Optik*, **32**, 205.
19. Niederpruem, C. J., Nelson, C. N. and Yule, J. A. C. (1966). Contrast index. *Photogr. Sci. Eng.*, **10**, 35.
20. See for example *American National Standard PH. 2.5*: American National Standard method for determining the speed of photographic negative materials.
21. Simonds, J. L. (1963). Application of characteristic vector analysis to photographic and optical response data. *J. Opt. Soc. Amer.*, **53**, 968.
22. Berg, W. F. (1969) The photographic emulsion layer as a three-dimensional recording medium. *Appl. Opt.*, **8**, 2407.
23. Burton, P. C. (1951). Interpretation of the characteristic curves of photographic materials. *In* "Fundamental Mechanisms of Photographic Sensitivity", J. W. Mitchell (ed.), Butterworth Press, London.
24. Farnell, G. C. (1954). The variation of gamma with wavelength in the visible region. *J. Photogr. Sci.*, **2**, 145.
25. Farnell, G. C. (1959). Relationship between sensitometric and optical properties of photographic emulsion layers with particular reference to the wavelength variation of sensitivity. *J. Photogr. Sci.*, **7**, 83.

26. Barrows, R. S. and Wolfe, R. N. (1971). Review of adjacency effects in silver photographic images. *Photogr. Sci. Eng.*, **15**, 472.
27. Nelson, C. N. (1971). Prediction of densities in fine detail in photographic images. *Photogr. Sci. Eng.*, **15**, 82.
28. Perrin, F. H. (1961). What is the sensitivity of a photographic system? *J. Soc. Motion Picture and Tel. Engrs.*, **70**: 515.
29. Selwyn, E. W. H. (1935). A theory of graininess. *Photogr. J.*, **75**, 571.
30. Higgins, G. C. and Stultz, K. F. (1959). Experimental study of *rms* granularity as a function of scanning spot size. *J. Opt. Soc. Amer.*, **49**, 925.
31. Ericson, R. H. and Marchant, J. C. (1972). RMS granularity of monodisperse photographic emulsions. *Photogr. Sci. Eng.*, **16**, 253.
32. Siedentopf, H. (1937). Concerning granularity, resolution and the enlargement of photographic negatives. *Physik Zeit.*, **38**, 454.
33. Jones, R. C. (1958). On the quantum efficiency of photographic negatives. *Photogr. Sci. Eng.*, **2**, 57.
34. Fellgett, P. B. (1958) Equivalent quantum efficiency of photographic emulsions. *Mon. Not. Roy. Astr. Soc.*, **118**, 224.
35. Shaw, R. (1965). The equivalent quantum efficiency of aerial films. *J. Photogr. Sci.*, **13**, 308.
36. Shaw, R. and Shipman, R. (1969). Practical factors influencing the signal-to-noise ratio of photographic images. *J. Photogr. Sci.*, **17**, 205.
37. Vendrovsky, K. V., Veitzman, A. I. and Ptashenchuk, V. M. (1972). The relationship of signal-to-noise ratio and quantum efficiency to resolution in photographic layers. *Zh. Nauch. Prikl. Fotogr. Kinematogr.*, **17**, 426.
38. Burton, W. M., Hatter, A. T. and Ridgeley, A. (1973). Photographic sensitivity measurements in the vacuum ultra-violet, *Appl. Opt.*, **12**, 1851.
39. Altman, J. H., Grum, F. and Nelson, C. N. (1973). Photographic speeds based on radiant energy units. *Photogr. Sci. Eng.* **17**, 513.
40. Salib, S. K., DePalma, J. J. and Gaspar, J. (1974). Specular optical density of diffusing media. *Photogr. Sci. Eng.*, **18**, 145.

Recommended Reading

Brown, F. M., Hall, H. J. and Kosar, J. (eds), "Photographic Systems for Engineers". Chapters 1, 2, 9, 10. (2nd Ed., 1969). Soc. of Photographic Scientists and Engineers, Washington D.C.

James, T. H. and Higgins, G. C. (1960). "Fundamentals of Photographic Theory". (2nd Ed., 1960), Morgen and Morgen, New York. Chapters 1–4, 9–11, 13.

Katz, J. and Fogel, S. J. (1971). "Photographic Analysis". Chapter 2.2–2.4. Morgen and Morgen, New York.

Hamilton, J. F. (1972). The Photographic Grain. *Appl. Opt.*, **11**, 13.

Kowaliski, P. (1972). "Applied Photographic Theory", Chapter 2. Wiley, London.

Exercises

Planck's constant, $h = 6 \cdot 62 \times 10^{-34} J s$;
velocity of light, $c = 3 \cdot 00 \times 10^8 \, m \, s^{-1}$.

(1) The ASA Speed of a certain negative film was measured with a source of appropriate wavelength distribution and found to be 200 (consult reference 20). Calculate the number of photons incident per unit film area, within the spectral region 400 nm–600 nm, that were required to yield a density of 0·1 above base and fog.

(2) A developed emulsion layer consists of grains of transmittance t and mean area $\bar{a}$. Calculate the density resulting from N such grains per unit image area. A certain emulsion is coated to thicknesses of 5, 10 and 25 μm on a suitable base. Predict any differences that may exist between the sensitometric and microimage properties of the coatings.

Suppose now that the same basic emulsion is diluted with gelatin solution in such a way that the three coatings each contain the same total mass of silver halide, (i.e., unequal thickness, equal coating weight); what differences might now be expected?

(3) A laser of wavelength 633 nm and 10 mW power is used to expose 5 cm² of a photographic film. It is known that 40 photons/μm² are required to give the appropriate density. Assuming that only 1% of the laser power reaches the film, calculate the necessary exposure time (neglecting reciprocity law failure).

(4) If a photographic film is exposed to a 10 Watt lamp emitting visible light of wavelength 600 nm for 10^{-2} s, calculate the number of silver atoms formed in an isolated silver halide grain 10^{-5} cm in diameter at a distance of 100 cm from the light source, assuming a quantum efficiency of one and that all the incident light is absorbed.

(5) Suppose that a developed layer consists of spherical grains of silver of radius 0·5 μm, and that the Nutting formula for image density is applicable. Plot a graph of density against mass of silver per unit image area (assume $\rho_{Ag} = 2\cdot6$ g cm^{-3}).

(6) The results of granularity measurements made on uniformly exposed and processed samples of a photographic plate using a scanning aperture of 25 μm radius are as shown:

Density	RMS Density
0·14	0·012
0·31	0·018
0·58	0·025
0·78	0·030
0·89	0·032

Use a graphical method to estimate the size of the photographic grains, and state clearly whatever assumptions are involved. What *rms* values would you expect to measure using a scanning aperture of 40 μm radius?

(7) For films A and E of Fig. 19, estimate the number of picture points in a 35 mm frame (36 × 24 mm). If a letter can be imaged legibly inside an area of 100 × 100 picture points, estimate for both films the image area required to store the contents of this book.

2. EXPERIMENTAL OBSERVABLES 67

(8) Use the curves of Fig. 25 to estimate the grain sizes of the range of films. For films (a) and (f) plot a graph of *rms* density fluctuation against scanning aperture diameter at a mean density of 1·0.

(9) Plot DQE as a function of $\log E$ for film A of those whose characteristic curves are given in Fig. 11, and whose noise characteristics are defined in F·g. 24. Note that the exposure axis in Fig. 11 is for monochromatic light of wavelength 420 nm. How would you expect DQE to vary with exposure wavelength?

(10) A system consists of an image intensifier tube plus photographic film. The system has a linear $D - E$ curve up to $D = 6·0$, with slope $g = 0·15$ when the exposure is expressed in terms of quanta/μm^2. The $G - D$ curve is also linear with unit slope if G is in terms of $\mu m^2 \ D^2$. Re-write equation (19) for DQE in terms of g, G and the exposure q, and hence find the DQE of the system as a function of exposure.

3. Input/Output Relationships for Conventional Photographic Processes: Analytical Models.

3.1 Models for Single-level Grains

Introduction

The hypothetical ideal imaging processes which were analysed in Chapter 1 were judged to be ideal by the fact that under certain conditions they had DQE values up to the theoretical limit of 100%. This limit could be attained over a wide exposure latitude for a uniform array of identical receptors, each having a gradual transition of image state from a threshold quantum exposure up to a saturation level. Spatial randomization of the receptor array introduced a fluctuation into the input/output relationships for an output involving simultaneous sampling of the image states of a large number of receptors. Although this fluctuation augmented the recorded quantum noise in the image, the calculated DQE values were still in the region of 50%.

The measured DQE values shown in Chapter 2 for conventional photographic processes had DQE values only up to the order of 2%. But a significant fact to emerge from Chapter 2 was the essential on/off nature of individual grains, which restricts them to a single output recording level. This fact alone can explain much of the difference between the DQE values calculated for the ideal arrays and experimental values in the practical case, although it was noted in Chapter 2 that there are two other important sources of input/output fluctuation: the variation in the number of quanta required to produce the single output level; and the variation in the size of the grains.

In this section we shall first investigate the appropriate DQE modifications for single-level grains, ignoring the two other variations. The influence of variable quantum sensitivity will then be introduced, and finally changes will be made in the model to include the distribution of grain sizes. Because the model then approaches closely to the practical case for conventional silver halide photography, we shall then develop in some detail the general input/output relationships including DQE throughout the rest of this chapter.

For single-level grains the functions f_1, f_2 and f_3—in terms of which the input/output characteristics were expressed in Chapter 1—reduce to a much simpler form than the general expressions. The threshold and saturation levels now coincide, and both can be set equal to the quantum level Q, (i.e., $Q = T = S$), representing the required number of exposure quanta per grain

3. ANALYTICAL MODELS

to achieve the single level of recording ($L = 1$). Equations (20), (21) and (22) of Chapter 1 then reduce to:

$$f_1 = f_3 = \sum_0^{Q-1} \frac{q^r}{r!}; \qquad (1)$$

$$f_2 = \frac{q^{Q-1}}{(Q-1)!} \qquad (2)$$

It is recalled that q denotes the average number of exposure quanta per grain.

DQE for Uniform Grain Array

Substitution of equations (1) and (2) into equation (37) of Chapter 1 for the DQE of a uniform array of receptors leads to

$$\varepsilon = \frac{q \left(\frac{q^{Q-1}}{(Q-1)!}\right)^2 e^{-2q}}{\left(1 - e^{-q} \sum_0^{Q-1} \frac{q^r}{r!}\right) e^{-q} \sum_0^{Q-1} \frac{q^r}{r!}}. \qquad (3)$$

This can be abbreviated by defining new functions:

$$s = e^{-q} \sum_0^{Q-1} \frac{q^r}{r!}; \quad t = e^{-q} \frac{q^{Q-1}}{(Q-1)!}, \qquad (4)$$

leading to

$$\varepsilon = \frac{qt^2}{s(1-s)}. \qquad (5)$$

This expression for DQE is equivalent to that for the Zweig checkerboard model[1], which was used in one of the earliest attempts to interpret the DQE of the photographic process. For any fixed value of the quantum requirements of grains, Q, the DQE can be calculated as a function of the average quantum exposure per grain, q.

In the simplest case of only $Q = 1$ required quantum,

$$\varepsilon = \frac{q\,e^{-q}}{1 - e^{-q}}. \qquad (6)$$

Similarly, when $Q = 2$, the expression reduces to

$$\varepsilon = \frac{q^3 e^{-q}}{(1 - (1+q)e^{-q})} (1 + q)^{-1}. \qquad (7)$$

Using expressions such as these DQE can be computed as a function of exposure for any Q-value of interest. Figure 1 shows the cases for $Q = 1, 2, 4, 8, 16, 32$ and 64 quanta, where DQE is plotted against the logarithm of the average quantum exposure level per grain. On the top scale the exposure levels are indicated corresponding to each $q = Q$ value.

An interesting feature of Fig. 1 is that as Q becomes large the maximum DQE value becomes independent of Q. This limiting value is $2/\pi \times 100\%$, or approximately 64%, and occurs at an exposure level of $q = Q - \tfrac{1}{2}$. The proof of this follows from equation (3), but is a rather difficult mathematical problem and is left as an exercise at the end of this chapter. For high values of Q the difference between the DQE curves lies mainly in their latitude along the log exposure scale, the curves becoming narrower with higher Q-values.

These curves again emphasize the difference between detective quantum efficiency (DQE) and other concepts of quantum efficiency as discussed in previous chapters. For constant quantum requirements of $Q = 100$ we might say that the grains had a quantum yield of 1 %. On the other hand the responsive quantum efficiency (RQE) defined as the number of (output) grains per (input) quantum will be a function of exposure level, and although for $Q = 100$

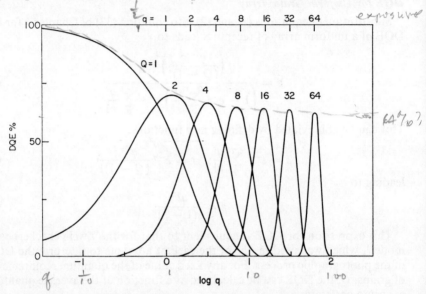

FIG. 1. *DQE* characteristics for a uniform array of grains with identical quantum requirements (Q). The exposure q is in terms of the average number of quanta per grain.

it will be low at all exposure levels, it will not be identical with the quantum yield, and it will be much lower than the DQE at exposure levels where DQE is near its maximum value of 64%. There is no anomaly between a high DQE and low RQE (or vice versa) when we remember that DQE relates input and output fluctuations as opposed to input and output numbers.

DQE for Random Grain Array

Substitution of equations (1) and (2) for single-level grains into equation (52) of Chapter 1 for the DQE of a random array of receptors leads to

$$\varepsilon = \frac{q \left(\dfrac{q^{Q-1}}{(Q-1)!} \right)^2 e^{-2q}}{1 - e^{-q} \sum_0^{Q-1} \dfrac{q^r}{r!}}. \tag{8}$$

3. ANALYTICAL MODELS

Abbreviated in terms of the functions s and t of equation (4), equation (8) reduces to:

$$\varepsilon = \frac{qt^2}{1-s}. \qquad (9)$$

This expression for randomly-distributed single-level grains was derived by Shaw[2] as a next approximation from the Zweig model towards the practical

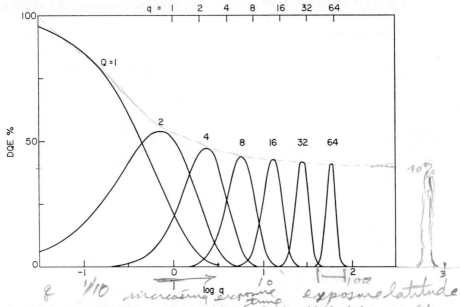

FIG. 2. DQE characteristics for a random spatial array of grains with identical quantum requirements (Q). The exposure q is in terms of the average number of quanta per grain.

case for conventional photographic layers. According to this model the individual grains are still identical in size, and it is assumed that the grains all have equal statistical access to exposure quanta in spite of their random distribution in the three-dimensional photographic layer.

In the simplest case of only $Q = 1$ required quantum,

$$\varepsilon = \frac{qe^{-2q}}{1-e^{-q}}, \qquad (10)$$

and similarly for $Q = 2$:

$$\varepsilon = \frac{q^3 e^{-2q}}{1-(1+q)e^{-q}}. \qquad (11)$$

The DQE curves for the same Q-values as for the uniform grain array are shown in Fig. 2 for the random grains array. There is again a limiting maximum DQE value as Q increases, although this value (which is $(2/\pi)^2 \times 100\%$

or approximately 40%) is approached rather more slowly as Q increases than for the uniform array. There is a similar decrease in log exposure latitude as Q increases.

The quantitative relationship between the DQE curves for the uniform and random grain arrays emerges more clearly by combining equations (5) and (9):

$$\varepsilon_r = s\,\varepsilon_u, \tag{12}$$

where the subscripts denote random and uniform. Inspection of s as defined

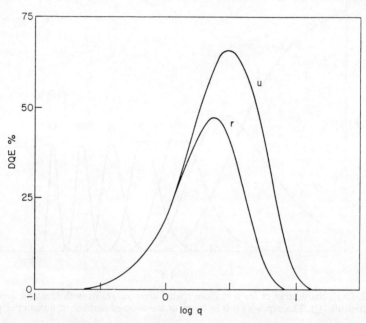

FIG. 3. DQE—log q characteristics for mono-sensitive grains ($Q = 4$). The curves for uniform (u) and random (r) spatial arrays are shown.

in equation (4) shows that it has an upper limit of unity when q is small compared with Q. Figure 3 shows the DQE curves for the uniform and random grain distributions for the individual case of $Q = 4$. At low average quantum exposure levels per grain, s is very close to unity and the two curves coincide. At the higher exposure end of the scale s decreases and this reflects in an increasing divergence as expressed by the ratio of the two curves. This ratio s of the two DQE curves at any exposure level provides a measure of the influence on DQE of the spatial randomness of the grain positions.

Influence of Sensitivity Distribution

The introduction of a single-level constraint on the grains only reduces the maximum DQE values from 100% down to around 40% for random grain arrays with constant Q-values greater than one. However it is necessary to introduce the effect of a spread of quantum sensitivities, since in Chapter 2 it was noted that the spread may be appreciable for conventional silver halide grains exposed to visible light. It is relatively simple to proceed to the appropriate DQE expressions when there is such a distribution of Q values rather than a constant value.[2,3] If as previously α_Q is the proportion of grains requiring Q quanta, where by definition $\sum_Q \alpha_Q = 1$, then the definitions of s and t

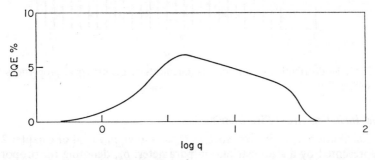

FIG. 4. DQE—log q characteristics for random grain array. The grains have identical size, and there are equal proportions of all quantum requirements from $Q = 4$ to 32 inclusive

according to equation (4) must now be replaced with weighted averages over the range of Q values, i.e.:

$$s = \sum_Q \alpha_Q \left(e^{-q} \sum_0^{Q-1} \frac{q^r}{r!} \right); \qquad (13)$$

$$t = \sum_Q \alpha_Q \left(e^{-q} \frac{q^{Q-1}}{(Q-1)!} \right). \qquad (14)$$

With s and t now redefined in this way, equations (5) and (9) remain valid for DQE when there is a spread of sensitivities. The justification for this simple change will become apparent during the course of this chapter.

An example of the influence of the spread in quantum sensitivity on DQE is shown in Fig. 4. This DQE curve is for a random grain array with an arbitrary spread of sensitivities from 4–32 required quanta inclusive and in equal proportions as represented in Fig. 5. The maximum DQE is now in the region of 6% as opposed to 40% for the mono-sensitivity cases of Fig. 2, although the latitude on the log exposure scale is now greater than that for any of these individual curves.

By use of the random grain model, Spencer[4] calculated DQE curves with maxima of around 8–10% for more practical types of quantum sensitivity distribution than the arbitrary one used in the above example. With the influence of the sensitivity distribution included, we conclude that the computed DQE values are nearer those measured experimentally than are those computed for constant sensitivity. However, the influence of the size distribution of the grains has been ignored in all these calculations, and we shall see that this can bring the computed DQE values even nearer to the practical case.

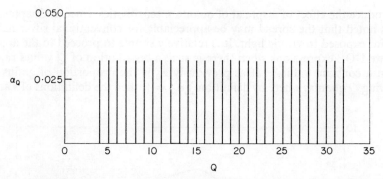

FIG. 5. Model distribution of quantum requirements of grains: equal proportions from 4–32 quanta inclusive.

Model Including Grain Size Distribution

A distribution of grain sizes such as those shown in Fig. 4 of Chapter 2 can be represented by a size-distribution parameter, β_a, denoting the proportion of grains having projection area a. Although these limiting statistical distributions are continuous in nature (unlike the quantum sensitivity distribution which is quantized by definition), discrete sampling is necessary when computing model characteristics, and we shall characterize the size distribution by a summation such that for the appropriate number of sampling intervals

$$\sum_a \beta_a = 1. \tag{15}$$

The mean and mean-square values will be given by the first and second moments of the distribution:

$$\bar{a} = \sum_a \beta_a a; \quad \overline{a^2} = \sum_a \beta_a a^2. \tag{16}$$

The spatial structure of this model with a distribution of sizes is illustrated in Fig. 6, along with the checkerboard model and that for a random array of mono-sized grains. In each case the grain array is shown on the extreme left, then three levels of density up to maximum density are indicated on the right. The structural nature of the intermediate density levels for the top two models depends only on the assumption of equal statistical access to exposure quanta. For the new model with size distribution it will also depend on the relationship between grain size and sensitivity.

The initial assumptions which will be made as a basis for this model are summarized as follows:

(a) all exposure quanta are absorbed by grains;
(b) all grains have equal statistical access to exposure quanta, independent of position within the photographic layer;
(c) the capture cross-section of grains for exposure quanta is equal to the geometrical projection area;

FIG. 6. Illustration of the spatial structure according to three models. The unexposed grain arrays are shown on the left and three stages of image density (including maximum density) are shown on the right.

(d) the sensitivity and size of grains are statistically independent, such that the proportion of grains with sensitivity Q and size a is defined by the product $\alpha_Q \beta_a$;
(e) there is a one-to-one size transform between unexposed grains and image grains;
(f) all image grains are completely opaque.

It remains convenient to define the exposure in terms of the local quantum exposure at the grains. To do this we define the absolute exposure as $E_q = \frac{N_A}{A} \bar{a} q$ quanta per unit image area, where N_A is the average value of the total number of grains included in the area A. In this case assumptions (a) to (c) imply that at the absolute exposure level E_q a grain of unit area will absorb q quanta on average, while a grain of projection area a will absorb aq quanta on average.

Corrections can be made when assumption (a) is not applicable and a random fraction of the total exposure quanta are not absorbed by grains, since there will only be a multiplicative scale change in absolute exposure. For example if only half the quanta are absorbed on average by grains, the total exposure must be twice that defined by $E_q = \frac{N_A}{A} \bar{a} q$ to maintain the same average quantum exposure level, q, per unit grain area. In practical layers there is a relatively low probability for quantum absorption by a grain during a single "collision", but due to the high degree of scattering within the layer during image formation there are many collisions. Scattering in the layer will be investigated in more detail in Chapter 7, and here we are only concerned with the final destination of the quanta and the way this final destination is distributed among the grains of various sizes.

With these assumptions the probability $p_{Q,a}$ that a grain of class (Q, a) will absorb sufficient quanta for latent image formation will now be defined by the Poisson statistics as

$$p_{Q,a} = 1 - e^{-aq} \sum_0^{Q-1} \frac{(aq)^r}{r!}. \tag{17}$$

On the basis of equation (17) it is possible to model the input/output characteristics for conventional photographic processes in quite a realistic manner. In the remainder of this chapter we shall investigate the characteristic curve, image noise and detective quantum efficiency according to this model.

3.2 The Characteristic Curve

Basic Theory

Following the model of Nutting which was outlined in Chapter 2 and which dates back to 1913, characteristic curve theory for silver halide processes has been given much attention. The investigations of Silberstein, Toy, Trivelli, Webb, Burton, Frieser and Klein are outstanding in this respect, and cover the period 1922 to 1960 (see references 5 to 17). The Nutting model only relates the image density to the number and geometry of the image grains, and

3. ANALYTICAL MODELS

the model to be developed here[18,19] essentially defines the exposure scale of the Nutting density formula by considering the probability of grains contributing to the image in terms of the quantum exposure.

The number of grains of class (Q, a) per unit image area which contribute to the image will be equal to the total number of this class times the probability of a contribution:

$$n_{Q,a} = \frac{N_A}{A} \alpha_Q \beta_a p_{Q,a},$$

and the total area of these will thus be

$$a\, n_{Q,a} = \frac{N_A}{A} \alpha_Q \beta_a\, a\, p_{Q,a}.$$

The Nutting formula as in equation (6) of Chapter 2 can now be written in a more general form as the summation of the image contributions from all classes of grains, i.e.:

$$D = \frac{\log_{10} e}{A} \sum_Q \sum_a a\, n_{Q,a}$$

$$= \log_{10} e \frac{N_A}{A} \sum_Q \sum_a \alpha_Q \beta_a\, a\, p_{Q,a}. \tag{18}$$

Substitution of the probability $p_{Q,a}$ from equation (17) then gives

$$D = \log_{10} e \frac{N_A}{A} \sum_Q \sum_a \alpha_Q \beta_a\, a \left(1 - e^{-aq} \sum_0^{Q-1} \frac{(aq)^r}{r!}\right)$$

$$= \log_{10} e \frac{N_A \bar{a}}{A} \left(1 - \frac{1}{\bar{a}} \sum_a \beta_a\, a\, s_a\right), \tag{19}$$

where s_a is defined by

$$s_a = \sum_Q \alpha_Q \left(e^{-aq} \sum_0^{Q-1} \frac{(aq)^r}{r!}\right). \tag{20}$$

Note that for constant grain size equation (20) reduces to that of equation (13).

The ratio $\frac{N_A \bar{a}}{A}$ will be a constant for a given type of photographic layer, and will be denoted by the layer parameter k. This parameter will provide a measure of the amount of silver per unit area or the extent to which the grains are packed within the layer. In terms of k equation (20) becomes

$$D = k \log_{10} e \left(1 - \frac{1}{\bar{a}} \sum_a \beta_a\, a\, s_a\right). \tag{21}$$

The slope of the characteristic curve follows from differentiation of equation (21):

$$\frac{dD}{dq} = -\frac{k \log_{10} e}{\bar{a}} \sum_a \left(\beta_a\, a\, \frac{ds_a}{dq}\right). \tag{22}$$

From equation (20):

$$\frac{ds_a}{dq} = \frac{d}{dq}\left(\sum_Q \alpha_Q \left(e^{-aq} \sum_0^{Q-1} \frac{(aq)^r}{r!}\right)\right)$$

$$= -\sum_Q \alpha_Q e^{-aq}\left(a \sum_0^{Q-1} \frac{(aq)^r}{r!} - \frac{d}{dq}\left(\sum_0^{Q-1} \frac{(aq)^r}{r!}\right)\right)$$

$$= -a \sum_Q \alpha_Q e^{-aq}\left(\sum_0^{Q-1} \frac{(aq)^r}{r!} - \sum_0^{Q-2} \frac{(aq)^r}{r!}\right). \quad (23)$$

Substituting this value of the differential in equation (22);

$$\frac{dD}{dq} = \frac{k \log_{10} e}{\bar{a}} \sum_a \beta_a a^2 (s_a - s_a'). \quad (24)$$

where s_a' is defined by

$$s_a' = \sum_Q \alpha_Q \left(e^{-aq} \sum_0^{Q-2} \frac{(aq)^r}{r!}\right). \quad (25)$$

The definition of s_a' is thus identical to that of s_a, except for truncation one term earlier of the inner summation.

The slope, gamma, of the characteristic curve can now be expressed as

$$\gamma = \frac{dD}{d(\log_{10} q)} = \frac{q}{\log_{10} e} \frac{dD}{dq}$$

$$= \frac{qk}{\bar{a}} \sum_a \beta_a a^2 (s_a - s_a'). \quad (26)$$

If for convenience we introduce the new functions h_1 and h_2, defined as

$$h_1 = \frac{1}{\bar{a}} \sum_a \beta_a a s_a; \quad (27)$$

$$h_2 = \frac{1}{\bar{a}} \sum_a \beta_a a^2 (s_a - s_a'); \quad (28)$$

then equations (21) and (26) reduce to

$$D = k \log_{10} e (1 - h_1); \quad (29)$$

$$\gamma = q k h_2. \quad (30)$$

Both density and gamma are seen to be directly proportional to the layer constant, k. As discussed in Chapter 1 in relation to the image characteristics of ideal photoreceptors, a convenient scale of density results when $k = (\log_{10} e)^{-1} = 2\cdot 30$, since in this case the maximum density is unity when all the grains contribute to the image. Such a value of k would imply that the number and sizes of grains were such that if the grains were packed closely together in a single layer they would occupy an area $2\cdot 30$ times that of the equivalent surface area of the image.

3. ANALYTICAL MODELS

The same density scale of 0–1 emerges by expressing density in terms of the ratio D/D_{max}, where D_{max} denotes the maximum density appropriate to the actual value of k. However, the exact value of k affects the relationship between the absolute exposure and the local quantum exposure per grain. But if the characteristics are expressed on a normalized density scale with the log exposure scale in terms of the quantum exposure at a grain rather than the absolute exposure, these characteristics will be independent of the number of grains, and hence independent of the assumptions made about the value of k. The actual density scale and the absolute log exposure scale may then be derived from these normalized characteristics by scaling according to the relevant layer parameters.

With these assumptions of normalization, equations (29) and (30) reduce to

$$D = 1 - h_1; \tag{31}$$

$$\gamma = 2\cdot 30 \, q \, h_2. \tag{32}$$

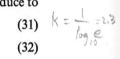

Constant Grain Size and Sensitivity

We have already investigated the DQE characteristics for the random grain model in the special case of identical grain size ($a = 1$) and sensitivity ($Q = $ constant), and we can now study the characteristic curve. From equations (20) and (27):

$$h_1 = s_a = e^{-q} \sum_{0}^{Q-1} \frac{q^r}{r!}; \tag{33}$$

and from equations (25) and (28):

$$h_2 = s_a - s_a' = e^{-q} \frac{q^{Q-1}}{(Q-1)!}. \tag{34}$$

The functions h_1 and h_2 have now reduced to the same expressions as those for s and t of equation (4), as might be anticipated.

Equations (31) and (32) for density and gamma now reduce to:

$$D = 1 - e^{-q} \sum_{0}^{Q-1} \frac{q^r}{r!}; \tag{35}$$

$$\gamma = 2\cdot 30 \frac{q^Q e^{-q}}{(Q-1)!}. \tag{36}$$

The maximum value of gamma follows from differentiation of equation (36):

$$\frac{d\gamma}{dq} = \frac{2\cdot 30 \, e^{-q}}{(Q-1)!} (Qq^{Q-1} - q^Q)$$

$$= \frac{2\cdot 30 \, e^{-q}}{(Q-1)!} q^{Q-1} (Q - q). \tag{37}$$

Inspection of equation (37) reveals that the maximum value of gamma will occur at an exposure level corresponding to $q = Q$; i.e., when the average exposure level per grain matches the fixed quantum requirements of the grain. At this level

$$\gamma_{max} = \frac{2 \cdot 30 \, Q^Q \, e^{-Q}}{(Q-1)!}. \tag{38}$$

This maximum value of gamma will increase as Q increases, as indicated by the examples shown below:

Q	γ_{max}	
1	$2 \cdot 30 \, e^{-1}$	$= 0 \cdot 85$
2	$2 \cdot 30 \, 2^2 \, e^{-2}$	$= 1 \cdot 25$
4	$2 \cdot 30 \, \dfrac{4^4}{3!} \, e^{-4}$	$= 1 \cdot 80$
8	$2 \cdot 30 \, \dfrac{8^8}{7!} \, e^{-8}$	$= 2 \cdot 58$

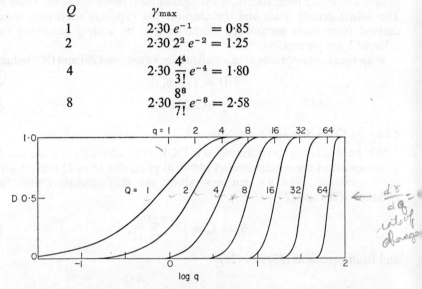

FIG. 7. Density-log exposure curves for mono-sized randomly-distributed grains with constant quantum requirements as shown. The exposure is in terms of the average quantum exposure level per grain, q.

The numerical values of γ_{max} for higher values of Q follow from an approximation formula known as the Stirling formula,

$$x! = \sqrt{2\pi x} \, x^x \, e^{-x},$$

for the factorials of large numbers. There is a further correction factor to this formula, but it is satisfactory in this form for values of $x > 10$ if an error of less than 1% is acceptable. By use of this formula equation (38) becomes

$$\gamma_{max} = 2 \cdot 30 \sqrt{\frac{Q}{2\pi}}.$$

For example, when $Q = 100$, $\gamma_{max} \simeq 9$.

Figure 7 shows the characteristic curves for constant Q-values of 1, 2, 4, 8, 16, 32 and 64, plotted in terms of the normalized density against the logarithm of the average quantum exposure per grain. The increase in slope with increase in Q shows clearly, with the curve tending to a step function for high Q-values. The log exposure spread of these curves is entirely due to the relative spread of the Poisson statistics of the exposure quanta, and in the limit the curve tends to the all-or-nothing step function which would result from a continuous rather than quantized absorption of exposure energy. The log exposure latitude of these curves explains that of the corresponding DQE curves which were shown in Fig. 2.

The curves of Fig. 7 form the basis of characteristic curves for practical layers with distributions of grain sizes and sensitivities.

Variable Grain Sensitivity

When there is a distribution of Q-values, but the grain size is still constant ($a = 1$), then

$$h_1 = s_a = \sum_Q \alpha_Q \left(e^{-q} \sum_0^{Q-1} \frac{q^r}{r!} \right), \tag{39}$$

$$h_2 = s_a - s_a' = \sum_Q \alpha_Q \left(e^{-q} \frac{q^{Q-1}}{(Q-1)!} \right), \tag{40}$$

which have reduced to identical expressions as those for the functions s and t of equations (13) and (14).

If the density and gamma for constant quantum requirements are denoted by D_Q and γ_Q, it follows from equations (31) and (32), and (39) and (40) that density and gamma for a distribution of grain sensitivities may be expressed in terms of the component Q-values as

$$D = \sum_Q \alpha_Q D_Q; \quad \gamma = \sum_Q \alpha_Q \gamma_Q. \tag{41}$$

The density level and gamma at any exposure level can thus be calculated as the sum of the weighted contributions from each component Q-value.

These relationships between the characteristics for constant and variable quantum sensitivity can be illustrated with the aid of a well-defined model sensitivity distribution such as that of Fig. 5. Since this has equal proportions of all Q-values from 4–32 quanta inclusive, the average value of Q is therefore $\bar{Q} = \frac{1}{2}(4 + 32) = 18$ quanta. In Fig. 8 the curves are shown for constant quantum requirements equal to the first, last, and average values of the distribution (i.e., $Q = 4$, 18 and 32). The composite curve for the distribution is seen to be spread out between the first and last Q-values and less steep than that for the average Q-value.

Since for this model distribution all alpha values are identical and equal to $(32 - 4 + 1)^{-1} = 0.0345$; for this case equations (41) become

$$D = 0.0345 \sum_{Q=4}^{32} D_Q; \quad \gamma = 0.0345 \sum_{Q=4}^{32} \gamma_Q.$$

Due to the averaging process, at any exposure level the slope of the curve for the sensitivity distribution is less than that of any of the component curves at the same exposure level. Compared with the $Q = 18$ curve there is a bigger difference at low exposure levels, due to the log exposure spread being greatest for the lowest values of Q.

An instructive way of illustrating the influence of the component curves on the summation curve is shown in Fig. 9. Each component curve is shown normalized to its proportional contribution of $\alpha = 0\cdot0345$ to the maximum density of unity. The $Q = 4$ curve is plotted at the bottom of the density scale, with a density range from 0–0·0345; next the $Q = 5$ curve with a density range from 0·0345 to 0·0690, and so on up to $Q = 32$. The latitude of the overall curve shows up clearly in terms of the latitudes of the component curves. Also, the overall curve fits almost exactly through the locus of half the

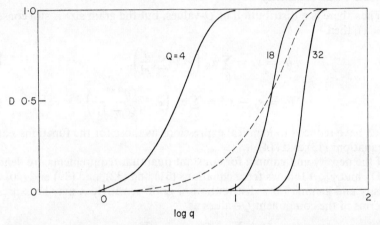

FIG. 8. Characteristic curves for constant quantum requirements of $Q = 4$, 18 and 32 quanta per grain, and for a uniform sensitivity distribution from 4–32 quanta inclusive (dashed curve).

difference between minimum and maximum density values for each component curve, with the exception of the first and last few Q-values.

Further relationships between the spread of the sensitivity distribution, the average sensitivity level, and the latitude of the characteristic curve, are illustrated in Fig. 10. The curve for the model distribution of sensitivities from 4–32 quanta is now shown with that for a uniform distribution from 33–61 quanta. The latter has the same number of successive Q-values, but is shifted along the Q-scale and with average sensitivity $\bar{Q} = \frac{1}{2}(33 + 61) = 47$ quanta, as opposed to 18 quanta previously. As a result of this increase in average quantum requirements the characteristic curve is now steeper, in spite of the same absolute spread in Q-values.

The curve is also shown in Fig. 10 for a uniform distribution of sensitivities which spans both the other two, i.e., from 4–61 quanta inclusive, with $61 - 4 + 1 = 58$ successive Q-values, as opposed to 29, and with $\bar{Q} = 37\cdot5$ quanta. The greater spread in sensitivities results in this curve having less slope and greater latitude than either of the two other curves. A more detailed examination of the relationship between sensitivity spread and curve shape

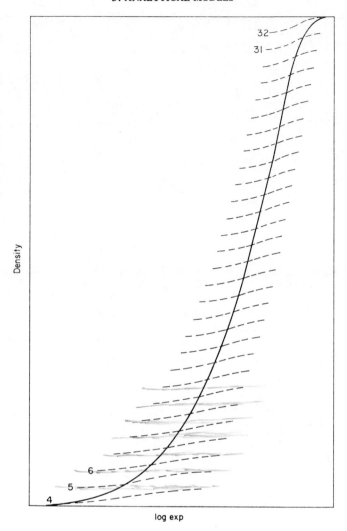

FIG. 9. Characteristic curve for uniform distribution of quantum requirements from $Q = 4$ to $Q = 32$ quanta inclusive, and the separate characteristics in individual proportions.

will be carried out in Chapter 4 in terms of a simple model of latent image formation.

Variable Grain Size

When there is a distribution of grain sizes as well as a distribution of quantum sensitivities, both will play a role in determining the shape of the characteristic curve. The function h_1 in equation (29) for the density will now contain a double summation, as defined by equations (20) and (27), over both the

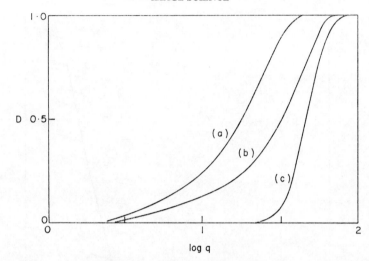

FIG. 10. Characteristic curves for uniform distribution of quantum requirements from: (a) 4–32 quanta; (b) 4–61 quanta; (c) 33–61 quanta.

size and sensitivity distributions. Such double summations can be carried out by computer, and characteristic curves calculated for any combination of the two distributions which is of practical or theoretical interest[19]. Figure 11

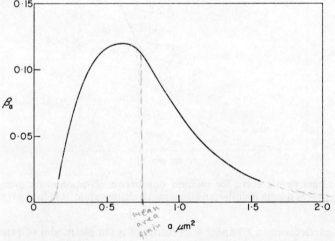

FIG. 11. Typical grain size distribution for a medium-speed film, used as a model for characteristic curve examples[19].

shows a typical distribution of grain sizes for a medium speed film, which will serve to exemplify the influence of the size distribution on the characteristic curve. The cross-section area of the grains is expressed in terms of μm^2, and the mean area is equal to 0.73 μm^2.

Since the average quantum exposure per grain will be a function of grain size, it is now convenient to relate the exposure scale to the absolute exposure level. For example, the exposure scales here have been defined by assuming that $E_q = kq = 2.30\,q$, giving a density scale of 0–1 as discussed earlier, where E_q is the absolute exposure per unit image area and q is the average quantum exposure for a grain of unit cross-section area.

Figure 12 shows the characteristic curves for constant grain size and Q-values of 1, 8 and 64, as in Fig. 7, and also the curves for the model grain size

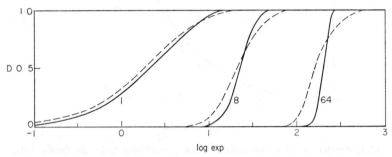

FIG. 12. The characteristic curves for grains with constant quantum requirements: mono-sized grains (solid curves), and grains with a size distribution of Fig. 11 (dashed curves).

distribution. In all three cases the maximum slope is decreased and the latitude is increased with the introduction of the size distribution. The change in curve shape is particularly marked for higher Q-values. Since the slope and latitude of the mono-sensitivity curves are entirely determined by the quantum fluctuations in the exposure when the grain size is constant, it is deduced from Fig. 12 that these exposure fluctuations still dominate the characteristic curve for low Q-values, but that the size distribution dominates for high Q-values.

In Fig. 13 the characteristic curve is now shown when both the grain size and sensitivity are distributed, the size according to the model distribution of Fig. 11 and the sensitivity according to that of Fig. 5. This curve is compared with that for the same sensitivity distribution but with a constant size. The difference between the two curves on removal of the size spread is not as great as that on removal of the sensitivity spread which was shown in Fig. 8. If these examples are realistic it would appear that the sensitivity spread has more influence on the shape of the characteristic curve than the size spread. However, in Chapter 4 we shall see that it is possible to argue that the sensitivity spread itself results from a random selection process operating by the grains on the exposure quanta during latent image formation. In this case the curve shape will relate closely to the quantum fluctuations in the exposure.

As a final observation on the interaction between the grain size and sensitivity spreads, Fig. 14 shows the curves for the same sensitivity distributions as for Fig. 10, but now with the size distribution included. In all three cases the maximum slope is now less and the latitiude is increased.

It was observed in Chapter 2 that the quantum sensitivity of commercial emulsion grains is only a weak function of grain size. Thus the major difference

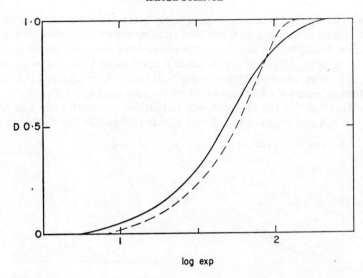

FIG. 13. Characteristic curves for variable size and sensitivity, and when the size distribution is removed (dashed curve).

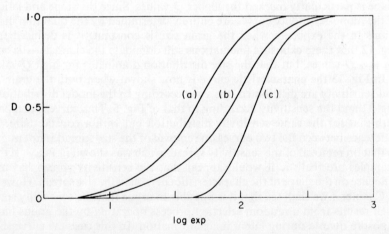

FIG. 14. Characteristic curves for the same sensitivity distributions as those of Fig. 10, but with the size now distributed.

between emulsions of widely different speeds is in grain size: bigger grains have a greater capture cross-section for exposure quanta and also make a greater unit contribution to image density. Since conventional photographic layers have similar requirements of characteristic curve shape for normal use, it is interesting to study the way in which similar curve shapes can be obtained over a wide speed range due to changes only in the size distribution.

The grain size distributions of Fig. 15 consist of the model distribution of Fig. 11, and those obtained by change of the size scale by factors of ½ and 2. Such changes are of practical relevance, since it was noted in Chapter 2 that

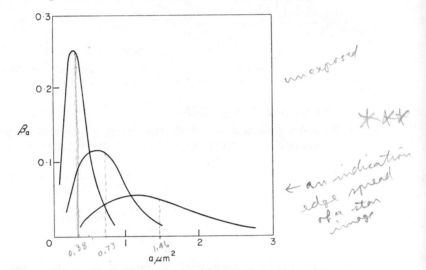

FIG. 15. Model grain size distribution, and those obtained by scaling the area up and down by a factor of 2.

size distributions may be approximately log normal, different distributions thus being related in the style of those of Fig. 15.

The three distributions have mean and root-mean-square grain areas as below:

$\bar{a}$	$(\overline{\Delta a^2})^{\frac{1}{2}}$
μm^2	μm^2
0·363	0·167
0·725	0·335
1·45	0·670

The corresponding changes in the characteristic curve are shown in Fig. 16. The original curve is translated up or down the log exposure axis as the grain size scale is halved or doubled. There is no change in the shape of the curve, so the only difference is one of speed. The log exposure shift will be $\log_{10} 2 = 0·30$ for a change of 2 in size scale: a change of size scale by a factor of 10 would be required for a curve shift of 1 log$E$ unit.

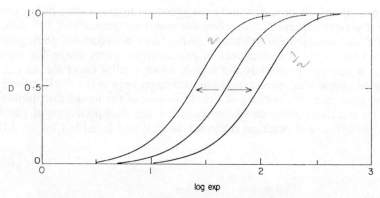

FIG. 16. Characteristic curve for variable grain size and sensitivity, and those corresponding to an increase and decrease in grain size-scale by a factor of 2.

Number and Size of Image Grains

When the grain size is variable the size distribution of the grains in the image will be a function of the exposure level, and not the same as the original distribution of unexposed grains. It follows that the average area, $\bar{a}_D$ of the image grains at the exposure level where the (Q, a) class of grains has probability $p_{Q,a}$ of contributing to the image, will be defined by

$$\bar{a}_D = \frac{\sum_Q \sum_a \alpha_Q \beta_a \, a \, p_{Q,a}}{\sum_Q \sum_a \alpha_Q \beta_a \, p_{Q,a}}. \qquad (42)$$

The ratio of the average area of the image grains to the average area of the unexposed grains will therefore be:

$$\frac{\bar{a}_D}{\bar{a}} = \frac{\sum_Q \sum_a \alpha_Q \beta_a \, a \, p_{Q,a}}{\sum_a \beta_a \, a \sum_Q \sum_a \alpha_Q \beta_a \, p_{Q,a}}. \qquad (43)$$

Similarly, the ratio of the image density D, expressed as a fraction of the maximum density, to the number of image grains, n_I, expressed as a fraction of the total number, will be defined by

$$\frac{D}{n_I} = \frac{\frac{1}{\bar{a}} \sum_Q \sum_a \alpha_Q \beta_a \, a \, p_{Q,a}}{\sum_Q \sum_a \alpha_Q \beta_a \, p_{Q,a}}. \qquad (44)$$

and hence from equation (43),

$$\frac{D}{n_I} = \frac{\bar{a}_D}{\bar{a}} \qquad (45)$$

Since at any exposure level bigger grains have a greater probability, $p_{Q,a}$, of contributing to the image than smaller ones of the same Q-value, it follows that the ratio in equation (45) will always be greater than unity, except at such high exposure levels that grains of all (Q, a) classes contribute to the image. This follows from intuition, since the two size distributions will coincide at maximum density (it is recalled that we have assumed a one-to-one size transform between unexposed grains and image grains). Of course, in the special case where there is no size distribution the ratio of equation (45)

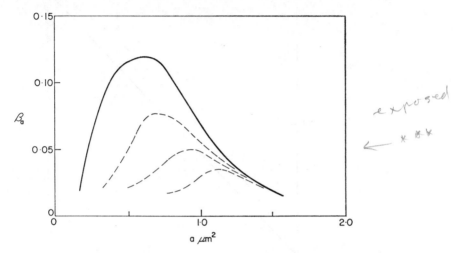

FIG. 17. The size distribution of the unexposed and undeveloped grains, and that of the developed image grains at three levels of image density (dashed curves).

will be equal to unity at all image densities. When the size is variable the mean-square fluctuation of the size distribution of image grains will also be a function of the density level.

Figure 17 shows the model size distribution of unexposed grains, which will also correspond to the size distribution of the image grains at maximum density. Also shown are the size distributions of the image grains at three lower density levels. In Fig. 18 the image density is plotted against the fractional number of image grains, the straightline relationship being that which would exist for mono-sized grains. The curve for the model distribution of grain sizes shows that as predicted a greater image density is associated with any given fraction of image grains: the bigger grains show up earlier in the image and make a bigger contribution to the image density.

Grain-Volume Absorption of Quanta

If analytical models are to predict the image properties with accuracy, it is necessary to relate the capture cross-section of the grains for exposure quanta to the physical properties of the grains. So far this has been assumed to depend only on the geometrical cross-section area. As mentioned in Chapter 2, the evidence indicates that this may be fairly realistic in the wavelength region of spectral sensitization, but that a volume dependence may be more realistic in the blue region. A general solution to include the laws governing the absorption of quanta by grains of all shapes and sizes is not feasible analytically, even if these laws were all known, and there are additional complications such as the dependence of the grain sensitivity itself on the size and shape of the

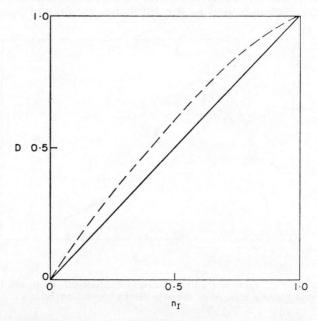

FIG. 18. Density plotted against fractional number of image grains for mono-sized grains and for the size distribution of Fig. 11 (dashed curve).

grains. However the following simple modification to the previous analysis leads to some interesting conclusions.

We now replace the probability $p_{Q,a}$ in equation (17) by a probability $p_{Q,v}$ which is based on grain volume:

$$p_{Q,v} = 1 - e^{-vq} \sum_0^{Q-1} \frac{(vq)^r}{r!}. \qquad (46)$$

The exposure q is now the average quantum exposure per unit grain volume, rather than per unit cross-section area as previously. With q redefined in this way, the absolute exposure is defined by

$$E_q = \frac{N_A \bar{v}}{A} q = k \frac{\bar{v}}{\bar{a}} q, \qquad (47)$$

where k is defined as the ratio $\frac{N_A \bar{a}}{A}$, as before.

Equations (20) and (25) for s_a and s'_a are now replaced by:

$$s_v = \sum_Q \alpha_Q \left(e^{-vq} \sum_0^{Q-1} \frac{(vq)^r}{r!} \right); \tag{48}$$

$$s_v' = \sum_Q \alpha_Q \left(e^{-vq} \sum_0^{Q-2} \frac{(vq)^r}{r!} \right). \tag{49}$$

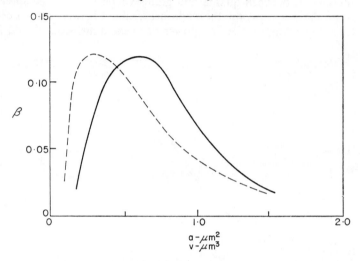

FIG. 19. Relationship between area and volume distributions for model distribution of grain sizes, assuming spherical grains. Dashed curve is the volume distribution.

Similarly, equations (27) and (28) now become

$$h_1 = \frac{1}{\bar{a}} \sum_v \beta_v \, a \, s_v \tag{50}$$

$$h_2 = \frac{1}{\bar{a}} \sum_v \beta_v \, a \, v \, (s_v - s_v') \tag{51}$$

where β_v denotes the distribution of grain volume, with $\sum_v \beta_v = 1$. Note that the cross-section area still appears in equations (50) and (51), since it appears in all products $ap_{Q,v}$ contributing towards image density. This is because we have retained the assumption that it is the cross-section area of an image grain which makes the contribution to image density. With the functions h_1 and h_2 redefined in this way, equations (29) and (30) remain valid for density and gamma, or equations (31) and (32) in the normalized case.

As a simple example it is supposed that all the grains are spherical, and hence that the relationship between volume and cross-section area is simply $v = \frac{4\pi}{3} \left(\frac{a}{\pi} \right)^{\frac{3}{2}}$ for all grains. For the model distribution of Fig. 11 this would produce a volume distribution related to area distribution as shown in Fig. 19.

Figure 20 shows the comparative characteristic curves for these cross-section and volume absorption laws. According to the volume law the curve has slightly lower maximum slope and more latitude. The direction of these changes follows from general reasoning. For a given size distribution the spread of grain volumes (cube of linear dimensions) will be greater than the spread of cross-section areas (square of linear dimensions). It is this spread in the interaction between quanta and grains which governs the spread, or latitude, of the response on the exposure scale. Likewise an absorption mechanism depending on the first power of linear dimensions would imply

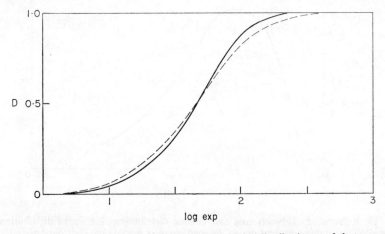

FIG. 20. Characteristic curve for model size and sensitivity distribution, and that assuming volume-absorption of quanta (dashed curve).

less latitude than either, and one depending on the fourth power of linear dimensions would imply more latitude. However it must be remembered that in practice the problem cannot be divorced from the actual grain shapes and sizes and the sensitivity and absorption laws. For example, if the grains were all flat tablets of uniform thickness parallel to the layer surface, the characteristic curves would be the same according to either a volume or cross-section hypothesis.

Grain-Size Amplifications

So far it has been assumed that there is a one-to-one relationship between unexposed emulsion grains and exposed and developed image grains (or more precisely, between the capture cross-section for quanta during image formation and during image densitometry). We suppose now that each grain of area a in the emulsion has area za in the image, where z is a size-amplification constant. From the remarks of Chapter 2, z might be anticipated to be of the order of 2 or 3 for conventional silver images.

Since the area a will remain as previously in the equations expressing the exposure/grain interaction during image formation, equations (20) and (25) for the functions s_a and s_a' will remain the same. Both the numerator and denominator in equations (27) and (28) for the functions h_1 and h_2 will be multiplied by a factor z, and so these will also remain the same. Thus equations (29) and (30) for density and gamma are unchanged. However if the definition of the layer constant k is retained as $k = \dfrac{N_A \bar{a}}{A}$, then the new effective constant in the image density equations will be given by the product zk. It is convenient to retain k as previously, since then the defined relationship between the absolute exposure and the local exposure at a grain remains unchanged. So finally we may write

$$D_z = zk \log_{10} e\, (1 - h_1) = zD; \qquad (52)$$

$$\gamma_z = qzk\, h_2 \qquad\qquad = z\gamma. \qquad (53)$$

These equations relate density and gamma to the values when there is no amplification ($z = 1$). If z were a random variable the average value, $\bar{z}$, would replace z in equations (52) and (53).

When $z = 2$, equation (52) predicts that at any exposure level the image density would be twice that corresponding to the $z = 1$ case for zero amplification. Figure 21 shows the characteristic curve for the model size and sensitivity distributions, and that for a grain-size amplification of $z = 2$. The maximum density will now be $D_{max} = 2$, and the whole curve will be scaled up by this factor of 2. However, we suppose now that the number of grains in the coated emulsion layer is halved, thus halving the parameter k. This would bring the maximum density back to unity, as before with zero amplification. The only change in the characteristic curve would then be that due to a shift of the $\log E$ scale. Remembering the assumption that all exposure quanta are absorbed by grains, less grains means more exposure quanta per grain, and reducing N_A by a factor of $\tfrac{1}{2}$ will also give a reduction of $\tfrac{1}{2}$ in the absolute quantum exposure necessary to produce the same image density. So the curve would be moved down the log exposure scale by $\log_{10} 2 = 0\cdot3$ $\log E$ units, as indicated in Fig. 21.

Another type of effective size amplification is that which is due to high-energy exposure quanta such as X-rays. In this case each exposure quantum may give rise to several developed image grains[20]. The photographic recording of electrons is an analogous case[21]. In order to model the input/output characteristics it is necessary to know the statistics governing the number of grains in one output image unit. Zweig[22] has carried out such an analysis for a Poisson distribution. However, an approximation can be made by assuming that each quantum gives rise to the same number, c, of output grains. It is as though the size of a grain has been amplified by a factor c. According to the Nutting relationship this will imply that a given image density will now be achieved by only a fraction $1/c$ of the quanta which were necessary in the $Q = 1$ quantum case with $c = 1$. So there is an effective exposure scale shift of $\log_{10} c$ between these curves.

Figure 21 shows the $Q = 1$ quantum curve for both $c = 1$ and $c = 5$, and both can be compared with the curves relating to the z-type amplifications and the standard sensitivity distribution. In all practical cases there will be an influence on the curve due to z-type amplifications, but only in special cases

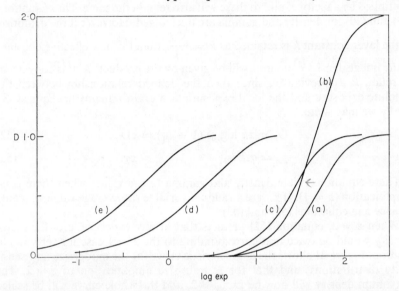

FIG. 21. Characteristic curves for: (a) standard size and sensitivity distributions; (b) with grain area amplification of $z = 2$; (c) with same amplification but half grain population; (d) the $Q = 1$ quantum curve; (e) the $Q = 1$ quantum curve, but with $c = 5$ grains per quantum.

such as for X-rays and electrons will there be a c-type amplification. Both will normally be variables.

Exposure Distribution in the Layer

The initial assumption was that all grains have equal statistical access to exposure quanta, independent of their position in the emulsion layer. The necessary correction to the characteristic curve can be made using the approach discussed in Chapter 2, based on the undeveloped density, Δ, of the emulsion layer.

A simple example is shown in Fig. 22, which is based on a value of $\Delta = 0.5$, and the assumption of an exponential decay in the exposure through the layer. As well as a movement of the overall curve towards higher exposure levels, there is a slight decrease of slope and increase in latitude due to the layer-depth exposure-spread effect. To make a more exact transformation from the thin-layer characteristic curve to that for the composite layer it is of course necessary to know the practical form of the exposure distribution in the layer.

Influence of Fog Grains

In practice there are grains which appear in the image due to chemical or thermal fog, rather than to the action of exposure quanta. A model which allows for the influence of such grains on the image characteristics follows as a simple modification of the present model[23].

When there are fog grains present these can be represented as $Q = 0$ grains, and the sensitivity distribution normalized according to $\sum_{Q=0} \alpha_Q = 1$; the summation now starts at zero rather than at unity as before. The Nutting relationship can now be expressed as the sum of proportional contributions from fog

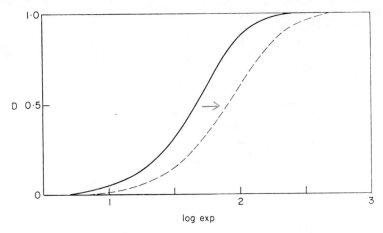

FIG. 22. Standard characteristic curve, and a composite curve assuming exponential exposure decay through the emulsion layer and $\Delta = 0.5$ (dashed curve).

grains, n_f, and from grains due to exposure quanta, n_q, where the respective numbers are related according to

$$n_I = n_f + n_q.$$

If α_f denotes the proportion of fog grains, then

$$\sum_{Q=1} \alpha_Q = 1 - \alpha_f,$$

and in terms of a density scale normalized to unity the fog density will be defined by

$$D_f = \alpha_f.$$

The density will now have a non-zero lower limit of α_f at zero exposure. The other changes in the characteristic curve can be calculated by renormalizing the old alpha distribution with no fog grains present, (i.e., from $Q = 1$ upwards) to $1 - \alpha_f$ rather than 1, and then computing the new curve. For comparative purposes it is necessary to specify the proportion of grains of the various sensitivities which become fog grains, and which are thus removed from the original alpha distribution and in effect placed at $Q = 0$.

To serve as an example two extreme cases were considered by Shaw[23]: equal proportions of grains of all sensitivities becoming fog grains, and the most sensitive grains with lowest Q-values preferentially becoming fog grains. Figure 23 shows results calculated according to these two assumptions for a typical distribution of sensitivities. In the former case the characteristic curves are identical with those for zero fog except for a re-scaling of the density between D_f and D_{max} instead of between 0 and D_{max}. However in the latter case the removal of the most sensitive grains with lowest Q-values to appear as fog grains is equivalent to a loss of the low exposure end of the scale, with

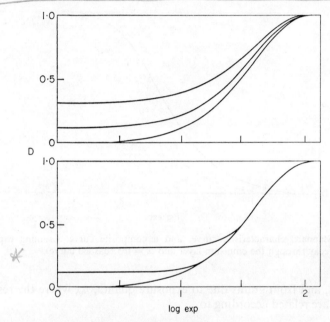

FIG. 23. Influence of fog grains on the characteristic curve[23]. Upper curves: equal proportions of grains of all sensitivities becomes fog grains. Lower curves: grains with lowest Q-values become fog grains. The fraction of fog grains is given by the intercept with the density axis.

little change at the high exposure end of the scale. This comparative behaviour is readily predicted from the curves for individual Q-values as represented in Fig. 9.

3.3 Image Noise Characteristics

Basic Theory

Attempts to model the noise characteristics of conventional silver images date back to the 1930s, with the work of Selwyn[24] and Siedentopf[25] prominent in this respect. Examples of more recent studies can be found in references 26–33. The analysis of the noise characteristics to be developed here follows as an extension[18,19] of the analysis of Siedentopf. Just as the Nutting formula relates the image density to the number and geometry of the image grains, so

3. ANALYTICAL MODELS

the Siedentopf formula relates the image density fluctuations to the fluctuations in numbers and geometry of the image grains. The extension here is thus again essentially that of relating the image grains to the quantum exposure.

In this noise analysis it proves convenient to work in terms of m_A, which denotes the average number of image grains of class (Q,a) per image area A. This can be related to other parameters used to specify the number of grains in the layer. For example, in terms of the total number, N_A, in the unexposed layer:

$$m_A = N_A \alpha_Q \beta_a p_{Qa}. \tag{54}$$

Similarly, in terms of the total number of image grains, n_A, as used in equation (6) of Chapter 2 for the Nutting formula:

$$n_A = \sum_Q \sum_a m_A.$$

From equation (18) the density can be expressed in terms of m_A by

$$D = \frac{\log_{10} e}{A} \sum_Q \sum_a m_A a. \tag{55}$$

Since we wish to analyse the fluctuations in density as recorded by an aperture A scanning a uniformly exposed and developed image area, it is necessary to relate changes in density to changes in m_A from one scanning aperture position to another. From equation (55) we can write

$$dD = \frac{\log_{10} e}{A} \sum_Q \sum_a a \, dm_A,$$

and squaring both sides of this equation:

$$(dD)^2 = \left(\frac{\log_{10} e}{A}\right)^2 \left(\sum_Q \sum_a a \, dm_A\right)^2. \tag{56}$$

Hence the relationship between the mean-square fluctuation in density, $\overline{\Delta D_A^2} = \sigma_A^2$, and in the numbers m_A will be

$$\sigma_A^2 = \left(\frac{\log_{10} e}{A}\right)^2 \sum_Q \sum_a a^2 \, \overline{\Delta m_A^2}. \tag{57}$$

Equation (57) follows from equation (56) by assuming that the image grains have a random spatial distribution which is independent of their size and sensitivity. Assuming Poisson statistics for the spatial distribution of the image grains,

$$\overline{\Delta m_A^2} = m_A, \tag{58}$$

and hence

$$\sigma_A^2 = \left(\frac{\log_{10} e}{A}\right)^2 \sum_Q \sum_a a^2 \, m_A. \tag{59}$$

Substitution of m_A from equation (54), with the probability p_{Qa} defined by equation (17), and using the relationship $k = \dfrac{N_A \bar{a}}{A}$, finally leads to

$$A\sigma_A^2 = G = (\log_{10}e)^2 \frac{k}{\bar{a}} \left(\overline{a^2} - \sum_a \beta_a \, a^2 \, s_a \right), \qquad (60)$$

where s_a is defined by equation (20).

If we introduce the new function defined by

$$h_3 = \frac{1}{\overline{a^2}} \sum_a \beta_a \, a^2 \, s_a, \qquad (61)$$

the noise may be expressed as

$$G = (\log_{10}e)^2 \, k \, \frac{\overline{a^2}}{\bar{a}} (1 - h_3). \qquad (62)$$

The Siedentopf Relationship

There is a close relationship between the image noise as defined by G and the image density level. This emerges after combining equations (55) and (59) in the form

$$G = \log_{10}e \; D \; \frac{\sum_Q \sum_a a^2 \, m_A}{\sum_Q \sum_a a \, m_A}. \qquad (63)$$

The mean and mean-square values of the areas of the image grains at image density level D will be defined by:

$$\bar{a}_D = \frac{1}{n_A} \sum_Q \sum_a a \, m_A; \quad \overline{a_D^2} = \frac{1}{n_A} \sum_Q \sum_a a^2 \, m_A, \qquad (64)$$

and so equation (63) can be expressed as

$$G = \log_{10}e \; D \; \frac{\overline{a_D^2}}{\bar{a}_D}. \qquad (65)$$

As discussed earlier in this chapter, both $\bar{a}_D$ and $\overline{a_D^2}$ will vary with the average image density level, except for an emulsion layer of mono-sized grains.

Since the mean-square fluctuation can be written as

$$\overline{a_D^2} = (\bar{a}_D)^2 + \overline{a_D^2} - (\bar{a}_D)^2 = (\bar{a}_D)^2 + \overline{\Delta a_D^2},$$

where $\overline{\Delta a_D^2}$ denotes the mean-square fluctuation about the mean, it then follows from equation (65) that

$$G = \log_{10}e \; \bar{a}_D \, D \left(1 + \frac{\overline{\Delta a_D^2}}{(\bar{a}_D)^2} \right). \qquad (66)$$

3. ANALYTICAL MODELS

This equation is now identical to the one quoted in Chapter 2, and is known as the Siedentopf relationship.

In the special case where all the grains are identical in size, $a_D = a$, independent of image density level, and then

$$G = a \log_{10} e\, D. \tag{67}$$

This predicts a linear relationship when G is plotted as a function of density, with a slope of $0\cdot 434\, a$. Due to the nature of equations (66) and (67), and since the image density level is the common output parameter, the image noise characteristics are often plotted in terms of a G vs. D curve, and we shall now investigate the influence of some of the model parameters on such a form of plot.

Influence of Layer Parameters

If the image grains were identical in size and of area 1 μm^2, then G would vary between 0 and $0\cdot 434\ \mu m^2\ D$. It is useful to express the units of G as $\mu m^2\ D$ rather than simply as μm^2, even though density is a ratio and not a unit as such. This reminds that the numbers are according to the density scale appropriate to the practical measuring system, and we have already discussed reasons why this can vary from one system to another (diffuse and specular density, etc.).

When the size is distributed it follows from equation (66) that the noise at any image density level will be higher than that predicted by the mean size of either the total grains or of the image grains. At maximum density the mean size and the size distributions are the same for the image grains and the unexposed grains. For example, for the size distributions of Fig. 11:

$$\bar{a} = 0\cdot 725;\quad (\overline{\Delta a^2})^{\frac{1}{2}} = 0\cdot 335;$$

and so at $D_{max} = 1$:

$$G = 0\cdot 434 \times 0\cdot 725 \times \left(1 + \frac{(0\cdot 335)^2}{(0\cdot 725)}\right)$$

$$= 0\cdot 315 \times (1 + 0\cdot 214) = 0\cdot 382.$$

Figure 24 shows the noise characteristics for the three size distributions of Fig. 15. Also shown are the characteristics which would result for monosized grains with area equal to the average values for the respective distributions. The scaling-up of the noise with increase in grain size shows clearly from these plots. When the grain size is constant there is a straight-line plot of G against D, as predicted. There is no such straight-line relationship when the size is distributed because then both the mean and mean-square size of the image grains are functions of image density level.

The influence of the sensitivity distribution on the G vs. D plot of the noise characteristics is less obvious than that of the size distribution. When the grains are mono-sized it follows from equation (67) that the sensitivity distribution has no influence on a G vs. D plot. However it may influence the characteristics when the size is distributed if, according to equation (66), it influences the mean or mean-square size of the image grains at any image density level.

If equations (62) and (29) are combined,

$$G = \log_{10}e \, D \, \frac{\overline{a^2}}{\bar{a}} \frac{(1 - h_3)}{(1 - h_1)}. \tag{68}$$

The functions h_1 and h_3 are both expressed in terms of s_a, which contains the sensitivity distribution α_Q. The sensitivity distribution will thus influence the G vs. D characteristics except when the grain size is constant; in which case

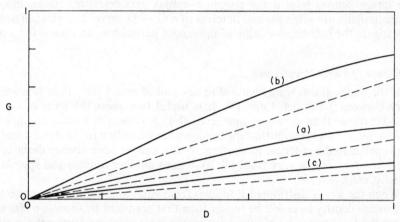

FIG. 24. G vs. D characteristics for: (a) standard size distribution; (b) that scaled up by a factor of two; (c) that scaled down by a factor of two. The straight lines (dashed) are for the cases of constant grain size equal to the respective average sizes.

inspection of the definitions shows that $h_1 = h_3 = s_a$, and hence $(1 - h_3)/(1 - h_1) = 1$.

In general the influence of the sensitivity distribution on the G − D plot is not a large one. Figure 25 shows the noise characteristics for the standard size distribution, and for constant quantum requirements of $Q = 1, 4, 16$ and 64 quanta. It is seen that for higher quantum requirements the noise at any density level will be slightly increased. For $Q = 64$ a limiting upper G-value is approached.

This influence of the sensitivity can be given a general interpretation in terms of the Siedentopf relationship as follows. When the grain quantum requirements are low, the relative fluctuations in the appropriate quantum exposures from grain to grain will be high. This will imply a relatively high spread in the size of the image grains. Thus greater proportions of small grains will contribute to the image density at low density levels than when the quantum requirements are high. So there will be a reduction in the average image grain area, $\bar{a}_D$, of the Siedentopf relationship when Q is low. This more than compensates for an increase in the correction factor due to the ratio of mean-square to squared-mean sizes of the image grains. In the limit when Q is high this is equivalent to a continuous absorption of energy by the grains, the quantum fluctuations are negligible in relative amount, and the G — D curve also takes its limiting form.

Figure 26 shows the G — D characteristics for the model size distribution and for model sensitivity distribution of Fig. 5, and also that for the constant sensitivity of $Q = 18$ quanta, equal to the average value of this distribution. There is only a small difference between the two curves, and in this sense the influence of the sensitivity spread has only a small effect on the noise characteristics. Most other changes in the emulsion layer properties and parameters, such as the grain-volume rather than grain-area assumption for quantum absorption, also have only a small effect on the size distribution of the image

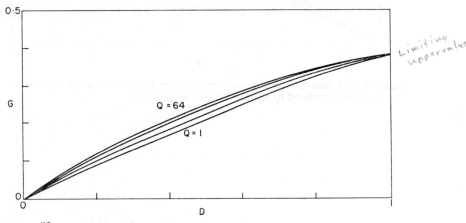

FIG. 25. G vs. D characteristics for standard size distribution with grains having constant quantum requirements of $Q = 1, 4, 16$ and 64 quanta.

grains, and hence on the G — D curve. The exceptions are due to those factors which have an amplifying effect on the output size unit, and these will be considered shortly.

There is an interesting comparison between the G — D curve for this random grain model and that for a uniform array of grains, when both the same constant grain size. The G — D characteristics for a uniform array (studied by several authors[3,30,34]), follows from the analysis of Chapter 1, if the equations are reduced to those for single-level grains.

Figure 27 shows the comparative G — D characteristics for the random and uniform grain arrays with grains of constant size. The intermediate curves follow from Langer[30], who specified intermediate states of the spatial grain distribution by a randomness coefficient varying from 0 for the uniform array to 1 for the random array. When the spatial distribution is completely uniform, G tends to a limiting value of zero at maximum density, as opposed to a limiting value of $(a \log_{10} e)$ for complete randomness. This is illustrated by the respective grain structures of the images at maximum density in Fig. 6.

Trabka[33] has analysed the influence on the noise of the degree of randomness which is lost when the grain population of the layer is high and grains are

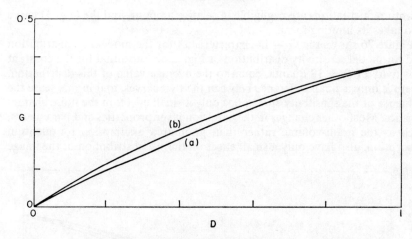

FIG. 26. G − D characteristics for: (a) standard size and sensitivity distributions; (b) sensitivity distribution replaced by mean value of $Q = 18$ quanta.

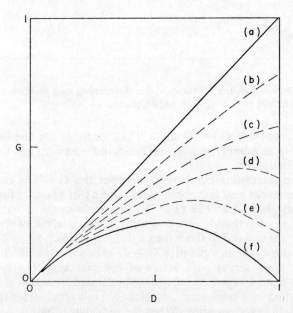

FIG. 27. G − D characteristics for mono-sized grains: (a) distributed randomly; (f) in a uniform spatial array; (b)–(e), for increasing degrees of spatial uniformity[30].

"crowded" into positions which are correlated with those of adjacent grains. Due to such crowding effects some images are found to have G — D curves which reach a maximum value at an intermediate density level before reducing at higher density levels, typically as curve *d* of Fig. 27. Some types of colour images are prone to such effects, since the diffuse dye clouds which form the output units may tend to blend into a continuum. More implications of these and other noise effects will be studied in Chapter 8.

Size Amplification Factors

The influence on the noise characteristics of a z-type amplification of grain size between unexposed grain and image grain will depend on whether z is constant or variable. If z is constant it follows from equation (57) that

$$\sigma_A^2(z) = z^2\, \sigma_A^2(z = 1). \qquad (69)$$

Hence equation (62) now becomes

$$G_z = z^2\, (\log_{10}e)^2\, k\, \frac{\overline{a^2}}{\bar{a}}\, (1 - h_3), \qquad (70)$$

where the layer-constant k and the function h_3 remain defined in terms of the undeveloped grain size, as previously. The noise is thus amplified directly by a factor z^2. However by equation (52) it was seen that the density is directly amplified by a factor z, leading to a simple change in the Siedentopf relationship. For grains of constant size equation (67) now becomes:

$$G_z = z\, a\, \log_{10}e\, D_z. \qquad (71)$$

This equation denotes that the lowest random image unit is now the amplified grain area, za, rather than the area a as before. The slope of the G — D curve will thus change directly with z.

If there is a variable amplification the new G — D curve will depend on the nature of the variation in z. The following simple example will serve to indicate the type of change to be expected. If the grain size is constant at the exposure stage, but z is a random variable with mean value $\bar{z}$, then the mean-value density relationship will depend only on $\bar{z}$ and not on the fluctuation in z, and in relation to the density without amplification, we can write

$$D_z = \bar{z}D. \qquad (72)$$

However it follows from the Siedentopf relationship that

$$G_z = \bar{z}a\, \log_{10}e\, D_z \left(1 + \frac{\overline{\Delta z^2}}{(\bar{z})^2}\right)$$

$$= (\bar{z})^2\, G \left(1 + \frac{\overline{\Delta z^2}}{(\bar{z})^2}\right). \qquad (73)$$

Compared with constant z-amplification as in equation (71), this type of variable amplification is seen to be entirely harmful in the noise sense. The output density level is only amplified by the same amount as for constant amplification, but the noise is increased by the additional factor which depends on the variance in z. As a general principle any spurious source of image fluctuation will increase the noise level in a way such as this.

Amplifications of the c-type, where the output unit is a clump of grains rather than a single grain, will modify the noise characteristics in a similar but not identical way to z-type amplifications. Again, a general analysis is complicated, but a few simplifying assumptions allow insight into the general nature of the associated noise characteristics.

Suppose that the exposure is to X-rays of sufficient energy to produce c grains per X-ray quantum, where c is a random variable with mean value $\bar{c}$. It is assumed that the average number of image grains, n_A, per area A is related to the average quantum exposure by

$$n_A = \bar{c}\, q_A. \tag{74}$$

We are now dealing with a primary process which produces a secondary cascade of events. If the primary-secondary interactions are statistically independent, it has been shown by Mandel[35] that

$$\overline{\Delta n_A^2} = (\bar{c})^2\, \overline{\Delta q_A^2} + q_A\, \overline{\Delta c^2}. \tag{75}$$

In our case the assumption of statistical independence implies that the exposure level is low enough for there to be only a small probability of "overlapping" of adjacent output clumps of image grains.

Assuming that both the X-ray quanta and the number of output grains per quantum are Poisson-distributed,

$$\overline{\Delta q_A^2} = q_A; \quad \overline{\Delta c^2} = \bar{c}, \tag{76}$$

and from equations (75) and (76):

$$\overline{\Delta n_A^2} = q_A \bar{c}\,(1 + \bar{c}), \tag{77}$$

If $\bar{c}$ is large, equation (77) approximates to:

$$\overline{\Delta n_A^2} = q_A\,(\bar{c})^2, \tag{78}$$

From equation (57) it follows that for grains of constant size:

$$G_c = (\log_{10} e)^2\, \frac{a^2}{A}\, \overline{\Delta n_A^2} = (\log_{10} e)^2\, \frac{a^2}{A}\,(\bar{c})^2 q_A = (\bar{c})^2 G; \tag{79}$$

and from the Nutting formula for density:

$$D_c = \log_{10} e\, \frac{a}{A}\, n_A = \log_{10} e\, \frac{a}{A}\, \bar{c} q_A = \bar{c} D. \tag{80}$$

G and D signify the noise and density at the equivalent exposure levels where only one grain per quantum is produced. From equations (79) and (80),

$$G_c = \bar{c} a\, \log_{10} e\, D_c. \tag{81}$$

Equation (81) can be compared with equation (71) for a z-type amplification.

Equation (81) is based on the assumption that the image clumps of grains are non-overlapping. At high exposure levels this will not be true, and at sufficiently high exposure levels to produce maximum image density, the noise will revert to the value it would have without any c-type amplification. There is no such restriction for a z-type amplification, and these differences are illustrated in Fig. 28. In this example $z = \bar{c} = 5$, implying the same degree of amplification in both cases, i.e., the grains appear five times as big in the

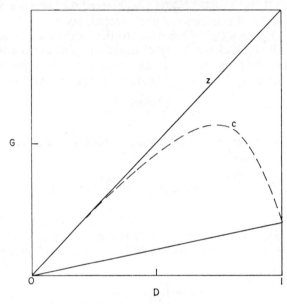

FIG. 28. G − D characteristics for randomly-distributed mono-sized grains, and also for z-type and c-type amplifications of $z = c = 5$.

image, or of the same size but five of them per quantum. The z-type amplification has a straight-line G − D relationship as for the case with no amplification, but with slope five times as great. At low densities the c-curve is identical with the z-curve, but returns to the original value for no amplification when maximum density is reached. Curves such as the c-type case of Fig. 28 are commonly found for X-ray images[36].

Amplifications of the z-type are in principle independent of quantum efficiency and occur for exposures to photons where this may be very low, but c-type amplifications will generally occur only when each quantum produces more than one developable grain, as for X-ray quanta.

In general our model predicts that only amplification changes make a significant change in the G − D characteristics. Other emulsion parameters such as the exposure distribution through the layer have only a minor influence. Fog grains make no basic modification[23], but imply that the G − D curve commences at some finite density level, D_f, and no longer extends back to $G = 0$ at $D = 0$.

3.4 Detective Quantum Efficiency

Formulation

Following the analysis by Zweig[1] in 1961 for the DQE of the checkerboard model, further attempts to interpret DQE according to more realistic models has led to a close understanding of the limitations of conventional photographic processes. In this respect the studies of Bird *et al*, Shaw, Hamilton, and others (see references 2, 3, 18, 23, 34, 37–40) have explored the influence of the most significant properties of conventional photographic layers.

The general DQE expression according to the model used in this chapter in terms of grain size and sensitivity distributions, was derived by Shaw[18], and follows from the expressions already obtained for the constituent parameters. Using the basic definition of DQE, as in equation (19) of Chapter 2,

$$\varepsilon = \frac{\gamma^2 (\log_{10}e)^2}{GE_q},$$

where all the parameters refer to the same exposure level. From equations (30) and (62) for the model:

$$\gamma = qkh_2; \quad G = (\log_{10}e)^2 k \frac{\overline{a^2}}{\bar{a}} (1 - h_3);$$

and the absolute exposure has been defined in terms of the local quantum exposure at a grain by $E_q = kq$, where q is the average number of quanta absorbed by a grain of unit cross-section area. Substitution of these parameters leads to

$$\varepsilon = \frac{q\bar{a}}{\overline{a^2}} \frac{h_2^2}{1 - h_3}. \tag{82}$$

The functions h_2 and h_3 are defined by equations (28) and (61). Since the definition of the absolute exposure assumes that all exposure quanta are absorbed by grains, equation (82) represents an upper limit in this sense.

It is straightforward to show that in the special case of mono-sized grains the expression becomes identical with equation (9) for the simple random grain model, and this has been suggested as an exercise at the end of the chapter.

It is interesting to note that γ, G and E_q are all directly proportional to the layer constant $k = \dfrac{N_A \bar{a}}{A}$, but DQE is independent of this constant. However it is unreasonable to assume that the equation $E_q = kq$ will hold for all values of k. We have already indicated that this value of Eq will represent a lower limit to the absolute exposure which is necessary to produce the local quantum exposure used in the equations for the probability of grains becoming developable. As the number of grains per unit area decreases it will be likely that the number of quanta absorbed by grains will decrease. If we define

$$E_{\text{ABS}} = x \, E_q = xkq, \tag{83}$$

where $1/x$ represents the fraction of quanta absorbed by grains, then for given grain properties and layer thickness x will be a function of k. DQE will then depend on k, since E_{ABS} will be the exposure value relevant in equation (82). However it is unreasonable to specify x out of the exact practical context, and the following investigations of the influence of model parameters will be based on the assumption that $x = 1$.

Influence of Layer Parameters

DQE curves relating to the random grain array have already been considered for the special case of mono-sized grains with constant sensitivity (Fig. 2)

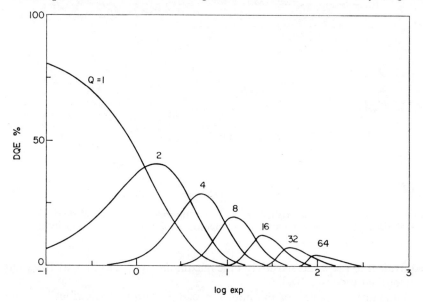

FIG. 29. DQE-log$E$ characteristics for random spatial array of mono-sensitive grains with size distribution of Figure 11.

and with a spread of sensitivities (Fig. 4). When there is a size distribution of grains the first new case of interest is that for constant quantum requirements.

Figure 29 shows the DQE-log$E$ curve for grains with constant quantum requirements, as indicated by the respective Q-values, and the size distribution of Fig. 11. Compared with the equivalent curves of Fig. 2 it is seen that the decrease in maximum DQE due to the size distribution is especially serious for high Q-values. For $Q = 64$ quanta DQE$_{max}$ is reduced from around 40% to around 5%, i.e., by a factor of 8. The general explanation for this is straightforward. High DQE values for identical sized grains with identical low quantum efficiency are mainly due to the high gamma associated with the step-function nature of the characteristic curve. However, the size distribution dominates the characteristic curve for high Q-values, and flattens out the step-like curve with substantial reduction in gamma. Comparison of

Fig. 2 and 29 shows that there is some compensation for this loss of DQE_{max} by an increase in $\log E$ latitude.

When the grain size and sensitivity are both distributed, the DQE-$\log E$ curve is such as that shown in Fig. 30, which relates to the model size and sensitivity distributions of Figs. 5 and 11, and has a maximum value of about 4%. Also shown are the curves for variable sensitivity and constant size, and vice versa. The size distribution now has little influence on DQE since the spread of sensitivities has become dominant. Without the spread of sensitivities, (i.e., by replacing the distribution with the constant value of $Q = 18$,

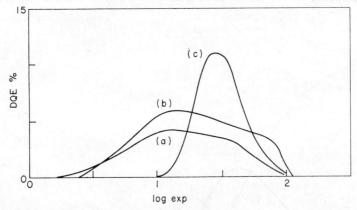

FIG. 30. DQE-$\log E$ characteristics for random grain model: (a) size and sensitivity both distributed; (b) size constant, sensitivity distributed; (c) size distributed, sensitivity constant.

equivalent to the mean sensitivity) the maximum DQE is as high as 12%, but the $\log E$ latitude is halved.

Some further aspects of the influence of the sensitivity distribution when the size is distributed are shown in Fig. 31. Here we see the changes in DQE when the quantum requirements are from 4-61 and 33-61 quanta inclusive, as opposed to 4-33 quanta for the standard distribution: these curves for DQE are thus the equivalent ones for the characteristic curves of Fig. 14. It is seen that there is no loss of maximum DQE when the quantum requirements have a constant amount added (33-61 quanta), but the latitude is halved; whereas when the spread is doubled the maximum DQE is halved but the latitude remains about the same.

A more exact relationship between average sensitivity, sensitivity spread, DQE and latitude will be investigated in Chapter 4. However, it is already clear that the sensitivity distribution has a major influence on DQE. Bird et al.[37] based their analysis not on the sensitivity distribution as such, but in terms of the separate factors such as recombination losses which determine the observed distribution of sensitivities. However, they concluded that increasing latitude by increasing sensitivity spread was a more costly way in DQE terms than by increasing the spread in grain sizes. Hamilton[40] calculated

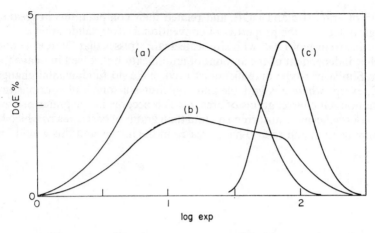

FIG. 31. DQE-log$E$ characteristics for random grain model with standard size distribution and sensitivity distribution: (a) from 4–32 quanta inclusive; (b) from 4–61 quanta inclusive; (c) from 33–61 quanta inclusive.

DQE curves based on a model directly in terms of the detailed mechanisms of latent image formation.

The role of grain size has other interesting implications in the DQE sense. We have already seen the direct influence on the density and noise characteristics brought about by a size scale-shift of the type of Fig. 15. Figure 32 shows

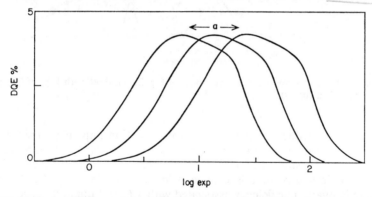

FIG. 32. DQE—log$E$ characteristics for standard size and sensitivity distribution (a), and those corresponding to an increase and decrease in grain size-scale by a factor of two.

that the only change in DQE is a translation of the DQE-log$E$ curve along the log$E$ axis. As discussed in Chapter 2, such changes in grain size distributions are typical of the way in which practical variations of speed are obtained, and hence these DQE curves imply a first-order invariance of DQE with speed, provided there is no appreciable change in quantum sensitivity with grain size. Farnell et al.[41] have computed the relationship between the spread of grain size and DQE, and related their computations in great technological detail to the properties of conventional silver halide grains.

The invariance of DQE with size scale again stresses that DQE is a concept which is independent of the amount of amplification involved in image formation. Similarly z-type amplifications need make no fundamental change in DQE, except where z is variable and constitutes a source of spurious image fluctuation. Of course, grains of area a which become image grains of area za may be less efficient in absorbing quanta than grains of basic area za which have no amplification and so also have area za in the image, and this would imply

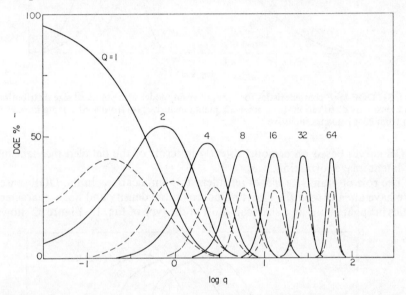

FIG. 33. DQE-log q curves identical with those of Fig. 2, and those with 10% of fog grains present[23] (dashed curves).

a comparative reduction in DQE. Image formation implying c-type amplifications may usually involve very high DQE values, sometimes approaching 100% as will be studied in Chapter 5. However, the amplification itself is not the basic source of this change, which comes from the fundamental increase in quantum efficiency associated with a $Q = 1$ quantum process.

The fundamental inter-relationship of image amplification is with image resolution, as discussed in Chapter 1. A modified and more rigorous definition of DQE which includes the resolution properties will be derived in Chapter 8.

The influence on DQE of the exposure distribution through the composite layer may be analysed as an additional effective spread in quantum sensitivity, and a factor of 2 reduction in maximum DQE might be typical[19]. The difference due to a grain-volume absorption of quanta might be anticipated to be small, as might other emulsion parameters of a similar nature[19]. The influence of fog grains might be expected to modify the DQE curves at low exposure levels where the number of fog grains is of the same order as the number of grains due to the exposure. A simple modification to the model to include fog grains confirms this conclusion[23]. Figure 33 shows the modification to the DQE-log$E$ curves for grains of constant size and sensitivity, under the assumption that 10% of all grains in the layer are fog grains.

3.5 Modelling Unconventional Processes

Although the model used throughout this chapter has been devised to interpret the input/output relationships for conventional photographic processes, the principles involved are quite general and can readily be adapted to any other type of imaging process whose essential imaging parameters are known.

To model the density characteristics (or whatever other form is relevant for the image output) it is necessary to know the way in which density is related to the quantum exposure level. For this we must specifiy the parameters of the imaging process which are effective in absorbing quanta, how many quanta are necessary to make a unit image contribution, and how this contribution relates to image density. The gamma of the process then follows by appropriate differentiation of the analytical expression. Hill and Griesmer[42] have modelled the gamma for a xerographic system in this way. Zweig[43] has modelled the characteristic curves of non-amplifying diazo-type processes.

Likewise to model the noise characteristics it is necessary to know the interaction between the statistics of the exposure quanta and the spatial statistics of the image units, and to translate these into fluctuations in image density. Marathay and Skinner[44] have modelled the noise characteristics for grains having a multilevel of response rather than a single-level with reference to processes such as vesicular photography. Goodman et al.[45,46] have calculated the noise characteristics of conventional photographic processes when used unconventionally, as in holography.

The DQE for any process follows by combining the gamma and noise characteristics with the exposure level in the appropriate manner. In this way Zweig[22] has modelled the DQE for detection systems with amplification, as with phosphorescent screens or photo-cathode detection. Shaw[34] has calculated in detail the DQE advantages of multilevel grain systems which would overcome the single-level limitations of conventional grains. During the course of defining the concept of the ideal noiseless chemical amplifier Bird[47] has analysed the potential DQE advantages of several unconventional photographic processes.

References

1. Zweig, H. J. (1961). Theoretical considerations on the quantum efficiency of photographic detectors. *J. Opt. Soc. Amer.*, **51**, 310.
2. Shaw, R. (1968). Image characteristics of model photodetectors II. *J. Photogr. Sci.*, **16**, 170.
3. Shaw, R. (1967). Image characteristics of model photodetectors. *J. Photogr. Sci.*, **15**, 78.

4. Spencer, H. E. (1971). Calculated sensitivity contributions to detective quantum efficiency in photographic emulsions. *Photogr. Sci. Eng.*, **15**, 468.
5. Silberstein, L. Quantum theory of photographic exposure. (I) (1922) *Phil. Mag.*, **44**, 257. (II) (1922) (co-author Trivelli, A. P. H.) *Phil. Mag.*, **44**, 956. (III) (1923) *Phil. Mag.*, **45**, 1062.
6. Toy, F. C. On the theory of the characteristic curve of a photographic emulsion. I. *Phil. Mag.*, **44**, 352. (1922); II. *Phil. Mag.*, **45**, 715. (1923).
7. Silberstein, L. (1928) Contribution to the theory of photographic exposure. *Phil. Mag.*, **5**, 464.
8. Silberstein, L. and Trivelli, A. P. H. (1938). Relations between the sensitometric and size-frequency characteristics of photographic emulsions. *J. Opt. Soc. Amer.*, **28**, 441.
9. Silberstein, L. The effect of gradual light absorption in photographic exposure. *J. Opt. Soc. Amer.*, **29**, 67, (1939); *J. Opt. Soc. Amer.*, **92**, 326 (1942).
10. Webb, J. H. (1939). Graphical analysis of photographic exposure and a new theoretical formulation on the H and D curve. *J. Opt. Soc. Amer.*, **29**, 314.
11. Silberstein, L. (1942). Uniform representation of the characteristic curves of pure silver bromide emulsions, unsensitized and sensitized. *J. Opt. Soc. Amer.*, **32**, 474.
12. Silberstein, L. and Trivelli, A. P. H. (1945). Quantum theory of exposure tested extensively on photographic emulsions, *J. Opt. Soc. Amer.*, **35**, 93.
13. Trivelli, A. P. H. (1946). Studies in the sensitivity of photographic materials. II. Effects on the shape of the characteristic curves of photographic emulsions. *J. Franklin Inst.*, **241**, 1.
14. Webb, J. H. (1948). Effect of light absorption on the shape of the photographic H and D curve. *J. Opt. Soc. Amer.*, **38**, 27.
15. Burton, P. C. (1951). Interpretation of the characteristic curves of photographic materials. *In* "Fundamental Mechanisms of Photographic Sensitivity". (Ed.) J. W. Mitchell. Butterworth, London.
16. Frieser, H. and Klein, E. (1960). Contribution to the theory of the characteristic curve. *Photogr. Sci. Eng.*, **4**, 264.
17. Klein, E. (1960). Theoretical considerations concerning the relationship between characteristic curves of elementary and thick layers. *Photogr. Sci. Eng.*, **4**, 341.
18. Shaw, R. (1969). Image characteristics of model photodetectors III. *J. Photogr. Sci.*, **17**, 141.
19. Shaw, R. (1972). Image evaluation as an aid to photographic emulsion design. *Photogr. Sci. Eng.*, **16**, 395.
20. Silberstein, L. and Trivelli, A. P. H. (1930). The quantum theory of X-ray exposures on photographic emulsions. *Phil. Mag.*, **9**, 787.
21. Hamilton, J. F. and Marchant, J. C. (1967). Image recording in electron microscopy. *J. Opt. Soc. Amer.*, **57**, 232.
22. Zweig, H. J. (1965). Detective quantum efficiency of photodetectors with some amplification mechanisms. *J. Opt. Soc. Amer.*, **55**, 525.
23. Shaw, R. (1973). The influence of fog grains on detective quantum efficiency. *Photogr. Sci. Eng.*, **17**, 495.
24. Selwyn, E. W. H. (1935). A theory of graininess. *Photogr. J.*, **75**, 571.
25. Siedentopf, H. (1937). Concerning granularity, density fluctuations and the enlargement of photographic negatives. *Physik Zeit.*, **38**, 454.
26. Picinbono, M. B. (1955). Statistical model for the distribution of silver grains in photographic films. *Comptes Rendus*, **240**, 2206.
27. Savelli, M. (1958). Practical results of the study of a three-parameter model for

the representation of the granularity of photographic films. *Comptes Rendus*, **246**, 3605.
28. Marchant, J. C. and Dillion, P. L. P. (1961). Correlation between random-dot samples and the photographic emulsion. *J. Opt. Soc. Amer.*, **51**, 641.
29. Haugh, E. F. (1963). A structural theory for the Selwyn granularity coefficient. *J. Photogr. Sci.*, **11**, 65.
30. Langner, G. (1963). Calculation of the granularity and contrast threshold of photographic layers. *Photogr. Korres.*, **99**, 177.
31. Bayer, B. E. (1964). Relation between granularity and density for a random-dot model. *J. Opt. Soc. Amer.*, **54**, 1485.
32. Schade, O. H. (1964). An evaluation of photographic image quality and resolving power. *J. Soc. Motion Picture and Tel. Engrs.*, **73**, 81.
33. Trabka, E. A. (1971). Crowded emulsions: granularity theory for monolayers. *J. Opt. Soc. Amer.*, **61**, 800.
34. Shaw, R. (1972) Multilevel grains and the ideal photographic detector. *Photogr. Sci. Eng.*, **16**, 192.
35. Mandel, L. (1959). Image fluctuations in cascade intensifiers. *Brit. J. Appl. Phys.*, **10**, 233.
36. Rossman, K. (1962). Modulation transfer functions of radiographic systems using fluorescent screens. *J. Opt. Soc. Amer.*, **52**, 774.
37. Bird, G. R., Jones, R. C. and Ames, A. E. (1969). The efficiency of radiation detection by photographic films: state-of-the-art and methods of improvement. *Appl. Opt.*, **8**, 2389.
38. Shaw, R. (1973). The influence of grain sensitivity on photographic image properties. *J. Photogr. Sci.*, **21**, 25.
39. Shaw, R. (1973). Some detector characteristics of the photographic process. *Optica Acta*, **20**, 749.
40. Hamilton, J. F. (1972). Simulated detective quantum efficiency of a model photographic emulsion. *Photogr. Sci. Eng.*, **16**, 126.
41. Farnell, G. C., Saunders, A. E. and Solman, L. R. The computation of the response of model emulsion layers. (a) Principles. *J. Photogr. Sci.*, **21**, 93 (1973). (b) Applications *ibid.*, **21**, 118 (1973).
42. Hill, E. R. and Griesmer, J. J. (1973). Macroscopic theoretical representation of the gamma for a xerographic system. *Photogr. Sci. Eng.*, **17**, 47.
43. Zweig, H. J. (1965). The behaviour of nonamplifying photographic detectors. *Photogr. Sci. Eng.*, **9**, 371.
44. Marathay, A. S. and Skinner, T. J. (1969). Multilevel-grain model for light-recording media. *J. Opt. Soc. Amer.*, **59**, 455.
45. Goodman, J. W. (1967). Film-grain noise in wavefront-reconstruction imaging. *J. Opt. Soc. Amer.*, **57**, 493.
46. Goodman, J. W., Miles, R. B. and Kimball, R. B. (1968). Comparative noise performance of photographic emulsions in holographic and conventional imagery. *J. Opt. Soc. Amer.*, **58**, 609.
47. Bird, G. R. (1973). Noiseless chemical amplifiers and the ultimate capabilities of organic imaging systems. *Photogr. Sci. Eng.*, **17**, 261.

Exercises

(1) Discuss in general terms why photographic image noise for X-ray recording may be significantly higher than that predicted by the Siedentopf relationship. Explain why in spite of this, the photographic process may be almost ideal in the DQE sense for recording X-rays.

If the number of grains made developable follows a Poisson distribution with average number $\bar{c}$ (noise characteristics defined by equation (77)), show that at low exposure levels the DQE approximates to $\dfrac{1}{1 + \dfrac{1}{\bar{c}}}$, and hence that $\bar{c}$ must be at least 4 if the DQE must not be less than 80%.

(2) A photographic process has mono-sized grains requiring 2, 3 and 4 quanta for latent image formation in equal proportions. Plot the characteristic curve for this process and hence find the maximum value of gamma. Express this value as a fraction of the maximum value appropriate for the three components.

(3) Two imaging processes with different spatial grain structures have density and noise defined by:

$$D_1 = \log_{10}\frac{1}{1-n}, \quad D_2 = n;$$

$$G_1 = c\,n\,(1-n), \quad G_2 = c\,n;$$

where n is the number of image grains as a fraction of the total number, and c is a constant. What conclusions would you draw about the respective spatial structures? Sketch the form of the $G - D$ relationship in each case. Prove that $G_1(\max) = \tfrac{1}{4} G_2(\max)$, and investigate the density levels at which these maximum noise levels occur.

(4) A film is exposed uniformly to light of wavelength 470 nm. The total exposure is equivalent to $1\cdot 45 \times 10^{-6}$ J m^{-2}, and a density level of 0·25 results. At this level the gradient of the characteristic curve is 0·42, while the *rms* granularity as measured with a circular aperture of 50 μm diameter is 0·021. It is known that the DQE is maximum at this exposure level. If it is assumed that the photographic layer consists of closely packed mono-sized grains all requiring 4 quanta for development and which is inefficient only in the sense of wasting a fraction of exposure quanta, calculate this fraction.

$(h = 6\cdot 62 \times 10^{-34}\,J\,s; \quad c = 3 \times 10^8\,m\,s^{-1})$

(5) The relationship between density D and average quantum exposure per grain, q, is found to be defined by $D = 1 - e^{-q^2}$. Prove that maximum gamma of 1·7 occurs at a density of 0·73, and plot the characteristic curve.

0·63

(6) Describe in general terms how the image noise G for mono-sized grains of area 1 μm^2 would compare with that for equal proportions of grains of 0·5 and 1·5 μm^2, assuming the grains are randomly distributed in the layer. For the latter case indicate the limiting relationship between G and D at low densities, and also calculate the value of G at $D_{\max}$.

3. ANALYTICAL MODELS 115

Explain why the quantum sensitivity of the grains has no influence on the G – D relationship for mono-sized grains, and only a minor influence for grains with practical size distributions.

(7) Prove that DQE_{max} for a checkerboard array of mono-sized grains each needing exactly Q quanta for image formation is independent of Q when Q is large, and has a limiting value of $200/\pi\%$ at an average exposure level of $Q-\frac{1}{2}$ quanta per grain.

Grains all requiring exactly 100 quanta for image formation might be said to have a quantum yield of 1%. Is the fact that such grains have a DQE as high as 64% in any way an anomaly?

(8) Show that for mono-sized grains the general DQE equation (82) reduces to that of equation (9): i.e., that with appropriate notation $h_2 = t$, $h_3 = s$. Discuss the relationship of t and s with parameters used in practical image measurements.

Prove that the ratio of DQEs according to the checkerboard and mono-sized random grain models is equal to $e^2/3$ when the average quantum exposure level is 2 quanta per grain in each case, and all grains require exactly 2 quanta.

(9) Write a flow diagram for a computer programme to calculate the DQE according to the random grain model. The basic input should consist of the grain size and sensitivity distributions, and other model parameters you think necessary if the output is to consist of DQE as a function of $\log E$.

Indicate how the relevant $\log E$ range and interval should be determined to give a satisfactory approximation to the complete DQE-$\log E$ curve. Suggest suitable test data.

(10) Discuss in general terms the influence of the size spread of the grains on the characteristic curve, the noise, and DQE associated with conventional photographic images.

In relating the absolute exposure to the average grain exposure the relationship $E_q = k\,q$ has been used in this chapter in calculating DQE. Since this represents a lower limit to the necessary exposure level, investigate the modification to DQE if the practical relationship is found to be $E_q = \dfrac{2k}{1 - e^{-k}} q$.

Discuss the physical implications of this equation.

4. Quantum Sensitivity and Ultimate Photographic Sensitivity

4.1 Measurement of Quantum Sensitivity

Experimental Techniques

There are fundamental experimental difficulties involved in the measurement of the quantum sensitivity of photographic grains. If direct methods of individual counting could be performed using a controlled number of quanta and observing the exact number necessary, grain by grain, the sensitivity distribution could then be built up as the limit after counting the requirements for many grains. However it is not possible to observe quanta themselves, but only their effect, and the image grains are the effect in this case. Due to the statistical fluctuations in the number of exposure quanta it is thus necessary to use indirect experimental methods based on the statistical interpretation of large numbers of grain counts.

In 1948 Webb[1] gave a clear survey of the theoretical and experimental problems involved, and made a first systematic attempt to deduce the quantum sensitivity. He pointed out that since the purpose is to study what happens in individual grains under controlled exposure conditions, it is desirable to have the grains spread out into a single layer and separated sufficiently so that they do not interfere with each other in any way during exposure and development and during subsequent grain-counting. Webb described the emulsion he used as being one of relatively uniform grain size, but in fact it had quite a size spread which introduced a degree of error in his deductions. He used monochromatic exposures of 360, 450 and 500 nm wavelength and several exposure steps, counting several hundred grains for each exposure step. The emulsion was in the form of single-grain-layer plates, sparsely populated such that only about 1/20th the plate area was covered by grains.

The exposure was obtained in terms of quanta per average grain by multiplying the exposure per unit area by the average grain size, which was 0.275 μm^2. Although the grains were not spectrally-sensitized, allowance was made in terms of the absorption of light at the three wavelengths used. An absorption of 20% was assumed at 360 nm, and the absorption at other wavelengths was measured relative to this. With this allowance the fractional grain counts were found to be largely independent of exposure wavelength, and only weakly dependent on development time. Figure 1 shows data for the fractional grain counts obtained by Webb for an exposure wavelength of 360 nm with 8 minutes development time.

4. QUANTUM SENSITIVITY AND ULTIMATE PHOTOGRAPHIC SENSITIVITY

An interesting point made by Webb has such important implications to be discussed later in the chapter that it is quoted in full:

> It cannot be supposed all quanta absorbed by a grain are actually utilized in the formation of the latent image speck which initiates development. Some of the quanta may go to forming internal specks of silver or in building up small specks of silver haphazardly distributed in the grain, neither of which necessarily contribute to that single speck of silver which first becomes large enough to initiate development of that grain. Hence the number of quanta absorbed would, of course, set an upper limit to the number of quanta forming the latent image, but would not uniquely determine this quantity.

Thus the quantum exposure of Fig. 1 relates to the number of quanta absorbed, though not necessarily those used to form the latent image.

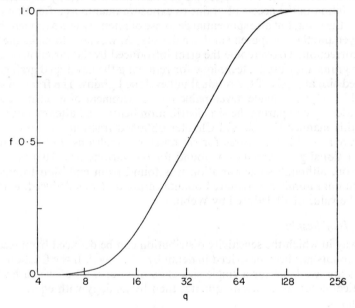

FIG. 1. Fractional grain counts due to Webb[1], for an experimental emulsion exposed to $\lambda = 360$ nm and developed for 8 minutes.

From curves such as those of Fig. 1, Webb argued that the most sensitive grains needed about 10 quanta, varying up to 100 quanta for the least sensitive grains. He also found that the curves fitted the mono-sized mono-sensitivity characteristics such as those of Fig. 7 of Chapter 3, for a value of around $Q = 10$. But the $Q = 10$ curve has an arithmetic exposure scale shift of about 4 towards lower exposures when compared to that of Fig. 1, and so corresponding to a $Q = 10$ quanta process with only 1 quantum in 4 of those absorbed being usefully employed.

Basic improvements in measuring technique were claimed by Farnell and Chanter[2] in 1961, during a description of their own method of measuring fractional grain counts. They defined the techniques needed in obtaining fractional grain counts in terms of absorbed quanta per grain as falling into two operations: firstly, the derivation of the response curves in terms of the incident exposure for the separate size classes, all grains being identically situated with respect to the incident radiation; secondly, the derivation of a factor which enables the incident exposure to be converted into absorbed exposure. In terms of these they discussed the shortcomings of the techniques of Webb, namely due to the size variations and the accuracy of the calculations of the very low degree of absorption of the thin layers.

To overcome the difficulties of calculating the absorbed exposure, Farnell and Chanter used ordinary emulsion layers rather than the sparse single-grain layers of Webb, since the absorption for these is quite high and can be measured accurately. This of course introduces the difficulty of determining the variation in the access of grains to exposure quanta as a function of position in the layer, but an exponential decrease of intensity was assumed through the layer and the unexposed emulsion density, Δ, was used to make the necessary correction. To overcome the error introduced by the variation of the size of the grains they used a technique for removing the developed grains, which allowed size analysis of the residual undeveloped grains. The fraction of each size class of grains made developable by an increment of exposure could be calculated by comparing the size distribution before and after exposure.

In this manner Farnell and Chanter obtained fractional grain counts for as many as eight size classes for six different emulsions. Little variation in the fractional grain counts was found for the various size classes of a given emulsion, although some variation was found from emulsion to emulsion. A minimum number of required quanta of about 4 was deduced, compared with the value of 10 deduced by Webb.

Analysis of Results

A way in which the sensitivity distribution can be deduced from fractional grain counts has been described in detail by Marriage[3]. If the fractional grain counts are available for a single size class of grains, all of which had equal statistical access to exposure quanta, then by analogy with equation (41) of Chapter 3,

$$f = \sum_Q \alpha_Q p_Q. \tag{1}$$

where f denotes the fractional grain count, and α is the distribution parameter for the Q-values, as previously.

The probability p_Q of an image contribution from grains requiring Q quanta follows from the Poisson statistics:

$$p_Q = 1 - e^{-q} \sum_0^{Q-1} \frac{q^r}{r!}. \tag{2}$$

From equations (1) and (2),

$$f = \sum_Q \alpha_Q \left(1 - e^{-q} \sum_0^{Q-1} \frac{q^r}{r!}\right). \tag{3}$$

4. QUANTUM SENSITIVITY AND ULTIMATE PHOTOGRAPHIC SENSITIVITY

The practical problem is that of deducing the set of alpha values, given the fraction of grains contributing to the image at a series of exposure levels per grain, q. The analysis of the experimental results can in principle be carried out by solving many simultaneous linear equations based on the complete range of measured $f - q$ values. However, as pointed out by Marriage, a much simpler method of approximation is possible due to the nature of equation (3).

For a fixed value of Q, equation (3) reduces to

$$f = 1 - e^{-q} \sum_{0}^{Q-1} \frac{q^r}{r!}, \tag{4}$$

and by differentiation,

$$\frac{df}{dq} = e^{-q} \sum_{0}^{Q-1} \frac{q^r}{r!} - e^{-q} \sum_{0}^{Q-2} \frac{q^r}{r!} = e^{-q} \frac{q^{Q-1}}{(Q-1)!}. \tag{5}$$

The derivative with respect to $\sqrt{q}$ will thus be

$$\frac{df}{d\sqrt{q}} = \frac{df}{dq} \frac{dq}{d\sqrt{q}} = 2\sqrt{q} \frac{df}{dq},$$

and so from equation (5)

$$\frac{df}{d\sqrt{q}} = \frac{2e^{-q} q^{+(Q-\frac{1}{2})}}{(Q-1)!}. \tag{6}$$

Substituting $y = \frac{df}{d\sqrt{q}}$, $x = \sqrt{q}$, enables equation (6) to be written as

$$y = \frac{2x^{2Q-1} e^{-x^2}}{(Q-1)!}. \tag{7}$$

The reason for this form of substitution will become apparent. When Q is large it can be shown that equation (7) approximates to

$$y = \sqrt{\frac{2}{\pi}} e^{-2(x-\sqrt{Q-\frac{1}{2}})^2}. \tag{8}$$

The proof of this is rather difficult and again involves the Stirling approximation for the factorial of large numbers. Equation (8) represents a Gaussian distribution with mean value at $x = \sqrt{Q - \frac{1}{2}}$. The rms of this distribution is $\sigma = \frac{1}{2}$, and hence is independent of Q. Plotted curves of y against x are shown in Fig. 2 for values of $Q = 1, 2, 4, 8, 16, 32,$ and 64 quanta. It is seen that with the minor exception of the $Q = 1$ curve the curves are almost exactly symmetrical, and identical except for progressive shifts along the x-axis as Q increases.

When Q is large the x-shift between the curves for successive Q-values is approximately $1/2\sqrt{Q - \frac{1}{2}}$; for example, the spacing between the $Q = 63$ and $Q = 64$ quanta curves will be approximately $1/16$ x-units. This fact, combined with the constant shape and size of the set of curves, leads to a simple approximate method of deconvoluting the separate contributions from combined curve for a mixture of alphas. This method assumes that the distribution of alphas varies slowly and smoothly with Q. In this case the y-value of the $y - x$ curve obtained from experimental grain counts is noted at the value $x = \sqrt{Q - \tfrac{1}{2}}$ and divided by the appropriate curve-spacing factor, which will be $2\sqrt{Q - \tfrac{1}{2}} = 2x$ at this point. The value so obtained will be a first

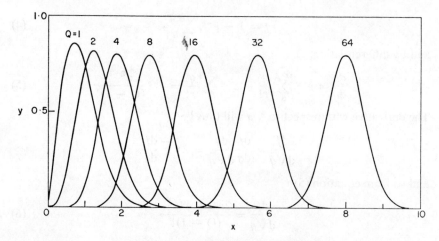

FIG. 2. $y - x$ curves for a range of Q-values.

estimate of α_Q. Suitable corrections can be made if the approximate form of the alpha distribution is known.

In his own analysis of the Farnell and Chanter grain counts, Marriage proceeded as follows. The first derivative of the experimental plots, $Y(x)$, was used to derive a curve

$$Z(x) = 2Y(x) - \tfrac{1}{2}(Y(x + \tfrac{1}{2}) + Y(x - \tfrac{1}{2})). \tag{9}$$

In this way allowance was made for parabolic variation in the alpha values. The alpha for a specific Q-value was then obtained by dividing the ordinates of this plot at the appropriate x-values by $2x$. The need for a correction to this first estimate was studied by summing the product α_Q and y for all the individual Q-values to see how near this led back to the original experimental curve. By noting the discrepancies and adjusting the alphas, a trial and error procedure quickly led to a more satisfactory set of alpha values. The experimental fractional grain counts and the corresponding gradients, both plotted against $x = \sqrt{q}$, are shown in Fig. 3 for three of the emulsions originally analysed in this way.

An interesting conclusion from this method of analysis is that if relatively thin layers of monosized grains are available, the need for tedious grain sizing and counting techniques might be eliminated. In this case, if the Nutting formula holds reasonable well,

$$f = \frac{D}{D_{max}}. \tag{10}$$

The only important assumption will then be that of equal statistical access to exposure quanta for all the grains. Obviously this will depend on the thickness

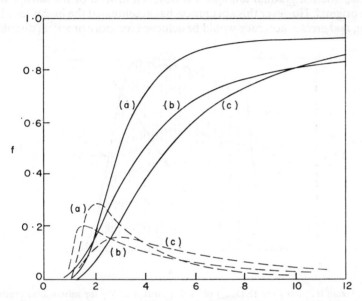

FIG. 3. Curves for fractional grain counts and their gradients (dashed curves) for[2,3]: (a) Screen X-ray; (b) Fine Grain Negative; (c) Fast Negative.

of the layer, or the possibility of working back from the thick-layer characteristic to the thin-layer characteristic if the exposure distribution through the layer is known. However, let us assume that the practical conditions are such that we may make use of equation (10).

An illustrative example[4] may be based on the model quantum sensitivity distribution of Fig. 5 of Chapter 3, which was used as the standard for computing image characteristics. This sensitivity distribution has equal values from $Q = 4$–32 quanta successively, and the computed characteristic curve for mono-sized grains having this sensitivity distribution is shown in Fig. 4. It is assumed that $D/D_{max} = f$, which is shown on the density scale, and the exposure scale is plotted as a function of $x = \sqrt{q}$ instead of the usual log q. The problem is whether it is possible to operate on this characteristic curve to deduce the constituent alpha values.

The necessary procedures of differentiation, parabolic correction, division of the value at $x = \sqrt{Q - \frac{1}{2}}$ by $2x$, etc., can all be performed by a computer taking digital values of the characteristic curve as input. The result according to a typical procedure is also shown for the present example, plotted in Fig. 4 in arbitrary units. Since no attempt has been made to smooth the intermediate Z-curve, the envelope of the resulting alpha values is irregular in shape. The intersection with the envelope of the vertical lines drawn at all relevant $x = \sqrt{Q - \frac{1}{2}}$ values gives the estimated alpha values for each value of Q. These are seen to be quite close to the original set, except at the high end of the Q-scale, where a gradual tailing-off is deduced instead of the abrupt halt as in the original. However this abruptness has accentuated the inaccuracy in this region, and greater accuracy would be achieved for more practical distribution

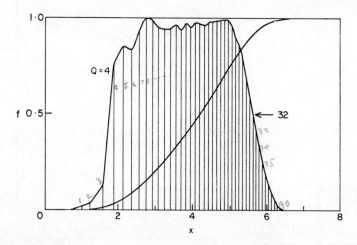

FIG. 4. Grain fraction f (or D/D_{max}) plotted against $x = \sqrt{q}$ for mono-sized grains with sensitivity from Q = 4 to 32 quanta inclusive and in equal proportions. Also shown, in arbitrary units, is the estimation of this sensitivity distribution from the $f - x$ curve, using a computer version[4] of the Marriage analysis.

shapes which diminish slowly and smoothly for high Q-values. At the lower end of the Q-scale there is a spurious predicted contribution at Q = 2 and 3, but this is small compared to those for Q = 4 upwards.

In practice the decisive error will be due to the determination of the average quantum exposure per grain. Farnell and Chanter estimated that in their experiments this might be as high as ± 30%. The upper plot of Fig. 5 shows a typical error spread of ± 20% in the D-log q curve after converting from absolute exposure to the average number, q, of quanta per grain. The lower plot shows the corresponding limits to the estimated sensitivity, the dashed line representing the original distribution. This demonstrates that it is unlikely that any further corrections to the first analytical estimates will produce much further useful information about the exact form of the sensitivity distribution unless there is a fundamental improvement in experimental accuracy.

4. QUANTUM SENSITIVITY AND ULTIMATE PHOTOGRAPHIC SENSITIVITY

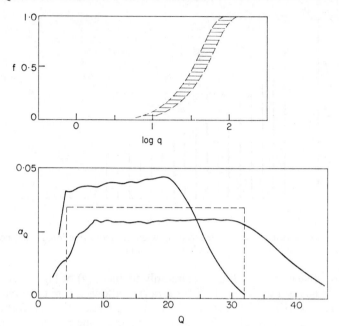

FIG. 5. Above: example of the error spread in the $f - \log q$ curve after converting the absolute exposure to the average number of quanta, q, absorbed per grain. Below: the corresponding spread in the estimate of the sensitivity distribution compared with the actual distribution (dashed).

Sensitivity Distributions

Two examples of the sensitivity distributions derived by Marriage from the fractional grain counts of Farnell and Chanter have already been considered in Chapter 2. A more recent one was deduced by Spencer[5], also based on a computer version of the analysis which has been described here. This distribution is shown in Fig. 6. The alpha values as listed by Spencer were only for Q-values up to $Q = 20$, covering 61% of the grains, and hence leaving a 39% contribution from grains needing 21 quanta or more. The extrapolation of Fig. 6 serves to remind that the distribution is not complete.

Spencer also related the sensitivity distribution to the procedure used in sensitization by giving fractional grain counts for the unsensitized emulsion and as sensitized by sulphur and sulphur-plus-gold. The distribution shown in Fig. 6 refers to the sulphur-plus-gold case. It is noted from the grain counts of Fig. 7 that there is a close similarity between the curve for sulphur sensitization and that shown in Fig. 1 due to Webb. It is thus possible that the shift of the curve from that of (b) to that of (c) gives a measure of the change in photographic speed due to technological improvements in the process between the years 1948 and 1971.

It is notable that all curves of Fig. 7 are similar in shape, the only significant change being in the position of the curve along the $\log q$ axis. Although the

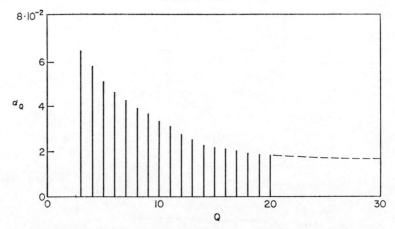

FIG. 6. Sensitivity distribution as deduced by Spencer[5] from fractional grain counts, for an experimental sulphur-plus-gold sensitized emulsion.

sensitivity distribution of Fig. 6 refers only to curve (a) of highest sensitivity, the invariance of the shape of all three curves allows a rapid estimate of the two other sensitivity distributions, following the reasoning of Webb in the analysis of his own results[1]. The q-values at the same f-value of curves (b)

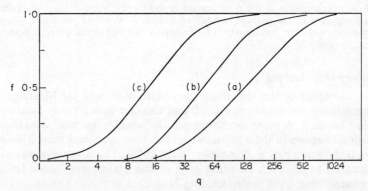

FIG. 7. The fractional grain counts for the experimental emulsion used by Spencer[5], as: (a) unsensitized; (b) sulphur-sensitized; (c) sulphur-plus-gold sensitized.

and (c) are approximately $4 \times$ and $8 \times$ those of curve (a), respectively. Thus if the sensitivity distribution relating to (a) ranges from $Q = 3$ to, say, $Q = 100$ quanta, then it follows that those relating to (b) and (c) range from $Q = 12$–400 quanta and $Q = 24$–800 quanta, respectively. The shape of the sensitivity distributions will be identical except for these $4 \times$ and $8 \times$ changes in Q-scale. More detailed implications of these curves will be considered in the following section.

Other measured sensitivity distributions and discussions of their significance are due to Haase[6] and Klein[7], among others.

4.2 Interpretation of Quantum Sensitivity
Problems of Interpretation

It is important that the exact definition of the sensitivity distribution as used here should be appreciated, since conceptual confusion exists in the literature. The sensitivity distributions so defined, are experimental observations—albeit indirect ones, due to experimental difficulties—related to the number of quanta absorbed by a grain before it becomes developable. It thus has a precise meaning to the experimentalist independent of its significance to the latent image theorist. As we shall see, with various assumptions made concerning the mechanisms of latent image formation it is possible to arrive at other definitions of the sensitivity distribution which relate to the utilization of absorbed quanta. However such definitions are complementary to that used here, and as indicated in the earlier quotation from Webb, the sensitivity distribution according to our definition will represent an upper limit to the number of quanta forming the latent image.

It can in fact be of great benefit to the latent image theorist if the observed sensitivity distributions are interpreted in terms of the various internal grain mechanisms which operate on the absorbed quanta, especially when the image characteristics are modelled in terms of these distributions as in the previous chapter. It is then possible to relate the mechanisms themselves to speed, curve shape, noise and DQE. Due to the complexity of the photographic process of latent image formation—as outlined in Chapter 2—this is not a straightforward task, and many of the details of latent image formation await further experimental and theoretical advances.

One of the most successful attempts to relate the mechanisms of latent image formation to the properties of the photographic image is due to a so-called Monte Carlo model devised by Bayer and Hamilton[8,9] in 1965. Such models are generally useful when the number of possible sequences of events and their conditional probabilities is such that it prohibits an analytical solution. In the case of latent image formation the model must take into account all the possible chance events following the absorption of a quantum, and by tracing the effective paths and fates for a large number of quanta, a limiting statistical picture can be drawn to represent the behaviour of the complete layer ensemble of grains. In Chapter 7 the use of such Monte Carlo methods will also be seen to be beneficial in tracing the chance events prior to quanta being absorbed by grains, and in this way the scattering and image resolution properties of photographic layers can be predicted. In principle the combination of these before-and-after-absorption Monte Carlo models would allow more or less all the important photographic input/output characteristics to be modelled in an overall manner.

Bayer and Hamilton included in their own computer simulation quantitative information on most of the known and theorized mechanisms, including: reversible electron-trapping; ionic neutralization of trapped electrons; thermal decay of single silver atoms; permanent hole-trapping; permanent electron-trapping at silver aggregates, etc. They did not generate sensitivity distribution as such, but were able to simulate characteristic curves in a satisfactory manner. They also concluded that the observed spread of quantum sensitivities could follow from the random statistical nature of the events occuring during latent image formation and not necessarily from any basic initial difference in grain properties. In 1972 Hamilton[10] applied this Monte Carlo model approach to the simulation of DQE-log$E$ curves.

Although such models are invaluable to those interested in the latent image *per se*, simpler models can perhaps provide more general insight into the relationships between speed, quantum sensitivity, curve shape and DQE. One such simple model which ties together these and other concepts in a neat manner has been described by Shaw[4], and the analysis that follows is based on this particular model.

Simple Model of Sensitivity Distribution

Latent image formation is assumed to be completely described by only two essential grain properties. First, it is assumed that all grains require the same fixed number, T, of exposure quanta for latent image formation, where T might be anticipated to be a small number in the region of 2, 3 or 4. Second, it is assumed that all absorbed quanta have a random probability of making a contribution to the number T. We shall denote this probability as $p = \dfrac{1}{m}$, and it applies equally for all grains for all absorbed quanta until the latent image is formed. Thus from a statistical viewpoint all grains have identical properties. The exact values of T and m will define the observed sensitivity distribution, which can be derived analytically using standard statistical arguments[11].

The problem can be translated into terms of a familiar problem of statistical sampling, and in particular to the total number of samples required before a certain number of these samples with some desired property is reached. It is as though the grains sample the absorbed quanta in this manner, only 1 in m of the quanta at random having the required property and constituting a partial success in the latent image sense. A total success will be achieved when T of these partial successes have been sampled. The question is how many of these total samples, Q, are needed to achieve T of these partial successes. The proportion of times that Q samples are needed will represent α_Q, the observed sensitivity corresponding to Q quanta. The statistical problem is thus to find the proportion, α_Q, of times that the Tth partial success is achieved at the Qth sample.

A simple method to obtain the probability distribution, is to define an overall state of "favourable failure," as the situation when $T - 1$ partial successes have been achieved. It is only when in this state that the next sample has any probability $(1/m)$ of producing the final partial success towards an overall total success. Furthermore, if we consider the probability of obtaining $T - 1$ partial successes after $Q - 1$ samples (absorbed quanta), then we can work out the probability that the next, Q^{th}, sample will bring the total success needed.

Since the probability of any sample providing a partial success is $1/m$, the probability associated with an individual failure will be $(1 - 1/m)$. Now

4. QUANTUM SENSITIVITY AND ULTIMATE PHOTOGRAPHIC SENSITIVITY

after $Q-1$ samples, $T-1$ partial successes will be accompanied by $(Q-1)-(T-1) = Q-T$ failures. The respective numbers and probabilities in an overall state of favourable failure will thus be:

	Probability of each
$T-1$ partial successes	$1/m$
$Q-T$ failures	$1-1/m$

The probability, written $p_{(T-1)/(Q-1)}$, of being in a state of favourable failure will be governed by binomial statistics[11], and can be written as

$$p_{(T-1)/(Q-1)} = {}_{(Q-1)}C_{(T-1)} \left(\frac{1}{m}\right)^{T-1} \left(1-\frac{1}{m}\right)^{Q-T}. \quad (11)$$

where,

$${}_{(Q-1)}C_{(T-1)} = {}_{(Q-1)}C_{(Q-T)} = \frac{(Q-1)!}{(Q-T)!\,(T-1)!}$$

signifies the number of ways the combination of $T-1$ partial successes and $Q-T$ failures can occur. The probability of a total success after the next, and Q^{th}, sample is thus the above probability multiplied by the probability of another, and hence conclusive, partial success:

$$\alpha_Q = \frac{1}{m}\, p_{(T-1)/(Q-1)}$$

$$= \frac{(Q-1)!}{(Q-T)!\,(T-1)!}\left(\frac{1}{m}\right)^{T}\left(1-\frac{1}{m}\right)^{Q-T}. \quad (12)$$

This type of distribution is known as a negative binomial distribution[11]. It may readily be shown that its properties are such that the average number of required quanta will be defined by

$$\bar{Q} = Tm. \quad (13)$$

and if m is an integer the distribution has maximum values at $Q = (T-1)m$ and $Q = (T-1)m + 1$ quanta.

Equation (12) can be used to generate the sensitivity distribution for any combination of T and m which is of practical and theoretical interest. Figure 8 shows a series of such distributions for values of $T = 2, 3$ and 4, each combined with values of $m = 1, 2, 4$ and 8. For convenience the continuous envelope curves are shown rather than a set of discrete spikes at the successive Q-values. However the $m = 1$ case represents constant quantum sensitivity at the respective $Q = T$ values, and hence is shown as a single spike which should extend to $\alpha = 1$.

The distributions of Fig. 8 demonstrate that this simple two-parameter model can account for a wide spread of observed sensitivities. This confirms the conclusion of Bayer and Hamilton[8], since the grains were all assumed to have identical properties, the spread in observed sensitivities being entirely due to the random events occurring during image formation. A parallel view was expressed by Baker[12], who in discussing the Farnell and Chanter grain counts pointed out that it is not unreasonable to argue that most of the quanta absorbed were not involved in latent image formation. His own analysis confirmed that nominally-identical grains would have a wide observed

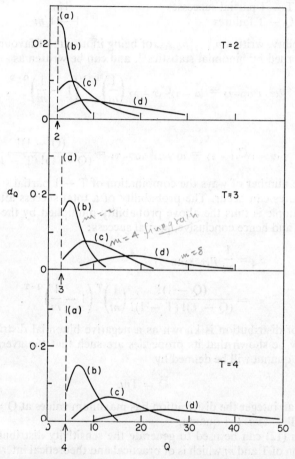

FIG. 8. Model sensitivity distributions based on a negative binomial distribution, for T = 2, 3 and 4. In each case: (a) $m = 1$; (b) $m = 2$; (c) $m = 4$; (d) $m = 8$.

spread of sensitivity. Following the arguments of Webb, Q is the upper limit to the number of quanta forming the latent image, which according to our simple model is exactly T. The larger the value of m, the greater will be the average difference, $Q - T = T(m - 1)$, between the actual number and this upper limit. However for any value of m there will always be a finite possibility that some grains will become developable after absorbing only T quanta.

The general shape of the model distributions of Fig. 8 are in reasonable agreement with measured distributions such as those of Fig. 2 in Chapter 2.

4. QUANTUM SENSITIVITY AND ULTIMATE PHOTOGRAPHIC SENSITIVITY

For example, the T = 3, m = 4 case corresponds quite closely with the practical distribution for a fine grain negative material. From equation (13) $\bar{Q} = 12$ quanta for this combination of T and m values, and the average "extrinsic" quantum yield of $100/\bar{Q}\%$ would be approximately 8% in this case. On the other hand, the "intrinsic" quantum yield related to the internal mechanism can be defined as $100/T\% = 25\%$, and these two definitions of quantum yield are entirely compatible. At the same time there are still some grains which would be observed experimentally to need to absorb more than 20 quanta for image formation, and it would also be in keeping to say that this particular class of grains had an extrinsic quantum yield of less than 5%.

The utility of a simple model depends on its ability to explain the pertinent facts. The justification of the present model is in the clarification of the interrelationships between some of the more important photographic input/output properties.

Quantum Sensitivity and the Characteristic Curve

Again making the assumption that the grains are mono-sized and that photographic density is proportional to the number of developed grains in the image, then equation (3) for the fractional grain counts, f, also represents the equation for the normalized density. In this case differentiation of (3) and substitution from equation (5) leads to

$$\gamma = \frac{q}{\log_{10} e} \frac{df}{dq} = \frac{q}{\log_{10} e} \sum_Q \alpha_Q \left(e^{-q} \frac{q^{Q-1}}{(Q-1)!} \right). \quad (14)$$

We can now make use of the analytical expression for α_Q from equation (12), and the summation on the RHS of equation (14) becomes:

$$\sum_Q \alpha_Q \left(e^{-q} \frac{q^{Q-1}}{(Q-1)!} \right) = e^{-q} \left(\frac{1}{m} \right)^T \frac{q^{T-1}}{(T-1)!} \sum_{Q=T}^{\infty} \frac{((1-1/m)q)^{Q-T}}{(Q-T)!}. \quad (15)$$

The summation which was over all contributory values of Q now extends by definition from T upwards.

Inspection of equation (15) shows that it is simply the natural expansion of e raised to the power of $(1 - 1/m)q$. The expression for gamma from equation (14) thus becomes

$$\gamma = \frac{e^{-q}}{\log_{10} e} \frac{(q/m)^T}{(T-1)!} e^{(1-\frac{1}{m})q} = \frac{1}{\log_{10} e} \frac{(q/m)^T}{(T-1)!} e^{-q/m}. \quad (16)$$

In the special case where $m = 1$ and all absorbed quanta contribute towards the number, T, equation (16) gives

$$\gamma = \frac{1}{\log_{10} e} q^T \frac{e^{-q}}{(T-1)!}, \quad (17)$$

which, since Q = T in this case, is identical with equation (36) of Chapter 3 for the gamma for mono-sensitive grains.

The only difference between equations (16) and (17) is that q in the special case of equation (17) is replaced by q/m in the general case of equation (16). Since

$$\log_{10}(q/m) = \log_{10}q - \log_{10}m,$$

it is concluded that the influence of the parameter m can be interpreted as a shift of the D-log$E$ curve along the log$E$ axis. The characteristic curve will move bodily by $\log_{10}m$ units towards higher exposure levels compared with the $m = 1$ case. However the shape of the characteristic curve will be independent of m, and completely determined by the T = Q curve.

Figure 9 shows characteristic curves for the same range of T and m values as for the sensitivity distributions of Fig. 8. The $m = 1$ cases will correspond exactly to the Q = 2, 3 and 4 mono-sensitivity curves. Since the characteristic curves shown for the other values of m correspond to successive changes of m by factors of 2, these curves will be displaced by $\log_{10}2 = 0.30$ units along the log$E$ axis.

The shape of the curves is determined entirely by the T = Q mono-sensitivity curves, and so the maximum slopes will be governed by equation (38) of Chapter 3. If the model assumptions are realistic, the slope of the characteristic curve can be said to be that determined by the single T = Q value, or that determined by the weighted average of the slopes for all Q-values of the sensitivity distribution: both interpretations will give exactly the same answer. If the model were not realistic then the above would still be true for sensitivity distributions which could be represented by negative binomial distributions, but T would no longer relate to anything of physical significance concerning latent image formation.

This dual way of interpreting the (low) observed slope of the characteristic curve as either the weighting of curves all of higher slope or as a single curve of the same low slope was at the centre of a controversy commencing in 1939, in which the justifications and merits of both ways were argued in detail[13,14], and these arguments have recently been reopened[15]. As we have seen, the viewpoint will depend on whether the interest is in the extrinsic or intrinsic quantum efficiency, and both approaches must be compatible if correctly analysed and related. However it must be stressed that the extrinsic sensitivity can always be given a precise definition by the experimentalist, whereas the definition of intrinsic sensitivity is dependent on assumptions concerning the mechanisms of latent image formation, the validity of some of these awaiting further scientific knowledge.

According to our simple model the grain counts relating to the sensitization procedure shown in Fig. 7 can be interpreted in terms of the changes in m. Since the three curves are very similar in shape, the total shift between unsensitized and sulphur-plus-gold sensitized curves representing an exposure change around $8x$, the overall sensitization procedure can be thought of as having produced an $8x$ reduction in m; i.e., in the efficiency of utilization of absorbed quanta. The model can thus explain the effects of sensitization in a satisfactory manner.

The reasons that the model may be inaccurate fall into two classes. First, the

4. QUANTUM SENSITIVITY AND ULTIMATE PHOTOGRAPHIC SENSITIVITY   131

assumptions may be fundamentally unrealistic. As analysed by Bayer and Hamilton there are certainly many conditional probabilities involved in latent image formation, and a random loss mechanism cannot represent

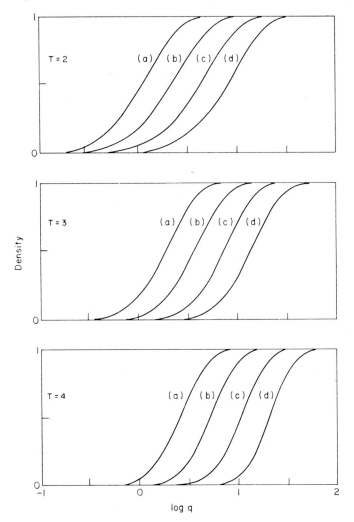

FIG. 9. The normalized density curves for mono-sized grains and sensitivity distributions of Figure 8.

these. However, just as the detailed Monte Carlo modelling of Bayer and Hamilton does not necessarily represent any actual path of events in any one grain—even though it may produce a limiting statistical picture—it is not unreasonable that the collective effect of many quanta absorbed by many grains with many conditional probabilities involved may approach an overall fixed-quantum-requirements plus random-loss situation in the limit. Second, even if the assumptions had a high degree of validity, it is still likely that T may be a variable–even if only over a small range of numbers as 2, 3 or 4–and m might also vary and be a function of this variable number for T. In such cases the alpha for a given Q-value would be obtained as the weighted mean of that for all the constituent T, m classes of grains. Models can be derived which allow for such variations, but are not necessarily of much further help in interpreting the photographic input/output relationships.

Quantum Sensitivity and DQE

Since we are concerned here with randomly-distributed mono-sized grains, the expression for DQE is defined as in Chapter 3:

$$\varepsilon = \frac{qt^2}{1-s}, \qquad (18)$$

where

$$s = \sum_Q \alpha_Q \left(e^{-q} \sum_0^{Q-1} \frac{q^r}{r!} \right); \quad t = \sum_Q \alpha_Q \left(e^{-q} \frac{q^{Q-1}}{(Q-1)!} \right).$$

From these equations it is possible to calculate DQE as a function of the average number of absorbed quanta, q, by inserting the appropriate distribution of alphas for any combination of T and m. However, it can be expressed more simply than this, due to the property of the negative binomial distribution whereby m is effectively only a direct multiplicative change in exposure scale. Because of this it is possible to define

$$\varepsilon_{T,m}(mq) = \frac{1}{m} \varepsilon_Q(q). \qquad (19)$$

This equation relates the DQE for the T, m model at the exposure level of mq absorbed quanta per grain to the DQE for the mono-sensitive case, T = Q, at the average exposure level of q absorbed quanta per grain[4]. ε_Q is still defined by equation (18), but now with

$$s = e^{-q} \sum_0^{Q-1} \frac{q^r}{r!}, \quad t = e^{-q} \frac{q^{Q-1}}{(Q-1)!}.$$

For any value of m the DQE-log$E$ curve will thus be the same as that for the appropriate T = Q curve of Fig. 2 of Chapter 3, but with the DQE scaled down in size by m, and the whole curve shifted along the log$E$ scale by $\log_{10}m$ units.

Figure 10 shows the DQE curves for the same range of T and m values as for the sensitivity distributions of Fig. 8 and the characteristic curves of Fig. 9. In each case the $m = 1$ curve follows the T = Q mono-sensitive curve. Increases in m by successive factors of 2 are seen to bring successive DQE decreases by this same amount, along with the predicted log$E$ shift.

4. QUANTUM SENSITIVITY AND ULTIMATE PHOTOGRAPHIC SENSITIVITY 133

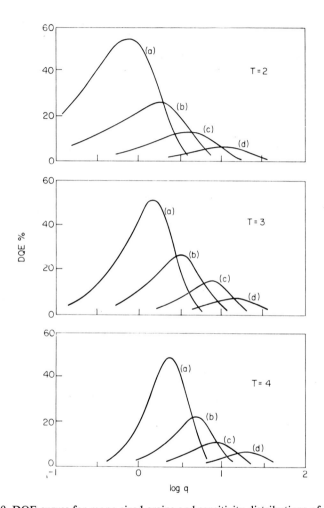

FIG. 10. DQE curves for mono-sized grains and sensitivity distributions of Fig. 8.

The relationship between maximum values of DQE also follows directly from equation (19) as

$$\varepsilon_{T,m}(\max) = \frac{1}{m}\varepsilon_Q(\max), \tag{20}$$

for the appropriate T = Q cases. As already noted in the previous chapter, the maximum DQE values are largely independent of Q when Q is high. The values given below summarize the influence of T and m:

T	DQE(max)
2	$\frac{1}{m} \times 54\%$
3	$\frac{1}{m} \times 50\%$
4	$\frac{1}{m} \times 48\%$
100	$\frac{1}{m} \times 41\%$

With this final DQE relationship the inter-relationships between speed, characteristic curve shape, quantum sensitivity distribution, quantum yield and DQE are now complete according to our simple T, m model for the sensitivity of randomly-distributed mono-sized grains with equal statistical access to exposure quanta. The characteristic curve shape and latitude are due entirely to the quantum fluctuations based on T quanta. The positions of the characteristic curves and DQE curves along the log$E$ axis (and hence the speed) are determined by m. The magnitude of the DQE is largely independent of T and inversely proportional to m.

The relationship between DQE and quantum efficiency is also a close one. The extrinsic quantum yield can be defined by

$$\eta_{\text{ext}} = \frac{100}{Tm}\%,$$

If T is restricted to low quantum numbers such that

$$\text{DQE}_{\max} \simeq \frac{1}{m} \times 50\%,$$

then

$$\text{DQE}_{\max} \simeq \frac{T}{2}\eta_{\text{ext}}. \tag{21}$$

Similarly, the intrinsic quantum efficiency may be defined by

$$\eta_{\text{int}} = \frac{100}{T}\%,$$

and hence

$$\mathrm{DQE}_{\max} \simeq \frac{T}{2m} \eta_{\mathrm{int}}. \tag{22}$$

4.3 Ultmiate Photographic Sensitivity

Some Estimates

The topic of ultimate sensitivity in photography and its relation to the existing state-of-the-art is one that has undergone much general discussion and speculation. Also of interest is the way in which progress in sensitivity has been made since the origins of photography. Several authors have plotted the logarithm of some measure of sensitivity against time, and have found a roughly straight-line relationship. In this way they have deduced that with respect to the Daguerreotype process, *circa* 1840, photographic sensitivity has increased by about a sixth power to the mid-twentieth century. Berg[16] and Kirillov[17] have given summaries of such surveys and discussed their significance.

On the basis of such surveys, extrapolations indicate that a further increase in photographic sensitivity by several powers might be anticipated by the mid-twenty-first century. However doubts have been cast on such extrapolations merely from the empirical basis of observing a slight tendency of the progress curve to flatten out. Empiricisms of this nature inevitably lead to speculations as to whether or not such flattening-out is due to fundamental limits having been reached; or, if not, where these fundamental limits lie and how they might be approached more closely by practical photographic processes. The problems of limiting sensitivity are distingushed here from those of the limits of spectral sensitivity, as discussed briefly in Chapter 2. The upper wave-length limit is set in practice by thermal fogging, and the sensitivity to low wavelengths (high-energy quanta) of the electromagnetic spectrum extends to X-rays and beyond. Here we are concerned with the magnitude of sensitivity, and not its spectral confines.

A Symposium organized by the Royal Photographic Society on "The Ultimate Sensitivity of Photography Today and Tomorrow" was held in London during December 1960[18]. As a part of this symposium, a challenge was made to a group of scientists—representative of most theoretical and practical aspects—to predict the future limits of photographic sensitivity. A set of rules accompanied this challenge, and since the answers provided have become classical in this respect, these rules are stated here in full:

(1) The photographic system can be optical, electronic, etc., but it must yield a reasonably permanent print, say 9 × 12 cm, of a normal pictorial subject.
(2) The system must have a usable latitude of 1·2 log$E$ units and the final copy must be capable of a full reflexion density range up to a maximum density = 1·2.
(3) The information content of the picture should not be inferior to that

given by a 625-line television system under optimum, closed-circuit conditions. A good television picture of this sort can be taken as the general guide for quality of the image.

(4) No practical restrictions whatsoever are placed on the sensitized materials or equipment used. For example, emulsions can be imagined of perfectly ordered grains of any desired size and quantum efficiency, and the pigment of each grain can have any desired covering power. Reciprocity failure, fading, etc., can be ignored.

(5) The maximum sensitivity of which the system is capable should be expressed in terms of the luminance required (meter-candles or lux, 3000°K) to obtain a satisfactory picture from a subject in which the highest light is a perfectly reflecting matt white surface. The exposure time is to be 0·02 second and the subject is to be imaged originally with a working aperture of f/2.

There were eight detailed responses to this challenge, these coming from Benarie, Berg, Fellgett, MacAdam, Meyer, Rose, Schade, and Vendrovsky and Sheberstov, and a summary and discussion of their estimates was provided by Selwyn[18]. Figure 11 gives an indication of the spread of the answers. Six of the eight estimates cluster quite closely together at an illumination level around that corresponding to starlight, while the other two estimates gave quite higher and lower values. This clustering of the majority of estimates is interesting, since the rules permitted a large degree of flexibility in the manner of making the estimates, which was reflected in the different approaches and details of the separate calculations. Two of the most sensitive existing photographic process are also charted in Fig. 11, along with a conventional negative film as used as the recording medium with a multi-stage electronic image amplifier. The latter represented the highest state-of-the-art sensitivity at the time of the symposium, corresponding to an illumination level of moonlight, and this type of pre-amplification of the exposure light prior to photographic recording will be discussed in more detail in this and the next chapter.

The highest estimate due to Vendrovsky and Sheberstov was equivalent to a checkerboard array of grains with area of 2000 μm^2. On the other hand, the most conservative estimate was due to MacAdam who pointed out that a free interpretation of the rules would lead to estimates corresponding to images of unacceptable quality. This is due to the fact that the visual impression of a rapid series of images, as in TV, is quite different from that of a single static picture. Setting the rules with reference to TV picture quality is not necessarily the ideal criterion. MacAdam discussed the potential improvements in terms of relative increases in quantum efficiency, and estimated that there was probably only a further factor of ten remaining before the ultimate sensitivity of conventional photographic grains was reached.

A further detailed review of the implications of the eight estimates was made by Kirillov[17], who also made further estimates concerning ultimate sensitivity. He analysed the problem in terms of the quantum sensitivity and size of ideal single-layers and multilayers. In this way he arrived at sets of characteristic curves similar to those of Chapter 3, and from these he deduced speed ratings for all combinations of grain size and quantum efficiency between those for grains of 2025 μm^2 needing only 1 quantum to grains of 0·0156 μm^2 needing as many as 128 quanta. He did not expressly calculate the corresponding image qualities, or define criteria of acceptability, but he was able to indicate that grains with appreciably bigger sizes than those currently available

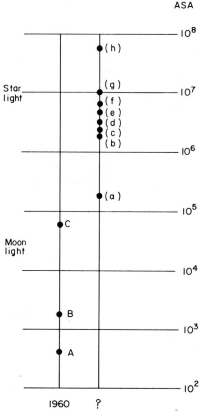

FIG. 11. Summary of estimates of ultimate photographic sensitivity made at the 1960 RPS Symposium. The state-of-the-art is shown for 1960: (a) Royal-X Pan; (b) Polaroid Type 47; (c) negative film as the recording medium with a multi-stage image amplifier. The estimates for the future (?) were made by: (a) MacAdam; (b) Schade; (c) Berg; (d) Fellgett; (e) Rose; (f) Benarie; (g) Meyer; (h) Vendrovsky and Sheberstov. Levels corresponding to moonlight and starlight, and ASA speed ratings, are indicated.

might give acceptable images, and suggested that finding the means of producing larger mono-sized grains might be as worthwhile as efforts to increase quantum sensitivity.

Speed and Resolution

The fundamental relationship between photographic sensitivity, or speed, and image resolution for ideal processes has already been investigated in

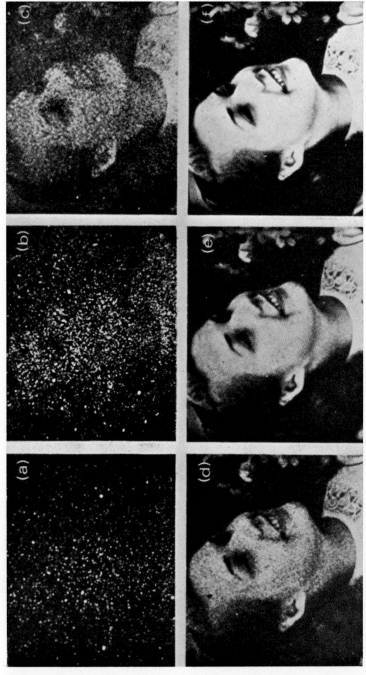

FIG. 12. Illustration of picture quality as a function of available photons[19]: (a) 3×10^3 photons; (b) $1 \cdot 2 \times 10^4$; (c) $9 \cdot 3 \times 10^4$; (d) $7 \cdot 6 \times 10^5$; (e) $3 \cdot 6 \times 10^6$; (f) $2 \cdot 8 \times 10^7$.

4. QUANTUM SENSITIVITY AND ULTIMATE PHOTOGRAPHIC SENSITIVITY 139

general terms in Chapter 1. A classical example of the implications of the quantum nature of light is due to Rose[19], and concerns the number of quanta required to produce an acceptable picture. This is shown in Fig. 12, consisting of a series of pictures made up of an increasing number of photons. These pictures were produced photo-electronically, and each photon (or photo-electron after making allowance for a quantum efficiency less than unity) shows up as a discrete visible speck. The numbers of photons vary from

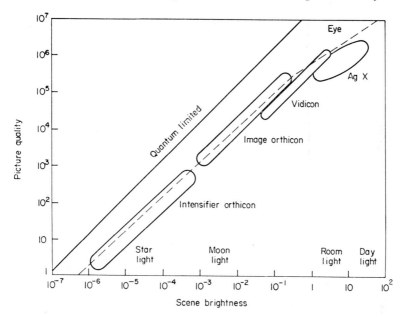

FIG. 13. Relationship between picture quality and scene brightness for several types of detector[20].

3×10^3 in picture (a), which shows no recognizable image, to 2.8×10^7 in picture (f), which shows a portrait of quite reasonable quality. The main features of a face become visible in (b), and in (c) it is possible to discern that it is the face of a woman. This series of pictures gives a clear illustration of the maxim that a photograph can be no better than the number of available quanta.

Also due to Rose[20] is an equivalent plot to that of Fig. 22 of Chapter 1, showing the working region of the resolution-exposure diagram for some practical types of detector in relation to the quantum-limited case. This is shown here in Fig. 13. The adaptability of the eye to available light-level shows clearly, and it is also seen that although conventional silver halide processes are capable of high picture quality, they are rather further displaced from the quantum-limited line than are the various electronic image tubes.

An elaboration of this type of diagram related entirely to photography and including many unconventional processes is due to Ooue[21]. A summary of his version, which included around fifty named processes and types, is shown in Fig. 14. Image resolution is plotted against ASA speed rating. Conventional silver halide processes are seen in general to have the most favourable balance of speed-resolution characteristics, although there is some deviation from a straight-line relationship at very high resolution approaching the wavelength of light. Other principal types of processes are indicated, including electrophotography, photographic papers, photo-polymerization etc. The area J of this diagram covers a wide range of high-resolution low-speed processes, including diazo materials, dichromated gelatin, light-sensitive glass,

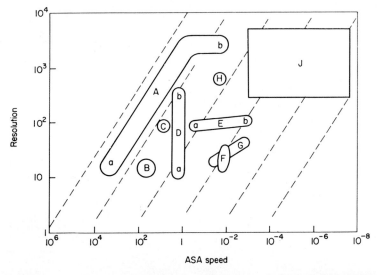

FIG. 14. Speed-resolution characteristics for various types of photographic processes, as calculated by Ooue[21]: (A) silver halide, from (a) Polaroid Type 47 to (b) Kodak Type 649F; (B) photo-polymerization based on charge transfer of N-vinyl-carbazole; (C) 3M Company dry silver process; (D) electrostatic photography from (a) cascade development to (b) liquid development; (E) photographic papers from (a) bromide to (b) chloride; (F) photo-polymerization based on radical initiation; (G) photosolubilization; (H) Itek RS process based on photo-activation of titanium dioxide; (J) a wide variety of low-speed high-resolution unconventional processes (see text).

lead iodide processes, aromatic azides, vesicular photography, etc. It should be stressed that all the areas represented on this diagram are only approximate state-of-the-art illustrations.

Comparative diagrams of this type are extremely useful, especially if they can be related via models such as those of Chapter 3 to the essential input/output mechanisms of each process. In this way the present and future possibilities can be established for each type of process. Schmidlin[35] has compared the factors limiting the ultimate sensitivity of two general types of electrophotographic processes.

Rosell[22] has made a detailed analysis of the sensitivity-resolution characteristics of electronic light amplifiers intended for use in night-vision systems.

4. QUANTUM SENSITIVITY AND ULTIMATE PHOTOGRAPHIC SENSITIVITY

Quantitative details of the illumination associated with moonlight and starlight, and the problems involved in low-light-level amplification have been given by Soule[23]. Some thirteen papers covering various aspects of low-light-level imaging systems were presented at a special seminar on this topic in 1970[24]. The practical factors limiting the resolution aspects of the speed-resolution characteristics for conventional processes will be studied in more detail in Chapter 7.

Ultimate Sensitivity and DQE

Although it is difficult to answer the question of ultimate sensitivity in terms of conventional speed ratings due to the necessity of specifying a criterion of acceptable image quality, the problem is relatively simple in terms of

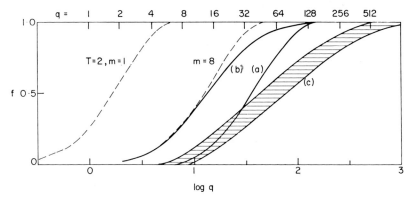

FIG. 15. Normalized density plotted against the logarithm of the average quantum exposure per grain, corresponding to: (a) fractional grain counts of Webb[1] (1948); (b) fractional grain counts of Spencer[5] (1971) for sulphur-plus-gold sensitized emulsion; (c) the band of curves representing commercial films (1960) as estimated by Frieser[25]. Also shown are the equivalent curves for the two-parameter sensitivity model with $T = 2$, $m = 1$ (monosensitivity), and $T = 2$, $m = 8$.

DQE. Without further investigation it is already possible to postulate that ultimate sensitivity corresponds to 100% DQE, with individual grains, or receptors, as big as are permissible according to whatever practical criterion is adopted. However it is instructive to translate this statement into grain properties of actual and hypothetical processes such those modelled in Chapters 1 and 3.

Frieser[25] discussed ultimate sensitivity in terms of quantum sensitivity and grain size, and in relation to existing conventional processes. The band of curves in Fig. 15 shows his estimates relating to 1960. Also shown are the fractional grain counts of Webb from Fig. 1, and those of Spencer from Fig. 7 for his sulphur-plus-gold sensitized emulsion. As discussed previously, it might be reasonable to take these latter two as state-of-the-art curves for 1948 and 1971. The band of curves due to Frieser has lower slope, but this may be due to the fact that they include the spread due to the grain size distribution. Without this they would probably align quite well with the 1948 curve.

The two other curves of Fig. 15 are those based on the two-parameter model for the sensitivity distribution, both curves being for T = 2, but with $m = 1$ and $m = 8$ respectively. The latter gives quite a close fit with the 1971 curve, especially from $f = 0$ to $f = 0.5$. Therefore it might be said that the most sensitive type of modern silver halide process behaves as though its grains each require 2 mechanistic quanta, with a random chance of 1 in 8 that any absorbed quantum will contribute to this number. For this 2-quanta case the limits to future increases in sensitivity would thus correspond to an eight-fold improvement in the utilization of absorbed quanta, the T = 2 mono-sensitivity curve shown in Fig. 15 then being attained. The log$E$ shift would be $\log_{10} 8 = 0.9$ units. The dangers in attempting to relate T and m to actual physical processes occurring in grains during latent image formation have already been stressed.

The extrinsic quantum yield would be increased from 1/16 to 1/2 during such a change, which would leave the intrinsic quantum yield unchanged at 1/2. From the curves of Fig. 10 it is deduced that the maximum DQE would be increased from 6·7% to 54%, (i.e., by a factor of 8), with no change in the recording latitude. For a further increase in sensitivity it would be necessary to achieve the 1-quantum curve, the intrinsic and extrinsic quantum yields then both reaching unity. In this case there would be an actual increase in useful log$E$ latitude, and from the DQE curves of Fig. 2 in Chapter 3 it is noted that the DQE would approach 100% when the exposure was low enough to avoid the possibility of any grain being exposed to more than 1 quantum. It is remembered that these conclusions are for randomly-distributed mono-sized grains with equal statistical access to exposure quanta, and all exposure quanta being absorbed.

Potential sensitivity and DQE improvements related to the quantum efficiency of conventional photographic grains can be compared with those for unconventional processes[26]. For example there might be a fundamental improvement in sensitivity made possible by grains with a multilevel of response rather than a single level, as analysed in Chapter 1. Shaw[27] has calculated some examples of the DQE—resolution—latitude relationships for such grains in comparison to conventional ones. He deduced that in principle a uniform mesh of multilevel grains having threshold and saturation levels of 4 and 32 quanta respectively would have 29 recording levels each corresponding to a change in opaque image area equivalent to the size of conventional silver halide grains, yet would still be capable of a DQE in excess of 90% over a log$E$ range of about one unit, and with a resolution down to about 3 μm.

While discussing quantum efficiency aspects of ultimate sensitivity it is recalled that a quantum efficiency in excess of unity may be relevant for high-energy exposure radiation. Thus photographic sensitivity for X-rays is observed to be higher than that for photons. However, as was demonstrated in Chapter 3, the real effect on the characteristic curve, speed and DQE, is that for a quantum yield of unity, but with an excess size-amplification. For X-rays the quantum efficiency limits of ultimate sensitivity are already attainable.

From the viewpoint of grain size it has been noted that Kirillov[17] concluded that further increases in size would still be capable of producing acceptable

4. QUANTUM SENSITIVITY AND ULTIMATE PHOTOGRAPHIC SENSITIVITY

images in pictorial photography. The highest-speed conventional processes have average grain diameters in the region of 1 μm. The limiting resolution is based on this dimension, but due to light scattering in the photographic layer during exposure the practical resolution is usually found to be a factor of about ten inferior to this. Thus there is no reason in principle why the resolution presently achieved should not be maintained for much bigger grains, so the associated increase in noise level is probably the pertinent factor so far as image quality is concerned. Translating the noise characteristics into the image quality observed in pictorial photography is a difficult task since it involves subjective factors. However there is probably a fairly close relationship between the visual impression and the noise as measured by $G = A\sigma_A^2$ at an image density of about 0.3[28].

Following the arguments of Kirillov, an increase by a factor of about 5 in cross-section area compared with existing high-speed emulsions might still give acceptable images. This would correspond to a characteristic curve shift of $\log_{10} 5 = 0.7 \log E$ units. Combined with the shift of 0.9 units calculated for a quantum efficiency increase associated with a change from $m = 8$ to $m = 1$, this represents an overall shift of $1.6 \log E$ units. A further factor of 5 increase might be possible for grains in a uniform array rather than randomly-distributed, due to the relative decrease in noise level, as analysed in the previous chapter. The technological possibility of improvements for either conventional or unconventional processes, is of course, a separate issue. For example the evidence is that for conventional grains larger sizes show some loss of quantum efficiency, and it is not known whether the causes of this trend are of a fundamental nature. Thus in practice benefits due to increase in size might be more than off-set by quantum efficiency drawbacks.

It is concluded that the whole question of ultimate sensitivity hinges on optimizing DQE. Further increases in speed can then be obtained only by the size-type amplifications which are permissible according to image quality criteria. A fairly detailed analysis of the factors influencing DQE for conventional processes has been given in this and the previous chapter. However, due to its importance to the question of ultimate sensitivity, one further overall DQE analysis is considered here. This is shown in Fig. 16 and is due to Shaw[29]. This model analysis starts with a typical set of parameters which give a DQE curve near that for existing silver halide processes, with a maximum DQE in the region of 2%, and all factors which have a deleterious effect on DQE are then successively removed. The order of removal, and hence the relative changes involved, do not necessarily have any practical or technological significance.

Initially it is assumed that the grains are randomly-distributed with sensitivity defined by $T = 4$, $m = 4$, a typical size distribution, only $\frac{1}{2}$ total quanta absorbed, and 10% fog grains present. The computed maximum DQE for this model is then around 2.5%. Assuming complete absorption, this increases to 5%. Elimination of fog grains shows an increase to 7%. Removal of the size distribution gives 12%. Doubling the utilization of absorbed quanta, (i.e., from $m = 4$ to $m = 2$) produces a maximum DQE of 24%, and a further doubling leads to the 4-quanta mono-sensitive curve and a DQE of 48%.

Moving to the 2-quanta curve gives 54% DQE, while then changing from a random to uniform grain array increases DQE to 70%. Removal of the saturation characteristics, (i.e., having a multilevel grain system from 2 quanta upwards) gives a DQE of 100% at the higher exposure end of the scale where the 2-quanta threshold has been overcome. The only final improvement would then be in reducing the threshold to 1 quantum. Other than defining the size scale of the uniform mesh of grains, it might then be said with some certainty

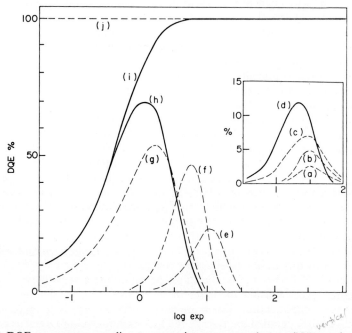

FIG. 16. DQE curves corresponding to successive parameter changes (N.B.: scale of inset is twice that of main diagram)[29]: (a) collection of parameters giving maximum DQE of 2·5% (see text); (b) complete quantum absorption; (c) fog eliminated; (d) size distribution removed; (e) utilization of absorbed quanta doubled; (f) ditto; (g) change from 4-quanta curve to 2-quanta curve; (h) grains in uniform array; (i) saturation removed; (j) 2-quanta threshold removed.

that ultimate photographic sensitivity had been achieved. Pictures would then be as good as the number of quanta available to form them, as illustrated in Fig. 12.

From the experimental evidence cited in Chapter 2, optimum DQE values for the photographic process have probably increased by a factor of at least two over the period 1958–1972, although by changes in which parameters represented in Fig. 16, or in other parameters, is not clear.

Exposure Addition and Multiplication

There are two types of method by which practical photographic processes can be made to operate with an effective increase in sensitivity, and to give

4. QUANTUM SENSITIVITY AND ULTIMATE PHOTOGRAPHIC SENSITIVITY

images at lower light levels than would be possible otherwise. These can be understood in general terms as follows. Suppose that grains require 100 photons for image formation, but that there is only available an average of 2 photons per grain. If we "added" an average of 98 quanta per grain or if we "multiplied" by a factor of 50, then image formation would be possible. Due to the statistical processes involved this is a very over-simplified picture, but serves to illustrate these two types of method.

It is not immediately clear in what way images formed as described would be images of the original photons or of the additive and multiplicative mechanisms. A detailed analysis is more appropriate in terms of the transfer of signal-to-noise ratio, rather than in terms of sensitivity or speed increases, and this will be left until Chapter 5.

Methods based on multiplication of the exposure usually consist of electronic pre-amplification of the available photons, and the amplification may be in terms either of numbers or energy. Such methods were described by Chibisov[30] as an effective means of increasing normal film speed-ratings, and it is recalled that the highest state-of-the-art speed-rating of Fig. 11 related to such a system. Early reports[31,32] of practical image intensifier systems used with the photographic process as recording element, have been followed by the development of systems which yield significant increases in photographic sensitivity[33]. As will be discussed in the next chapter, in DQE terms such systems may approach the ultimate limit of 100%.

The majority of image intensifier devices have been developed for special detection problems as in astronomy, and may involve bulky hardware. From the viewpoint of conventional photography one of the most recent and interesting intensifiers[34] consists of a wafer-thin device plus power supply in a self-contained unit for attachment in a standard type of camera. The film-back of the camera attaches directly to the unit, such that an output fibre optics window comes in direct contact with the film. Only one-thirtieth the amount of light is then required with normal use. Since $\log_{10} 30 = 1\cdot 48$, this increase in speed comes near to that calculated earlier of 1·6 for possible quantum efficiency improvements and acceptable size-amplifications.

Methods based on "addition" involve a simultaneous (or pre- or post-) exposure as well as the normal exposure. Non-additivities in latent image formation and the subsequent variations of quantum efficiency make it especially difficult to estimate the practical advantages in real terms. However a DQE analysis and description of successful applications will be given in Chapter 5.

4.4 Sensitivity and DQE for Two-stage Processes

Two-Stage Imaging

When imaging is a two-stage process, and when both stages are photographic rather than a combination of electronic and photographic, the transfer of image characteristics from one stage to the next has important implications to the question of ultimate sensitivity.

In order to acquire a final positive image in conventional photographic reproduction, it is usual to employ either a so-called reversal process which gives a direct positive image, or a two-stage process for which both stages are negative-working. The latter is nearly always the case when a final print is required rather than a slide for projection, with the exception of certain one-step camera processes. The combination of the characteristics of two negative processes has been widely analysed from the view-point of tone-reproduction (or characteristic curve theory), but of interest here are the speed and DQE relationships. In practice both negative systems employ the amplification factor associated with the development process of the order of 10^9 for silver halide grains, as discussed in Chapter 2. Processes which can be used as the first stage are usually referred to as having camera speed, and in general unconventional photographic processes with lower amplification factors are considered unsuitable for general photographic purposes due to their lack of camera speed. However, since we have seen the limited way in which useful sensitivity is defined by speed-rating, it is useful to consider the DQE transfer for two-stage processes in order to define the mutual requirements of both stages in a fundamental way.

DQE Transfer

No adequate theory exists in the literature for the transfer of DQE from one photographic stage to another, but the following simplified analysis will illustrate the general principles.

The DQE of the first stage is defined as $\varepsilon_1(q)$ at the absolute quantum exposure level, E_q quanta per unit area. By definition the DQE of this stage at the exposure level E_q is:

$$\varepsilon_1(q) = \frac{\gamma_1^2(q)(\log_{10}e)^2}{E_q G_1(q)}, \tag{23}$$

where the variables of the process, γ and G, both relate to this same exposure level.

It will be assumed that printing on to the second stage is at a one-to-one magnification, as for contact prints. Suppose the exposure level E_q of the first stage corresponds to a density D in the final print image of the second stage. The magnitude of the overall gamma of the two stages will be

$$\gamma_{1+2}(q) = \gamma_1(q)\gamma_2(D). \tag{24}$$

The overall noise can be obtained by assuming that the noise of the first stage is transferred through the characteristic curve of the second stage, and then adds to that of the second stage at density D:

$$G_{1+2}(q) = \gamma_2^2(D)G_1(q) + G_2(D). \tag{25}$$

The implicit assumptions and justifications for use of equation (25) will be considered in Chapter 8, where noise transfer is analysed more rigorously.

Replacing γ_1 and G_1 by γ_{1+2} and G_{1+2} in equation (23) leads to:

$$\varepsilon_{1+2}(q) = \frac{\varepsilon_1(q)}{\left(1 + \dfrac{G_2(D)}{G_1(q)\gamma_2^2(D)}\right)}. \tag{26}$$

4. QUANTUM SENSITIVITY AND ULTIMATE PHOTOGRAPHIC SENSITIVITY

The justification of defining the overall DQE in terms only of the exposure level, E_q, relating to the original exposure of the first stage, is that it is at this first camera stage where all available exposure light should be used efficiently, and so DQE is measured with respect to this amount of light. The amount of light available for the second, printing stage, is usually at our disposal and of negligible interest from a DQE criterion. This is why the characteristics of the second stage have been specified as functions of image density level.

When there is a linear magnification of x during printing, the area A in the first stage image will appear as $x^2 A$ in the final print. From the definition $G = A\sigma_A^2$ it is seen that if the product of A and σ_A^2 is a constant independent of A, then the noise of the second stage is effectively added as measured with an aperture $x^2 A$. Equation (25) for the $x = 1$ case is now replaced by the more general relationship:

$$G_{1+2} = \gamma_2^2(D) G_1(q) + \frac{G_2(D)}{x^2}. \tag{27}$$

The overall DQE expression now becomes

$$\varepsilon_{1+2}(q) = \frac{\varepsilon_1(q)}{1 + \frac{1}{x^2} \frac{G_2(D)}{G_1(q)\gamma_2^2(D)}}. \tag{28}$$

In those cases where

$$x^2 \frac{G_1(q)}{G_2(D)} \gamma_2^2(D) \gg 1,$$

the overall DQE will be that of the first stage with a negligible reduction due to the second stage. In other words, if both stages have DQE of, say, 1%, the overall DQE may remain very close to 1% if these conditions are fulfilled.

In order to define the overall DQE-log$E$ curve it would be necessary to know the DQE-log$E$ curve for the first stage, and γ and relative G as functions of density for the second stage. The matching of the exposure scale of the first stage with the density scale of the second stage would be as defined by that adopted for tone-reproduction purposes, and hence the overall DQE could be calculated. Of course it may be that better matching of characteristics could be made according to a DQE criterion, for example by optimizing the area under the overall DQE-log$E$ curve. However for normal pictorial purposes the overall characteristic curve is the first consideration.

Implications

From equation (28) it is concluded that there is no basic reason why there should be any appreciable loss of DQE during printing, since in practice it can usually be arranged that the noise G of the second stage is relatively small compared to that of the first stage. In the special case where the aim is to obtain a print with a maximum overall DQE relating to an original signal of low amplitude confined to a small exposure range, then whereas the first stage should be used at the exposure level which optimizes DQE, the second stage should be used at the density level which optimizes its ratio of γ^2/G. The significance of this ratio will be explored in Chapter 5.

From this analysis we see that the basic requirements of the first (camera) stage are not those of camera speed as such, but of maximum DQE. It is true that conventional silver halide processes which employ 10^9 amplification have normal speed ratings and produce a visible image at the camera stage. However there is no fundamental need for this. If the exposure has been recorded in the first stage with optimum DQE, then as long as the first "image" can be sensed (chemically, physically, electronically, etc.), amplified, processed, etc., to produce a final visible print, the real limits of ultimate sensitivity will be attained.

We conclude that the fundamental need of unconventional processes is not camera speed as such, but high DQE plus the means to read-out the image during what is traditionally the printing stage, where available light, time, techniques etc., are at our disposal. It is thus not always appropriate to view the large family of newer processes such as those included in area J of Fig. 14 merely from a speed viewpoint. In the sensitivity sense a high-DQE low-speed process is fundamentally superior as a first stage compared with a low-DQE high-speed process, so long as second-stage amplification is possible. In which way future moves in the direction of ultimate sensitivity will be achieved technologically, and whether unconventional processes can overtake conventional ones for normal pictorial photography, is open to speculation.

References

1. Webb, J. H. (1948). Absolute sensitivity measurements on single-grain-layer photographic plates for different wave-lengths. *J. Opt. Soc. Amer.*, **38**, 312.
2. Farnell, G. C. and Chanter, J. B. (1961). The quantum sensitivity of photographic emulsion grains. *J. Photogr. Sci.*, **9**, 73.
3. Marriage, A. (1961). How many quanta? *J. Photogr. Sci.*, **9**, 93.
4. Shaw, R. (1973). The influence of grain sensitivity on photographic image properties. *J. Photogr. Sci.*, **21**, 25.
5. Spencer, H. E. (1971). Calculated sensitivity contributions to detective quantum efficiency in photographic emulsions. *Photogr. Sci. Eng.*, **15**, 468.
6. Haase, G. (1960). The sensitivity distribution of photographic emulsions. *Naturwiss.*, **47**, 320.
7. Klein, E. (1962). The theoretical limits of photographic sensitivity. *J. Photogr. Sci.*, **10**, 26.
8. Bayer, B. E. and Hamilton, J. F. (1965). Computer investigation of a latent image model. *J. Opt. Soc. Amer.*, **55**, 439.
9. Hamilton, J. F. and Bayer, B. E. (1965). Investigation of latent-image model: recombination of trapped electrons and free holes. *J. Opt. Soc. Amer.*, **55**, 528.
10. Hamilton, J. F. (1972). Simulated detective quantum efficiency of a model photographic emulsion. *Photogr. Sci. Eng.*, **16**, 126.
11. Feller, W. (1965) "An Introduction to Probability Theory and its Applications; Volume I". Chapter 6. Wiley, New York.
12. Baker, E. A. (1969). Surface conduction theory and the Farnell-Chanter counts. *J. Photogr. Sci.*, **17**, 48.

13. Webb, J. H. Number of quanta required to form the photographic latent image as determined from mathematical analysis of the H and D curve. *J. Opt. Soc. Amer.*, **29**, 309 (1939); **31**, 348 (1941); **31**, 559 (1941); Selwyn, E. W. H. (1939). *J. Opt. Soc. Amer.*, **29**, 518.
14. Silberstein, L. (1941). On the number of quanta required for the developability of a silver halide grain. *J. Opt. Soc. Amer.*, **31**, 343.
15. Ames, A. E. (1973). How many quanta, revisited. *Photogr. Sci. Eng.*, **17**, 154.
16. Berg, W. F. (1961). The ultimate sensitivity in photography today and tomorrow. *Photogr. J.*, **101**, 61.
17. Kirillov, N. I. (1967). "Problems in Photographic Research". Chapter 1. Focal Press, London.
18. Towards the Ultimate Speed in Photography. (1961). *J. Photogr. Sci.*, **9**, 247. (Introduction by G. I. P. Levenson; the Rules; Contributions from M. M. Benarie, W. F. Berg, P. B. Fellgett, D. L. MacAdam, R. Meyer, A. Rose, O. H. Schade, K. V. Vendrovsky and V. I. Sheberstov; Summary by E. W. H. Selwyn.)
19. Rose, A. (1953). Quantum and noise limitations in the visual process. *J. Opt. Soc. Amer.*, **43**, 715.
20. Rose, A. (1970). Quantum limitations to vision at low light levels. *Image Technology*, **12**, No. 4: 13.
21. Ooue, S. (1972). Ultra-microimage. *J. Soc. Photogr. Sci. Tech. Japan*, **35**, 105.
22. Rosell, F. A. (1969). Limiting resolution of low-light-level imaging sensors. *J. Opt. Soc. Amer.*, **59**, 539.
23. Soule, H. V. (1968). "Electro-Optical Photography at Low Illumination Levels". Wiley, New York.
24. "Low Light Level Imaging Systems". (1970). Proceedings of a Seminar organized by the Society of Photographic Scientists and Engineers. An SPSE Publication.
25. Frieser, H. (1961). Limits and possibilities of the photographic process. *J. Photogr. Sci.*, **9**, 379.
26. Shaw, R. (1972). Image evaluation as an aid to photographic emulsion design. *Photogr. Sci. Eng.*, **16**, 395.
27. Shaw, R. (1972). Multilevel grains and the ideal photographic detector, *Photogr. Sci. Eng.*, **16**, 192.
28. Klein, E. and Langner, G. (1963). Relations between granularity, graininess and Wiener-spectrum of the density deviations. *J. Photogr. Sci.*, **11**, 177.
29. Shaw, R. (1973). Some detector characteristics of the photographic process. *Optica Acta*, **20**, 749.
30. Chibisov, K. B. (1961). Methods of raising the effective sensitivity of photographic systems. *J. Photogr. Sci.*, **9**, 26.
31. Wilcock, W. L., Emberson, D. L. and Weekley, B. (1960). An image intensifier with transmitted secondary electron multiplication. *Nature*, **185**, 370.
32. McGee, J. D. and Wheeler, B. E. (1961). Single stage photoelectronic image intensifiers. *J. Photogr. Sci.*, **9**, 106.
33. Reynolds, G. T. (1966). Sensitivity of an image intensifier—film system. *Appl. Opt.*, **5**, 577.
34. Brown, T. J. (1973). Proximity focused image intensifiers for photographic camera systems. *Image Technology*, **15**, 31.
35. Schmidlin, F. W. (1972). Physical theory of charged pigment electrophotography. *I.E.E.E. Trans. Electron Devices*, **ED19**, 448.

Exercises

(1) Fractional grain counts for mono-sized grains with equal statistical access to exposure quanta are as follows:

average quantum exposure per grain	fractional grain count
5	0·09
10	0·26
20	0·60
40	0·93

Making use of the characteristic curves of Fig. 7 of Chapter 3, derive and plot an estimate of the quantum sensitivity distribution, indicating the assumptions which are involved.

(2) Discuss the practical requirements of photographic images which limit the useful increase in image density obtained by increasing the size of the grains.

Make a rough estimate of whether practical photography would be feasible by starlight, assuming that on a clear moonless night the illumination at the earth's surface corresponds to 10^{12} photons $m^{-2} s^{-1}$.

(3) Assume that all emulsion grains contain two "sites" for latent image formation, and that either site needs exactly T absorbed quanta (or whatever proceeds them in latent image formation) to lead to the developability of the grain as a whole. If all absorbed quanta make a contribution to one of the two sites, every quantum having equal chance of contributing to either, derive an expression for the observed sensitivity distribution in terms of T, and plot the distribution for the special case of $T = 4$.

Discuss the DQE advantages of reducing T or having only one site.

(4) To what extent do the concepts of quantum yield, amplification associated with development, and detective quantum efficiency, provide a useful indication of the limits of ultimate photographic sensitivity?

In what way would having grains in a checkerboard array—as opposed to a random distribution in the photographic layer—contribute towards ultimate sensitivity?

(5) A thin photographic layer has mono-sized grains with an equal mixture of two sensitivity types according to the model of equation (12): the first are specified by $T = 2$, $m = 4$; the second by $T = 4$, $m = 2$. Calculate the sensitivity distribution for the grains as a whole which would be deduced from the analysis of fractional grain counts.

Plot the characteristic curve for each type of grain, and hence obtain the overall curve.

(6) Prove that the mean quantum requirements of grains defined by equation (12) is given by $\bar{Q} = Tm$, and that equal maximum values of alpha occur at $Q = (T-1)m$ and $Q = (T-1)m + 1$, where T and m are integers.

4. QUANTUM SENSITIVITY AND ULTIMATE PHOTOGRAPHIC SENSITIVITY

Explain why for this sensitivity model the extrinsic quantum yield is an average value, but the intrinsic quantum yield is constant. What are the extrinsic and intrinsic quantum yields of the class of grains having maximum alpha values?

(7) Give a detailed appraisal of the following statement: "The sensitivity distribution model for photographic sensitivity is questioned on the grounds that it is little more than fitting functions to data and has no direct relevance to photographic theory unless it can be shown independently that the utilization of absorbed quanta proceeds with 100% efficiency". (Reference 15.)

(8) Explain the practical implications of the relationship between speed and grain-size for grains of the same quantum sensitivity.

A photographic layer has mono-sized grains of area 1 μm^2 and is capable of a maximum density of 2. The grains all require 3 quanta for latent image formation, but only 1 in 4 quanta absorbed by a grain contribute to this number. Making reasonable assumptions estimate how many exposure quanta per μm^2 are necessary to give an image density of 0·1, and indicate how you would calculate the conventional ASA speed rating.

(9) Explain in general terms why a photographic image may be printed on a material with very low DQE without significant loss of overall DQE.

If a contact print is made on the same material as the negative, show that DQE_{max} might be expected to be reduced by a factor of around one-half.

(10) Plot the negative binomial distribution for sensitivity based on values $T = 2$, $m = 4$, and also plot the Poisson distribution which gives the same mean value of quantum requirements.

Are there any features of a negative binomial distribution which make it more satisfactory than a Poisson distribution for the representation of measured sensitivity distributions?

5. Detective Quantum Efficiency, Signal-to-Noise Ratio, and the Noise-Equivalent Number of Quanta

5.1 Detective Quantum Efficiency and Signal-to-noise Ratio

Introduction

The basic definition of DQE was established in Chapter 1 by comparing the noise level of a practical radiation detector to that of an ideal detector working at the same exposure level. This definition was made without reference to any particular type of detector, and hence is universal for any radiation detector independent of the technology involved. The basic assumption is that the output of the detector expressed in the relevant practical units, (e.g., density in the photographic case) can be related to the exposure by an operating characteristic (the characteristic curve in the photographic case), and hence the output fluctuations, or noise, can be referred back to the equivalent input fluctuation. In the ideal case the lower noise limit is set by the quantum fluctuations in the exposure, the size of which can be calculated from the quantum statistics. This lower noise limit defines the upper limit of 100% DQE.

In Chapters 2, 3 and 4 the practical values of DQE measured for conventional processes were interpreted in terms of models which included most of the important photographic parameters and inefficiencies. In this way it is possible to understand the limitations of the photographic process and to indicate the directions in which further improvements might be achieved. However, in applied photography the interest usually comes from the direction of incorporating the photographic process in an overall detection system. In such applications there is little concern with the photographic process *per se*, and it may be regarded as a "black-box" with prescribed DQE characteristics. Some implications of DQE in applied detection problems will be studied in this present chapter, but first it is necessary to analyse the transfer of signal-to-noise (S/N) ratio that takes place during photographic recording.

S/N Ratio Analysis

From equation (30) of Chapter 1, the mean-square fluctuation in image density, σ_A^2, as measured with a scanning aperture of area A, may be related back to the equivalent input exposure fluctuation by

$$\sigma_A^2 \left(\frac{dq}{dD}\right)^2 = \sigma_A^2 \left(\frac{q_A}{\gamma \log_{10} e}\right)^2, \tag{1}$$

5. DETECTIVE QUANTUM EFFICIENCY

where q_A denotes the absolute exposure in terms of the average number of exposure quanta per image area A. Assuming Poisson statistics for the exposure quanta,

$$\overline{\Delta q_A^2} = q_A. \tag{2}$$

By definition it follows from equations (1) and (2) that

$$\text{DQE} = \frac{\overline{\Delta q_A^2}}{\sigma_A^2 \left(\frac{dq}{dD}\right)^2} = \left(\frac{\gamma \log_{10} e}{\sigma_A}\right)^2 \frac{1}{q_A}. \tag{3}$$

Equation (3) can be given another interpretation in terms of S/N ratio, which is usually the important criterion in detection problems. When the input noise is only that due to the quantum fluctuations, the input S/N ratio associated with q_A exposure quanta may be defined as the ratio of this number to the rms fluctuation, i.e.;

$$(S/N)_{IN} = \frac{q_A}{\sqrt{q_A}} = \sqrt{q_A};$$

and hence

$$(S/N)_{IN}^2 = q_A. \tag{4}$$

The output S/N ratio will be defined by the input number of quanta divided by the output noise referred back in terms of the exposure, i.e.;

$$(S/N)_{OUT} = \frac{q_A}{\sigma_A \left(\frac{dq}{dD}\right)}, \tag{5}$$

q for quanta

and so from equations (1) and (5),

$$(S/N)_{OUT}^2 = \left(\frac{\gamma \log_{10} e}{\sigma_A}\right)^2. \tag{6}$$

From equations (4) and (6):

$$\frac{(S/N)_{OUT}^2}{(S/N)_{IN}^2} = \frac{1}{q_A} \left(\frac{\gamma \log_{10} e}{\sigma_A}\right)^2. \tag{7}$$

By comparison of equations (3) and (7) it is seen that DQE may also be expressed in the form:

$$\text{DQE} = \frac{(S/N)_{OUT}^2}{(S/N)_{IN}^2}. \tag{8}$$

This is an intuitively satisfying relationship, since the output S/N ratio will always be degraded in comparison with that of the input, except in the special case of an ideal detector with 100% DQE. The output S/N ratio will never exceed that of the input. Figure 1 shows this "black-box" representation of the photographic process in terms of DQE and S/N ratios.

The input and output fluctuations as measured with scanning aperture A might be as illustrated in Fig. 2(a). In the above analysis the signal has been

Fig. 1. Photographic process as a "black-box" DQE operator on the input S/N ratio, when the input noise is due to quantum fluctuations.

defined in terms of the mean input and output levels with respect to the origin. In practical detection problems the input and output S/N ratios often refer to discrimination between two exposure levels as illustrated in Fig. 2(b). If

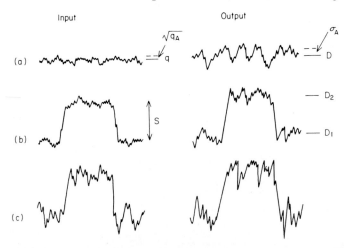

Fig. 2. Illustration of the input and output S/N ratios with waveforms recorded by a scanning aperture A: (a) input quanta with rms $\sqrt{q_A}$; output density level D with rms of σ_A; (b) as for (a), but with signal as distance between two levels; (c) as for (b), but with background noise as well as quantum noise.

we denote the average exposure levels by $q_A + \dfrac{S_A}{2}$ and $q_A - \dfrac{S_A}{2}$, the signal may be defined by the difference:

$$\text{signal} = q_A + \frac{S_A}{2} - \left(q_A - \frac{S_A}{2}\right) = S_A.$$

If the two levels are sufficiently close together that the photon noise at the upper and lower levels may be assumed to approximate to that at the average level between them, then:

$$\text{noise} = \sqrt{q_A},$$

and the input S/N ratio will be

$$(S/N)_{IN} = \frac{S_A}{\sqrt{q_A}}. \tag{9}$$

D for density

If the corresponding output density levels are denoted by D_1 and D_2, then from the characteristic curve it will be deduced that the average output difference D_2-D_1 corresponds to an input signal of S_A quanta, but the uncertainty, in this deduction will be greater than that at the input stage. If it is assumed that the output density range corresponding to the low input exposure range is small enough for the slope $\frac{dD}{dq}$ and the output noise σ_A to be taken as constant over this range, then the output noise referred back in terms of uncertainty in the exposure will be $\sigma_A \frac{dq}{dD}$, and the output S/N ratio will be

$$(S/N)_{OUT} = \frac{S_A}{\sigma_A \frac{dq}{dD}}. \tag{10}$$

From equations (9) and (10):

$$\frac{(S/N)_{OUT}^2}{(S/N)_{IN}^2} = \frac{q_A}{\sigma_A^2 \left(\frac{dq}{dD}\right)^2},$$

and thus by definition

$$\frac{(S/N)_{OUT}^2}{(S/N)_{IN}^2} = DQE. \tag{8}$$

Equation (8) is thus still applicable when the S/N ratio is interpreted in this more general sense. So far as S/N ratio transfer is concerned, photographic linearity can be defined as the exposure region over which DQE does not vary sufficiently to invalidate equation (8) for practical purposes.

In practice even more general considerations may have to be given to the definition of S/N ratio, since when attempting to discriminate between two exposure levels there may be so-called background noise (from other, unwanted signals, etc.), and this may be much bigger than the photon noise. This type of situation is illustrated in Fig. 2(c). Implications and definitions of S/N ratio in terms of DQE in such detection problems have been studied by Marchant and Millikan[1,2]. The exact form of the relationship between input and output S/N ratios will now depend on the statistics of the background noise, its additive or multiplicative nature in terms of the output density fluctuations, etc. Wolfe[3] has discussed why equation (8) should be used with caution in the general practical case, and we can summarize by saying that it refers only to the transfer of S/N ratio in quantum-limited cases.

Although in the general case, equation (8) may no longer hold such a simple form and DQE will no longer be a simple linear operator in the S/N ratio sense, the ratio of output to input S/N ratios will usually increase as DQE increases and be maximum when the image is recorded in the region of maximum DQE. When the signal and the background noise are so large that they are recorded in the image over a region within which there is wide variation in DQE, then the situation becomes even more complicated. The output S/N ratio must then take into account this variation of DQE as well as the nature of the background noise.

Noise-Equivalent Number of Quanta

Another useful way of interpreting DQE has its origins in the S/N ratio relationships above, and in the concepts explored in Chapter 1 concerning the ideal photographic process as an array of individual photon counters. In the practical case the photographic process may be treated as an inefficient photon counter, making a lesser number of counts than the quanta available to it[4]. The problem is to define this lesser number of counts in a satisfactory manner. From equation (4):

$$\frac{(S/N)_{OUT}^2}{(S/N)_{IN}^2} = \frac{(S/N)_{OUT}^2}{q_A}, \tag{11}$$

where $(S/N)_{OUT}^2$ is defined by equation (6). By definition both sides of equation (11) must be a pure ratio, and so it is possible to define

$$(S/N)_{OUT}^2 = q_A', \tag{12}$$

where by equation (6):

$$q_A' = \left(\frac{\gamma \log_{10} e}{\sigma_A}\right)^2. \tag{13}$$

The physical interpretation of q_A' is straightforward. If each quantum could be counted by an ideal detector, then at the exposure level q_A the S/N ratio would be that corresponding to q_A quanta. However due to inefficiencies of practical detectors such as the photographic process it is found that the S/N ratio is only equivalent to that associated with a lesser number q_A' quanta. It might therefore be said that the image associated with the practical detector is worth q_A' quanta, while that associated with an ideal detector working at the same exposure level would be worth q_A quanta. It is as though the photographic process were used as a photon counter, and in attempting to count q_A only q_A' of them were actually counted. Due to this manner of definition and interpretation q_A' is usually termed the noise-equivalent number of quanta (NEQ).

In spite of the straightforward nature of this definition of NEQ, caution must be taken in equating it with the physical number of events in the output, as for example the number of developed grains in the photographic image. The remarks of Chapters 1 and 4 distinguishing between responsive quantum efficiency (RQE) and DQE are recalled here. The concept of NEQ originates from that of DQE rather than RQE, and has a natural upper limit defined by $q_A' = q_A$. The number of output events has no such upper limit, and as already discussed may exceed the input numbers for image intensifier systems or the photographic process exposed to X-rays. Even though the number of grains in the photographic image is usually much less than the number of exposure photons, the numbers will still not generally coincide. We shall examine the implications of NEQ in more detail later in this chapter, but note here that NEQ would coincide with the number of image grains in the special (non-typical) case where all imaging inefficiencies acted on the exposure quanta as

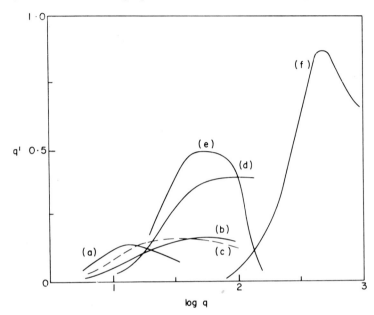

FIG. 3. Noise-equivalent quanta as a function of log q, for six aerial films[5]. Both q' and q are in terms of quanta μm^{-2}. The films were identified as: (a) AERIAL N; (b) HYPERPAN AERIAL; (c) HP3 AERIAL; (d) AERIAL A; (e) FP3 AERIAL; (f) HR AERIAL.

a series of random-loss filters and where the remaining quanta each caused exactly one grain to develop.

If the NEQ per unit image area is defined as q', then since quanta are additive in the sense of their contribution to the square of the S/N ratio, as in equation (4), then it follows that

$$q_A' = Aq', \tag{14}$$

and hence

$$q' = \frac{1}{A}\left(\frac{\gamma \log_{10} e}{\sigma_A}\right)^2 = \frac{(\gamma \log_{10} e)^2}{G}. \tag{15}$$

When the product $G = A\sigma_A^2$ is independent of the scanning aperture A, we arrive at the satisfying conclusion that the NEQ per unit image area is independent of the aperture used to measure it. A systematic series of measured NEQ curves[5] is shown in Fig. 3. The curves concern six aerial films whose noise characteristics were shown in Fig. 25 of Chapter 2. These six films cover a wide speed range, and the NEQ per unit area is shown plotted against the logarithm of the average quantum exposure per unit area. The maximum values fall within the range of 0·1–0·9 per μm^2, and in general they increase with decreasing film speed, (i.e., with decreasing noise, G). The not-surprising conclusion is that in a given image area fine-grained films can count more quanta than coarse-grained films. This conclusion has some interesting implications in the design of photographic systems for detection problems.

S/N Ratio Optimization

We consider first a simple numerical example[6] based on the curves of Fig. 3. This relates to the maximum NEQ values per unit image area for AERIAL N and HR AERIAL, and the assumption that there is available for exposure, say in some fixed experimental time interval, a total of 16 000 quanta. If it is required to record this exposure with maximum S/N ratio, the respective areas of film to be used to record all the signal at maximum NEQ per unit image area, and the respective numbers of quanta which will then be counted, are as shown below.

	NEQ_{max} per unit area. $q'\ \mu m^{-2}$	Exposure level per unit area. $q\ \mu m^{-2}$	Area to count 16 000 quanta $A\ \mu m^2$	Total quanta counted. $A q'\ \mu m^{-2}$
HR AERIAL	0·86	500	$\frac{16\ 000}{500} = 32$	$32 \times 0·86 = 27·5$
AERIAL N	0·13	16	$\frac{16\ 000}{16} = 1\ 000$	$1,000 \times 0·13 = 130$

It is concluded that it is important to match the total exposure quanta to the film area which will maximize the total NEQ. As we shall see shortly, DQE provides the criterion for this, and for both films in this example a greater total number of quanta could be counted at the exposure level for maximum DQE. Figure 4 shows the straight-line relationship between total NEQ and film area, and for a given film area total NEQ for HR AERIAL is always greater than that for AERIAL N. However for a fixed total number of exposure quanta it is possible to count more of them with HR AERIAL. The effective rate at which the counting progresses along the straight lines of Fig. 4 has been represented by the length of the dashes: the crosses indicate the count-level reached for a fixed number of 16 000 exposure quanta as in the above example.

The relationship between DQE and S/N ratio follows directly from equation (8), since for a fixed input S/N ratio the output S/N ratio will be proportional to DQE rather than NEQ, i.e.,

$$(S/N)_{OUT}^2 \propto DQE.$$

Thus for two films having DQE values DQE_1 and DQE_2 at the operative exposure levels per unit image area,

$$\frac{(S/N)_{OUT,1}^2}{(S/N)_{OUT,2}^2} = \frac{DQE_1}{DQE_2}. \qquad (16)$$

The relationship between NEQ, DQE, exposure level, available film area and magnification, have been considered in some detail with respect to typical detection problems where equation (8) can be applied[4]. Figure 5 depicts the

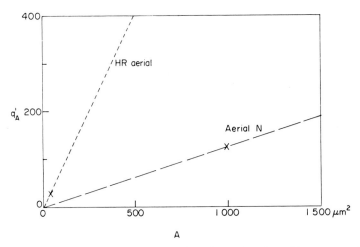

FIG. 4. Noise-equivalent number of quanta plotted as a function of image area, for two films[5] at exposure levels which optimize their NEQ values per unit image area. Lengths of dashes indicate the relative rates at which the total NEQs increase with available exposure, assuming the exposure is spread over the image area so as always to give optimum NEQ per unit area. The crosses indicate NEQ levels reached for a fixed exposure (see example in text).

comparison between the images obtained when using a fine-grained film with high NEQ, and a coarse-grained film with low NEQ but high DQE. These represent the images of a low-contrast signal on a uniform background, and at the respective magnifications which optimize the NEQ of the fine-grained film and the DQE of the coarse-grained film. Scanning apertures as shown—equal to the size of the signal in the image—would then be appropriate in achieving the comparative S/N ratios as calculated. Which of the two images yields the highest S/N ratio will then depend only on the respective DQE values.

This type of calculation is also important when deciding the optimum exposure level of detection for a given film, and whether this should be based on optimization of NEQ or DQE. Figure 6 shows both as a function of image density level as published for HP3 AERIAL film[5]. This illustrates the general observation that maximum DQE occurs at lower density level than NEQ. Above the level at which DQE is optimum the NEQ per unit image area can

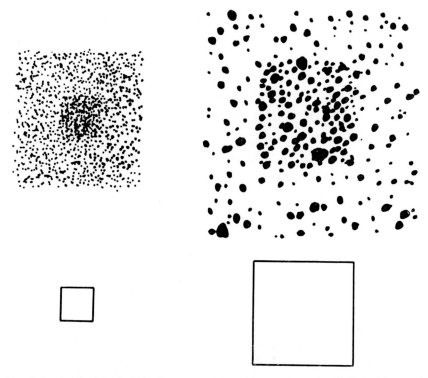

FIG. 5. Images for identical absolute exposures to a low-contrast signal[4], but with magnifications to optimize the NEQ for a fine-grained film and DQE of a coarse-grained film.

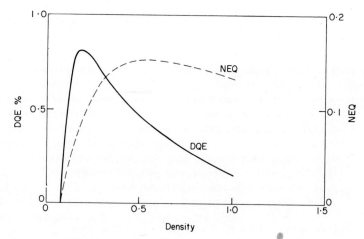

FIG. 6. DQE and NEQ for HP3 AERIAL film[5] as a function of image density level. NEQ is in terms of quanta μm^{-2}.

still be increased up to the level at which it itself is optimum, but at an ever-decreasing rate. Figure 7 shows the same example as Fig. 6, but now NEQ and DQE are plotted on the log$E$ scale. In general terms we may think of the input quanta and the quantum requirements of individual grains being ideally matched at optimum DQE in the S/N ratio sense. Increasing mis-match occurs above this level due to more and more of the individual grains being saturated. Beyond maximum NEQ the mis-match becomes unfavourable according to either a NEQ or DQE criterion.

The comparative DQE and NEQ characteristics as calculated according to the model described in Chapter 3 predict the behaviour which is observed in

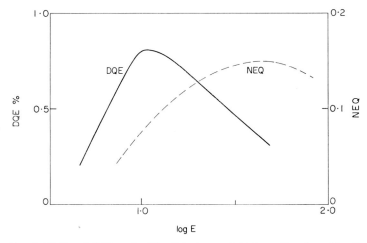

FIG. 7. DQE and NEQ as for Fig. 6, but now plotted as a function of log$E$.

practice. Figure 8 shows model characteristics calculated for a sensitivity distribution similar to those measured experimentally and a constant grain-size of 1 μm^2 cross-section area with 10% fog grains. The density is scaled to $D_{max} = 1$, and on this normalized scale the difference between the maximum values of NEQ and DQE corresponds to about 0·2 density units.

In principle equation (16) applies to any two comparative detection conditions, either for the same film at two different exposure levels or for two different films at the same or different exposure levels. So for the exposure levels which optimize NEQ and DQE for a given film:

$$\frac{(S/N)_{OUT,DQE_{max}}^2}{(S/N)_{OUT,NEQ_{max}}^2} = \frac{DQE_{DQE_{max}}}{DQE_{NEQ_{max}}}, \qquad (17)$$

and for the example of Figs 6 and 7,

$$\frac{DQE_{DQE_{max}}}{DQE_{NEQ_{max}}} = \frac{0·80}{0·37} = 2·16$$

The practical exploitation of optimum DQE values will depend on several factors. It is recalled that equation (17) only holds when the input noise is quantum-limited. It must also be noted that a DQE criterion is immaterial if there is no limit to the available exposure (although such cases cannot usually be classed as posing detection problems). In retrospect it is now easier to understand the result of Chapter 4, where it was concluded that the second, printing, stage of a two-stage process is optimum when used at the density level which optimises the ratio γ^2/G. We see now that this is the level which gives maximum NEQ, or S/N ratio, *per unit image area*, the criterion of DQE being the important one for the first stage where there is an exposure criterion.

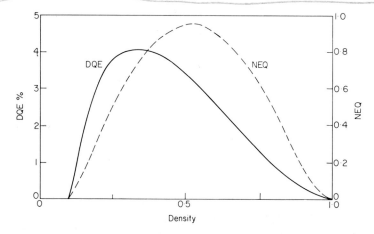

FIG. 8. DQE and NEQ as a function of density, calculated (random grain model of Chapter 3) for mono-sized grains (1 μm²), with typical sensitivity distribution and 10% fog grains.

According to the experimental constraints in any detection problem it may be necessary to devise special methods and techniques to achieve optimum S/N ratio. Some of these will be described in the following section, but the principle in common to all such methods is that the input should be matched statistically to the response of the detector in the DQE sense.

5.2 Detection Problems

Contrast and Energy Detectivity

Other useful concepts which are closely related to DQE and NEQ have been investigated—notably by Jones[7,8] and Zweig[9,10], concerning the ability of the photographic process to detect minimum amounts of signal energy or contrast. Jones defined the detectivity $\mathscr{D}$ as the output S/N ratio per unit input quantum, and hence in the present notation

$$\mathscr{D} = \frac{S/N_{\text{OUT}}}{q_A}. \tag{18}$$

5. DETECTIVE QUANTUM EFFICIENCY

From equations (4) and (8):

$$\mathscr{D}^2 = \frac{\text{DQE}}{q_A}. \tag{19}$$

Zweig denoted the reciprocal of the detectivity as the noise-equivalent energy, NEE:

$$\text{NEE}^2 = \frac{1}{\mathscr{D}^2} = \frac{q_A}{\text{DQE}}. \tag{20}$$

It is seen that this is a measure of the number of exposure quanta needed to make the output S/N ratio as high as that for an ideal detector with DQE = 1.

Zweig also used the term noise-equivalent contrast, NEC, defined as

$$\text{NEC} = \frac{\text{NEE}}{q_A}, \tag{21}$$

and hence

$$\text{NEC}^2 = \frac{\text{NEE}^2}{q_A^2} = \frac{1}{q_A \text{DQE}}, \tag{22}$$

while Jones termed the reciprocal of NEC the contrast detectivity:

$$\mathscr{D}_c = \frac{1}{\text{NEC}}, \tag{23}$$

and hence

$$\mathscr{D}_c^2 = \frac{1}{\text{NEC}^2} = q_A \text{DQE}. \tag{24}$$

In the previous section the NEQ corresponding to image area A has been defined as

$$q_A' = q_A \text{DQE}, \tag{25}$$

and hence

$$\mathscr{D}_c^2 = \text{NEQ}; \tag{26}$$

i.e., the square of the contrast detectivity is simply the noise-equivalent number of quanta. From equations (20) and (23):

$$\mathscr{D}\mathscr{D}_c = \frac{1}{\text{NEE}} \frac{1}{\text{NEC}} = \text{DQE}. \tag{27}$$

Zweig[10] calculated NEE and NEC for mono-sized grains in a uniform array with constant sensitivities, thus corresponding to the DQE curves of Fig. 1 in Chapter 3. Such NEE and NEC curves are shown in Fig. 9 as a function of exposure level for constant quantum requirements of Q = 1, 2, 4 and 8. Since both NEE and NEC depend on the image area, an area of $100 \times 100 \ \mu m^2$ has been assumed for these calculations.

By comparison of the DQE curves for the random and uniform grain arrays (Figs. 1 and 2 of Chapter 3) it is concluded that at all exposure levels the

NEE and NEC values would be greater for a random array of grains. In other words, more input energy or contrast would be necessary to produce the same effect in the image. Similarly, other photographic inefficiencies such as those considered in Chapters 3 and 4 will also increase both NEC and NEE. It is noted that NEE and NEC are "upside-down" concepts compared with NEQ, which is low when the detector efficiency is low. However it is often useful to

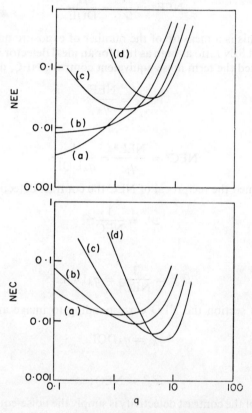

FIG. 9. Noise-equivalent energy NEE and contrast NEC for mono-sized grains in a uniform array[10], and with constant quantum requirements of: (a) $Q = 1$; (b) $Q = 2$; (c) $Q = 4$; (d) $Q = 8$.

specify a detector by the amount of input necessary to give a defined output effect, and this can be done in terms of quantities such as NEE and NEC.

Jones[7] calculated NEE as a function of exposure level based on the properties of four typical negative films. He found values in the region of 1×10^{-9} to 5×10^{-9} ergs for an area equivalent to one image resolution unit, based on the spread function.

By inspection of curves such as those of Fig. 9, Zweig concluded that for quantum requirements greater than $Q = 2$ the NEE could be reduced by use of a uniform additive exposure, and that NEC could be reduced even for quantum requirements of $Q = 1$ and 2. These conclusions are based on the existence or not of a negative slope for the curves of Fig. 9 in the low exposure region.

Pre-Exposure Advantages

The following analysis is based on the arguments of Jones[8]. We consider an arbitrary DQE-EXP curve as measured for some practical detector, plotted on a log-log scale as shown in Figure 10. Also shown are two straight lines

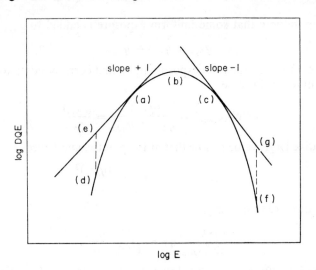

FIG. 10. A practical DQE-exposure curve plotted on a log-log scale, with tangents of slope $+1$ and -1.

with slopes of $+1$ and -1, being tangents to the curve at points a and c respectively. The maximum value of DQE lies at the point b.

By equation (19) the slope of the curve will be given by

$$\frac{d(\log \text{DQE})}{d(\log q_A)} = \frac{d(\log q_A + 2 \log \mathscr{D})}{d(\log q_A)} = 1 + 2 \frac{d(\log \mathscr{D})}{d(\log q_A)}.$$

At the point a on the curve the slope is $+1$, and so in this case:

$$\frac{d(\log \mathscr{D})}{d(\log q_A)} = 0, \text{ and hence } \frac{d\mathscr{D}}{dq_A} = 0.$$

A similar argument shows that when the slope is -1, as at point c on the curve, then

$$\frac{d\mathscr{D}_c}{dq_A} = 0.$$

These results lead to the possibility of increasing DQE when the exposure level is less than that at point a on the curve where the slope is $+1$. Such an increase would be possible in principle by "adding" a uniform background exposure. The amount to be added and the corresponding increase follow by considering some point such as d on the curve, at a lower exposure level than that of a. From equations (4) and (8):

$$(S/N)_{IN}^2 = q_{A,d}; \quad (S/N)_{OUT}^2 = DQE_d \, q_{A,d} \tag{28}$$

where the suffix d indicates the values of q_A and DQE at the corresponding point (d).

We suppose now that some uniform exposure is added to $q_{A,d}$ such that

$$q_{A,d} + q_{add} = q_{A,a},$$

i.e., sufficient to bring the total exposure up to that corresponding to the point a on the curve. In this case

$$(S/N)_{IN,add}^2 = \frac{(q_{A,d})^2}{q_{A,d} + q_{add}} = \frac{(q_{A,d})^2}{q_{A,a}}$$

The operative DQE value is now that at the point a, and hence

$$(S/N)_{OUT,add}^2 = DQE_a \frac{(q_{A,d})^2}{q_{A,a}}. \tag{29}$$

From equations (28) and (29):

$$\frac{(S/N)_{OUT,add}^2}{(S/N)_{OUT}^2} = \frac{DQE_a \, q_{A,d}}{DQE_d \, q_{A,a}}. \tag{30}$$

From equation (30) it is concluded that the effect of the uniform background exposure is as though the DQE has been increased from DQE_d to the value DQE_{add}, where

$$DQE_{add} = DQE_d \frac{DQE_a \, q_{A,d}}{DQE_d \, q_{A,a}} = DQE_a \frac{q_{A,d}}{q_{A,a}}. \tag{31}$$

If the DQE value in Fig. 10 corresponding to the intersection e of the vertical drawn from d with the tangent of slope $+1$ drawn from a, is denoted by DQE_e, then by the geometry of the figure:

$$\log DQE_a - \log DQE_e = \log q_{A,a} - \log q_{A,d},$$

and hence

$$DQE_e = DQE_a \frac{q_{A,d}}{q_{A,a}}. \tag{32}$$

By comparison of equations (31) and (32):

$$DQE_{add} = DQE_e. \tag{33}$$

In other words, it is possible in principle to make use of an additive background exposure to raise the DQE from that at point d to that of point e. It should be kept in mind that curves may exist for which there is no tangent of slope +1, for example for exposures to X-rays, but tangents usually exist for normal exposures with conventional photographic processes. Figure 11 shows curves plotted in the form of Fig. 10, but now relating to experimentally-measured DQE values for four film types[8]. The tangents of slope +1 are indicated by the dashed lines, and potential DQE increases are seen to be possible for a wide region of low exposure levels.

The above analysis has demonstrated how compensation may be made for so-called detector underloading by an additive exposure (although such methods are usually referred to as pre-exposure techniques, the uniform exposure might be added simultaneous with, or after, the signal exposure). Similarly, it may be shown that for detector overloading at the higher end of the exposure scale, the use of a neutral filter to reduce the average exposure

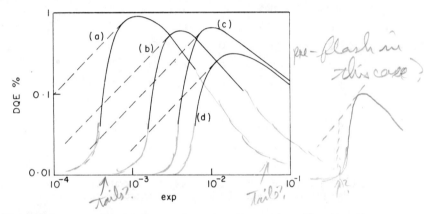

FIG. 11. DQE—exposure curves[8] and tangents of +1 for: (a) Royal-X; (b) Tri-X; (c) Plus-X; (d) Pan-X. The exposure is in terms of ergs cm^{-2}.

level will allow the DQE to be increased to that value associated with the tangent of slope −1, as illustrated by points f and g of Fig. 10. In general however, overloading is not such a serious practical problem as that posed by underloading, since it is usually those applications in which there is insufficient available exposure which raise detection, and hence DQE, problems.

A practical detection test object which is particularly useful for investigating the influence of a background exposure has been described by Zweig[10]. This is represented in Fig. 12, and consists of a matrix of transparent circular discs on an opaque background. The discs in a given column have a fixed diameter, the diameters of different columns ranging from 65 μm to 5 850 μm. The ratio between diameters in successive columns is $\sqrt{2}$, giving a ratio between areas of 2. Two neutral step wedges are placed over this matrix at right angles, with wedge density increasing by 0·3 in successive columns, (i.e., with transmittance decreasing by a factor of 2), and increasing by 0·15 in successive rows. Thus for a uniform overall exposure the total number of photons per second through a disc in a given row is constant, and the number in successive rows is increased by $\sqrt{2}$.

In one test with a coarse-grained film, representing one of the fastest processes then available, Zweig used pre-exposures of 1, 10 and 100 photons per μm^2. These led to visible images for 9, 10 and 7 rows of dots respectively, corresponding to totals of 2000, 1400 and 4000 photons (at 430 nm). Thus Zweig concluded that 1400 photons represented the lowest number of photons necessary to produce a visible photographic image, and that the pre-exposures

$\sqrt{area}$ (μm)	=	175	250	350	500	700	1000	1400	···
rel. area	=	1	2	4	8	16	32	64	···
rel. trans.	=	1	½	¼	⅛	1/16	1/32	1/64	···

rel. trans.	total photons
1	350
$\sqrt{2}$	500
2	700
$2\sqrt{2}$	1000
4	1400
$4\sqrt{2}$	2000

F.G. 12. Illustration of test matrix of exposure aperatures and transmittances devised by Zweig[10] for practical determination of pre-exposure advantages.

of 1 and 100 photons per μm^2 corresponded to regions of detector underloading and overloading.

Practical Detection Techniques

The systematic application of these S/N ratio relationships can be greatly beneficial in the design and optimization of photographic systems. The general rule is that the signal and detector should be matched statistically in the DQE sense. However, due to the complexities of the photographic process (for example, its non-linearities and non-additive latent image effects) and the diversity of practical detection problems (from those with complex instrumentation and automatic image read-out and analysis, to those involving simple camera systems with visual image read-out), the conversion of principles to practical techniques must be carried out with care, and each type of application will evolve its own set of practical rules.

By its nature, astronomy has provided many of the most fundamental problems in photographic recording and detection. Baum[11] and Hoag and Miller[12] have reviewed the application of photographic materials to detection problems in astronomy, and described some of the special techniques which are used. Baum[11] and Fellgett[13,14] have discussed the relevance of DQE and NEQ values to the design of telescope systems.

A study of the advantages of uniform background exposures has been made by Bird et al.[15], who interpreted the combined influence of signal and background on the processes of latent image formation. In this way they defined a general receipe for raising the DQE of an existing film, as follows. The main signal exposure is applied as a fast exposure which carries grains to or beyond a threshold of stable centres of two silver atoms, but does not render them developable. The film is then given a long, low-intensity post-exposure, which generally produces no effect in grains not already possessing a stable centre (due to low-intensity reciprocity failure), but carries some of those grains with a stable centre into a developable state. These conditions would be appropriate for normal high-speed photography, with a typical signal exposure of 10^{-5} seconds and a post exposure of 1 hour. In the event of the two exposures being insufficient to lead to the image density of maximum DQE, a uniform exposure would also be added simultaneous with the signal exposure.

For normal exposure times Bird et al. described how the first signal, exposure might still be made effectively "fast" by cooling the film, typically in the region of $-30°C$ to $-60°C$ for an exposure in astronomy of the order of 10^3 seconds. Miller[16] has given a general review of the practical application of pre-exposure techniques in astronomy.

A detailed practical report of the advantages of adding a background exposure has been given by Guttman[17]. In this case the uniform exposure was added simultaneous with the main exposure, and signal exposures through apertures in the range of 25 to 2000 μm^2 were tested. One of his results is illustrated in Fig. 13. This is for Tri-X film under specified development conditions, with a background exposure which produces an image density of 0·53. Without background exposure about four times signal intensity was necessary to obtain the same state of detectability compared with the use of background.

Reciprocity-law failure for the long exposure times sometimes used in astronomy has also been studied in the DQE sense by Porteous and Shaw[18]. Figure 14 shows the maximum DQE for a typical plate used in astronomy as a function of exposure time in the region of 1 to 1000 seconds. Maximum DQE decreases by a factor of almost two at the higher end of the scale. Shaw and Shipman[19] studied the influence of development time on DQE for a range of four films and three developers, and found that maximum DQEs are generally achieved with lower development times than those taken as standard for normal tone-reproduction purposes.

Smith et al.[20] have described how the response of spectroscopic plates was increased using a special technique of baking in a controlled nitrogen atmosphere, and subsequently it has been claimed[21] that a DQE of 4% has been achieved by this technique for detection problems in astronomy. Shaw[22] has demonstrated why practical techniques which suppress fog, may be as important as those which lead to real increases in sensitivity by quantum efficiency improvements.

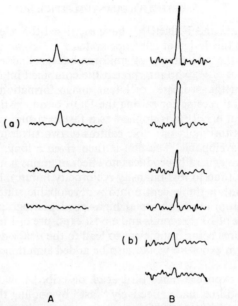

FIG. 13. Illustration of the influence of background exposure for Tri-X film[17]: (a) without background; (b) with background corresponding to a density of 0·53. Successive traces represent an increase of the main (signal) exposure. Equivalent levels of detection are shown at (a) and (b) respectively, the intensity required at (a) being 4× that at (b).

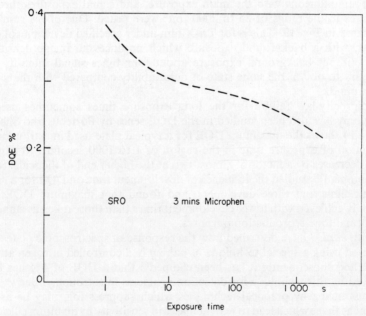

FIG. 14. Maximum DQE as a function of exposure time for a plate used in astronomy[18].

5. DETECTIVE QUANTUM EFFICIENCY

Many other investigations of the photographic process from a S/N ratio and detection viewpoint have been described in the literature. Courtney-Pratt[23] gave a detailed systems analysis of the problem of photographing a satellite in the region of the Moon. A comparison of the photographic process with standard photon-counting techniques in the detection of weak optical signals has been described by Amoss and Davidson[24]. Falconer[25] has considered image enhancement from the viewpoint of photographic S/N ratios and DQE. Other investigations of S/N ratio and photographic detection include those of Weinstein[26], Shepp and Kammerer[27], Yu[28] and Barnard[29]. Pryor[30] has given one of the most extensive descriptions of general image analysis in terms of DQE. Visual aspects of detection in terms of S/N ratio have been studied by Chambers and Courtney-Pratt[31], who also gave an extensive bibliography on visual properties. Hayen and Verbrugghe[32] used S/N ratio techniques to compare photographic and electronic systems for application in colour television. Several conferences have been held on topics closely related to the S/N ratio of imaging systems, including ones on computerized imaging techniques and image information recovery[33].

The principle of obtaining an increased S/N ratio by taking the composite S/N ratio for a series of nominally-identical exposures has been discussed by several authors. This is based on the observation that increased NEQ values per unit image area can be achieved at exposure levels above those for maximum DQE. However if n times the exposure is necessary to achieve NEQ_{max} compared with DQE_{max}, then as analysed in the previous section,

$$n\,\text{NEQ}_{\text{DQE}_{max}} > \text{NEQ}_{\text{NEQ}_{max}}.$$

It follows that n exposures at maximum DQE will produce a composite S/N ratio in excess of that at maximum NEQ. However the "splitting" of an exposure below maximum DQE could lead to a composite S/N ratio which is less than that of the single exposure[4].

Although multiple exposure techniques based on DQE optimization may prove easier to apply when more automatic image-processing methods are used to obtain the composite S/N ratio, Kohler and Howell[34] have described image enhancement by a physical method of superimposing multiple images.

It was noted in Chapter 4 that the highest state-of-the-art sensitivities had been achieved by signal intensification prior to photographic recording, and the application of intensifying devices has had remarkable success in detection problems in applications such as astronomy. The photographic DQE may then approach 100%, with the input NEQ identical to the photographic input number. The input quantum noise then appears in the image as the total image noise. An analysis of the DQE advantages of such systems will be carried out in the following section, which is concerned entirely with the role of quantum noise during imaging.

5.3 Noise-equivalent Quanta and Quantum Noise

X-Rays and Unconventional Exposures

When the photographic process is exposed to photons both the quantum efficiency of the grains and the measured DQE values are low, and NEQ is much less than the number of quanta in the exposure. However, as previously noted, for exposures to quanta with sufficient energy (such as X-rays) or in cases where the light is pre-amplified (as with image intensifiers), the photographic process may approach the ideal detector in so far as each quantum gives rise to one or a cluster of grains in the image. In such cases the NEQ may approach the number of exposure quanta, although this number may of course be less than the number of grains in the image when clusters of grains are produced.

A simple analysis for NEQ for these types of exposure follows from equations (74) and (75) of Chapter 3:

$$n_A = \bar{c}q_A; \quad \overline{\Delta n_A^2} = (\bar{c})^2 \overline{\Delta q_A^2} + q_A \overline{\Delta c^2}. \tag{34}$$

These equations relate the average number of image grains n_A and exposure quanta q_A, per image area A, in terms of the average number of grains per quantum $\bar{c}$, and also the mean-square fluctuations in these numbers. If as in Chapter 3 we assume that c and q are both Poisson-distributed:

$$\overline{\Delta q_A^2} = q_A; \quad \overline{\Delta c^2} = \bar{c}; \tag{35}$$

then DQE follows by definition:

$$\text{DQE} = \frac{\overline{\Delta q_A^2}}{\overline{\Delta n_A^2} \left(\frac{dq}{dn}\right)^2} = \frac{1}{1 + \frac{1}{\bar{c}}}. \tag{36}$$

This expression for DQE was derived by Zweig[32] for detection systems involving absorption and secondary-emission amplification as with phosphorescent screens or photocathode detection. Here we have considered the equations leading up to equation (36) as being applicable to the cascade process or processes relating to the production of photographic image grains. We have already noted that in the photographic case these equations will only hold where clumps of image grains have only a low probability of coinciding. However, for phosphorescent screens equation (36) should hold more exactly, independent of exposure level, limited only by the dead-time of the phosphors. In other words, equation (36) may hold without serious restriction for on-off-on receptors, but only under limited conditions for on-off receptors such as photographic grains.

In the case where the process of each quantum producing one or more image grains is preceded by a random loss of input quanta that can be described by an average quantum yield, η, where $\eta < 1$, then

$$\text{DQE} = \eta \frac{1}{1 + \frac{1}{\bar{c}}}; \quad \text{NEQ} = q_A \eta \frac{1}{1 + \frac{1}{\bar{c}}}. \tag{37}$$

Suppose for example that $\eta = 0.8$ and $\bar{c} = 10$. In this case $n_A = 10q_A$; DQE = 73%; NEQ = $0.73 \, q_A$. The NEQ will thus be 0.73 of its limiting value, even though it is less than one-tenth the number of image grains.

5. DETECTIVE QUANTUM EFFICIENCY

More detailed statistical analyses of such cascade processes have been given by Wilcock and Baum[36] and Catchpole[37]. Wilcock and Baum related the number of photoelectrons from a photomultiplier and amplifier to the eventual number of image grains, and included the number of fog grains in the output fluctuation. In this way they related the threshold magnitude of faint stars to the number of photoelectrons per grain, and compared this with unaided photography. Catchpole analysed the multistage amplification of X-rays and considered a further stage than that represented by equation (36), relating to an electronic multiplier with an exponential probability of multiplication. He deduced a relationship equivalent to:

$$\text{DQE} = \frac{1}{1 + \dfrac{2}{\bar{c}}}.$$

relating to the DQE after this additional stage.

When $\eta = 1$ DQE approaches 100% if $\bar{c}$ is large. It might seem that a process where there is one output unit (clump of grains) for every input unit (quanta) should have a DQE of 100%, independent of the value of c. This would be the case if c were constant and hence $\overline{\Delta c^2} = 0$ in equation (35). It is also remembered that the relationships between the fluctuations represent large-scale averages as carried out by a scanning aperture A. In this way the variation in c influences the total observed fluctuation. If on the other hand the image were analysed by counting individual output clumps of grains, the DQE would be 100% independent of c, as long as each input quantum produced at least one output grain and all output units were distinguishable in the image. In this context it is also relevant that the DQE will be even less than predicted by equation (36) when the output is in terms of density rather than grain numbers, unless the grains all make an identical contribution to the image density, as for example when the grains are mono-sized and the Nutting relationship holds. Otherwise fluctuations in grain size will introduce an additional fluctuation between grain numbers and density. An analysis of the NEQ relationships for grain counts and density measurements will be made shortly for conventional exposures to photons.

Practical details of the density and noise relationships which are implicit in the photographic recording of X-ray quanta, and the appearance of so-called quantum mottle, have been considered by Rossmann[38] and de Belder[39], among others. In radiography it is generally appreciated that in spite of the high observed noise compared with that for photon exposures, the photographic process may be an ideal recorder (NEQ → input quanta) for X-ray quanta, and that the problem is essentially one of limiting available quanta rather than one limiting photographic noise. Cain[40] has described a method of X-ray image enhancement, but of course such methods cannot fundamentally improve a system with DQE near its limiting value. Beyond this signal enhancement implies concomitant noise enhancement.

The direct photographic recording of electrons, as in electron microscopy, is analogous to that of X-rays. Nitka[41] gave one of the first detailed accounts of the comparative sensitivities of the photographic process to electrons and photons, and analysed the input/output relationships in terms of the factors limiting detail recognition of electronbeam images. Subsequently Valentine and Wrigley[42] published measured DQE values for 60 kV electron images with a large series of photographic plates and developers, finding that in some cases the DQE approached 100%. They concluded that the photographic process was almost a perfect detector in electron microscopy. Further DQE values for image recording in electron microscopy are due to Hamilton and Marchant[43], and Fig. 15 shows the results they obtained for an electron recording plate exposed to 100 kV electrons with three different development processes. From Fig. 15 it is seen that DQE values as high as 75% were measured. They also used a simple model to interpret these DQE results, and

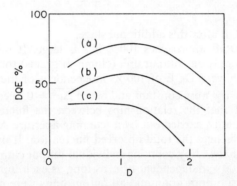

FIG. 15. DQE—density curves for Kodak Electron Image Plates exposed to 100 kV electrons and developed according to processes (a), (b) and (c), due to Hamilton and Marchant[43].

related the differences between the three curves to the differences in the variances of the number of developed grains per electron. A further noise analysis for electron recording, and a general survey of practical photographic requirements, is due to Valentine[44].

Some of the most successful types of image tubes used in astronomy involve photographic exposures to electrons, or so-called electronography. The Spectracon image tube for the photometry of faint stars and nebulae is capable of a systems DQE in the region of 40%[45]. The image to be recorded is focussed on a photocathode, and the emitted photoelectrons are accelerated and focussed through a mica window and recorded in a special photographic emulsion. Nearly all photoelectrons are recorded photographically and the DQE for the recording stage approaches 100%.

Other types of exposure which involve changes in DQE and NEQ values compared with those for conventional photon exposures, are those where the effective change is in the exposure statistics rather than due to energy amplification or cascade processes. Rosenblum[46] has studied the special case of laser light being passed through a rotating ground glass. For reasons based on the so-called coherence time of the radiation, Rosenblum discussed why this would lead to exposures with the characteristics of thermal radiation, and governed by Bose-Einstein rather than Poisson statistics. In this case, at the quantum exposure level of q quanta per grain on average, the Poisson terms

$$\frac{q^x}{x!} e^{-q},$$

in the probabilistic equations used in Chapter 3 to express grain contributions to the image will be replaced by Bose-Einstein terms:

$$\frac{q^x}{(q+1)^{x+1}}.$$

Rosenblum modelled the appropriate new D-log$E$ characteristics. For example, for mono-sized randomly-distributed grains all requiring exactly $Q = 4$ quanta:

$$\left(\frac{D}{D_{max}}\right)_{Poiss} = 1 - e^{-q}\left(1 + q + \frac{q^2}{2!} + \frac{q^3}{3!}\right); \qquad (38)$$

$$\left(\frac{D}{D_{max}}\right)_{B.E.} = 1 - \left(\frac{1}{(q+1)} + \frac{q}{(q+1)^2} + \frac{q^2}{(q+1)^3} + \frac{q^3}{(q+1)^4}\right). \qquad (39)$$

The comparative characteristic curves for $Q = 4$ quanta are shown in Fig. 16. The slope of the curve is generally lower for Bose-Einstein statistics, although image formation takes place at lower exposure levels due to "clumps" of quanta coinciding with grain quantum requirements.

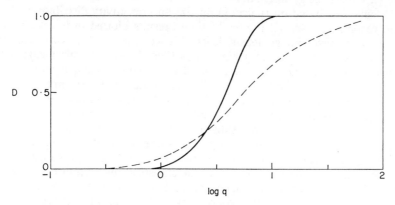

FIG. 16. The $Q = 4$ quanta characteristic curve for mono-sized randomly-distributed grains, for Poisson exposure statistics and for Bose-Einstein exposure statistics (dashed curve)[46].

Although Rosenblum did not model the noise characteristics, it is clear that the DQE and NEQ will also change. In particular, due to exposure correlations the fluctuations in quanta, Δq_A^2, in and image grains, Δn_A^2, will no longer be defined by the mean values, as for Poisson statistics.

Role of Photon Noise

For exposures to photons with low quantum efficiency, the relationship between NEQ and the number of exposure quanta is not as straightforward as for exposures with high quantum efficiencies and DQEs, as for X-rays. However prior to the investigation of the relationship between NEQ and photon noise, the variation of NEQ with the form of image read-out is of interest. For X-rays it has been argued that equation (36) for DQE relates to input/output numbers of events. Similarly for photon exposures the NEQ appropriate to grain counts will not generally be the same as density for measurements.

An analysis of this variation in NEQ is possible by use of the model used in Chapter 3, and especially by use of equation (44) of that chapter, since this equation compares the input/output characteristics in terms of grain numbers and density. In terms of image grain numbers per area A, the DQE will be defined as

$$\text{DQE} = \frac{\overline{\Delta q_A^2}}{\overline{\Delta n_A^2}\left(\frac{dq}{dn}\right)^2} = \frac{q_A}{n_A \left(\frac{dq}{dn}\right)^2}. \tag{40}$$

The probabilistic expression for n_A in Chapter 3, which differs from that for density by lacking a grain-area weighting factor, can be differentiated to obtain the slope $\frac{dn}{dq}$, and DQE can then be calculated from equation (40) with reference to image grain counts.

Calculations based on typical grain size and sensitivity distributions have been reported by Shaw[4]. Figure 17 shows results plotted in terms of NEQ against normalized image density level. It is seen that at all density levels the NEQ is higher in terms of grain counts than density measurements. Also shown is the NEQ curve for mono-sized grains equal to the average size of the size distribution. In this case NEQ for grain counts and image density measurements coincide, since the Nutting formula predicts an identical contribution to image density from all image grains. The reason that NEQ for grain counts when the size is distributed does not compensate completely back to the NEQ for mono-sized grains is due to the remaining statistical fluctuation relating to quantum capture varying with grain area during exposure: grain counting removes only the statistical fluctuation due to grain size variation at the output image stage, and not at the input exposure stage.

The first general arguments concerning the relationship between photon noise and NEQ for photographic images were outlined by Fellgett[47] in 1963. Fellgett discussed the basic laws of physics relating to the quantum nature of light, and argued why the output image noise is by necessity mainly an amplified manifestation of the input photon noise. To that date it had been assumed that on the evidence of low measured DQE values, photon noise contributed to the total noise as a negligible additive term. However, Fellgett outlined the amplying mechanisms—such as those studied quantitatively in Chapters 3 and 4—which reduce DQE from 100% to the order of 1%, and pointed out that there is nothing to suggest that noise mechanisms which act additively are of predominant importance. As will be studied shortly, fog grains provide the only significant additive noise source, and in general these produce an additive term which is independent of the photon noise, and which is small

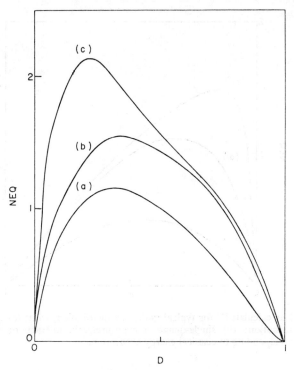

FIG. 17. NEQ per μm^2 image area, plotted against normalized density, as computed for typical grain size and sensitivity distributions[4]: (a) relates to density measurements; (b) to counting grains; (c) the corresponding curves for mono-sized grains. (coincident)

except at low image density levels where the number of fog grains compares with the number of photon-produced image grains.

A further quantitative investigation of the relationship between NEQ and photon noise along the lines suggested by Fellgett has led to more detailed conclusions concerning the influence of photon noise on characteristic curve shape, image density fluctuations and NEQ[48]. For this the procedure was as follows:

(a) derive an analytical expression for NEQ as a function of quantum exposure level (using the model of Chapter 3);
(b) substitute in this model typical grain size and sensitivity distributions and calculate NEQ;

(c) make changes in the sensitivity distribution equivalent in effect to extremes of possible changes in the exposure photon fluctuations, and re-calculate NEQ.

The comparative NEQ values obtained in this way are shown in Fig. 18. The NEQ values for the practical distribution of quantum sensitivities are seen

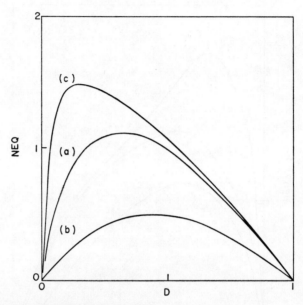

FIG. 18. NEQ as calculated[48] for typical grain size distribution, with: (a) practical grain sensitivity distribution; (b) single-quantum exposure/grain interaction; (c) sensitivity distribution corresponding to classical energy absorption.

to fall around half-way between those for a $Q = 1$ quantum/grain interaction and those corresponding to a continuous absorption of energy by the grains.

The NEQ values for the random grain model with constant grain size are shown in Fig. 19 for constant quantum requirements of $Q = 2, 4, 8, 16$ and 32 quanta. This set of NEQ-log q curves thus corresponds to the D-log q and DQE-log q curves of Fig. 2 and 7 in Chapter 3. The maximum NEQ values are directly proportional to Q. As Q increases the quantum fluctuations associated with the exposure level needed to yield this requisite number become bigger in absolute terms ($\sqrt{q}$) but smaller in relative terms ($\sqrt{q}/q$), and this relative decrease in the photon noise leads directly to the increase in NEQ with Q. Thus in these cases of constant quantum requirements, photon noise is the determining factor so far as NEQ is concerned even when Q is very large.

Photon Noise Amplification

We shall now investigate the conditions which would lead to a more general conclusion on the influence of photon noise in the practical case of a distribution

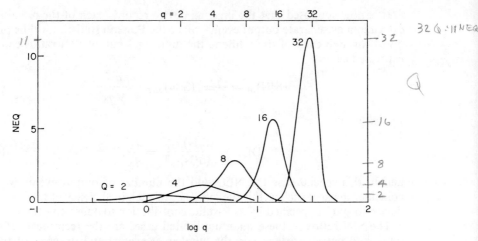

FIG. 19. NEQ per grain as a function of average quantum exposure per grain, for monosized randomly-distributed grains, with constant quantum requirements as indicated.

of sensitivities. In particular, we shall discuss the possibility of representing all photographic inefficiencies as linear amplifiers on the input photon noise.

Suppose that q_A input quanta are incident on area A of an imaging device, all of whose inefficiencies are such that they can be represented as a series of n filters, the effect of each one being to transmit only a fraction of the incident

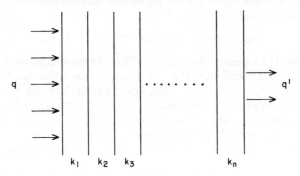

FIG. 20. Illustration of an imaging device whose inefficiencies act as a series of random quantum loss filters operating on the exposure.

quanta at random. If the number remaining at the end of the series of filters is denoted by q_A' then we may express this number as:

$$q_A' = \frac{1}{k_1}\frac{1}{k_2}\frac{1}{k_3}\cdots\frac{1}{k_n}q_A, \tag{41}$$

where there is a k-value greater than unity associated with each filter. This situation is shown schematically in Fig. 20.

It is now supposed that the imaging device records each of the remaining q_A' quanta as separate output events. Since the Poisson statistics will be preserved through each of the n filters, the input and output S/N ratios can be expressed as

$$(S/N)_{IN} = \frac{q_A}{\sqrt{q_A}}; (S/N)_{OUT} = \frac{q_A'}{\sqrt{q_A'}},\qquad(42)$$

By definition,

$$DQE = \frac{(S/N)_{OUT}^2}{(S/N)_{IN}^2} = \frac{q_A'}{q_A},\qquad(43)$$

and in this special case the NEQ and the number of output events would coincide. In cases such as this the influence of each inefficiency of the imaging device would be specified by its k-value contribution to equation (41).

The S/N ratios in these quantum-limited cases are the reciprocals of the input and output variances in the number of events, and in terms of these variances:

$$\sigma_{IN}^2 = \left(\frac{\sqrt{q_A}}{q_A}\right)^2 = \frac{1}{q_A} = \sigma_p^2;\qquad(44)$$

$$\sigma_{OUT}^2 = \left(\frac{\sqrt{q_A'}}{q_A'}\right)^2 = \frac{1}{q_A'}.\qquad(45)$$

The input noise is due entirely to the photon noise, represented by the variance σ_p^2. From equation (41) it follows that

$$\sigma_{OUT}^2 = k_1 \times k_2 \times k_3 \times \ldots \times k_n \sigma_p^2\qquad(46)$$

Equation (46) then defines the way in which the output noise is entirely an nth stage amplified manifestation of the input photon noise: the k-value is the amplifying factor relating to each stage. Levi[52] has analysed the noise of a series of linear electro-optical components in this way, and termed the noise-to-signal power ratio (i.e., σ^2 in the above equations) the "specific noise".

An extension of equation (46) is necessary if some of the n stages have k-values which are functions of the average exposure level, rather than constant. In this case

$$\sigma_{OUT}^2 = k_1 \times \ldots \times k_i \times k_j(q) \times \ldots \times k_n(q) \sigma_p^2,\qquad(47)$$

where factors k_1 to k_i are constants, and k_j to k_n are functions of exposure level. This is analogous with spectral transmission when some of the filters are neutral and some spectrally-selective.

When an equation such as (46) or (47) holds, then no matter how small the observed DQE value, the output noise is entirely an amplified manifestation of the input photon noise. For example, for an 8-stage process, each stage having k-value of $\frac{1}{2}$, the measured DQE would be 2^{-8}, or 0·4%, but entirely accounted for by amplifications of the photon noise. No additive noise mechanisms due either to the input photons or background effects have been

5. DETECTIVE QUANTUM EFFICIENCY

considered in arriving at these equations, which represent the simplest type of analytical basis for the reasoning of Fellgett[47]. An extreme type of imaging device would be one whereby each exposure quantum is recorded faithfully as one output event, but during recording a large background noise is added such that

$$\sigma_{OUT}^2 = \sigma_{ADD}^2 + \sigma_p^2. \tag{48}$$

DQE will now be defined by

$$DQE = \frac{\sigma_p^2}{\sigma_{ADD}^2 + \sigma_p^2}. \tag{49}$$

If $\sigma_{ADD}^2 \gg \sigma_p^2$; then $DQE \simeq \dfrac{\sigma_p^2}{\sigma_{ADD}^2}$; $\sigma_{OUT}^2 \simeq \sigma_{ADD}^2$.

For a measured DQE of 1% the photon noise would then have played a negligible role in determining the output noise.

Depending on whether the physical processes involved can be represented by an equation such as (47) or one such as (48), it is thus possible to account for DQE values in the region of 1% as amplified photon noise, or as almost completely uninfluenced by photon noise. It should be noted that even as an additive mechanism photon noise must make some contribution to the output noise except when DQE = 0; i.e., when there is no image except random noise internally-generated by the detector. As will be demonstrated, it is implicit from the model used in Chapters 3 and 4 that apart from fog grains the photographic process acts mainly as an amplifier in the sense of equation (47).

Fried[49] has analysed an analogous amplifying case concerning noise in photoemission currents. He showed that the fact that the secondary photoelectron-current obeys a Poisson distribution follows from the Poisson statistics of the primary photons, so that the noise calculated for the photoelectron current automatically contains the contribution due to the photons. The photoelectron noise is thus the total detector noise, just as the output noise of equation (47) represents the total image noise for a hypothetical imaging device acting in the multiplicative sense on the photon noise.

As an example Fried considered the case where incident photons produce either one or no photoelectrons at random, with average quantum efficiency of η electrons per photon. In this case,

$$(S/N)_p = \sqrt{q_A}; \quad (S/N)_{p.e.} = \sqrt{\eta q_A}. \tag{50}$$

It now follows that DQE and quantum efficiency are identical and independent of exposure level, and NEQ is identical with the number of output events:

$$DQE = \eta; \quad NEQ = \eta q_A. \tag{51}$$

Schade[50] represented the overall quantum efficiency involved in photographic latent image formation as a cascade of quantum efficiencies associated with the various stages. In this way he defined

$$\eta = \eta_1 \eta_2 \eta_3. \tag{52}$$

where:

η_1 = number of photoelectrons per incident photon;
η_2 = number of stored silver atoms per electron;
η_3 = number of developed grains per stored atom.

In this way Schade arrived at a concept of noise-equivalent grain number, identical to the concept of NEQ. However, there is no longer a simple relationship between quantum efficiency and NEQ as in equation (51), since equation (52) relates to average values, and to calculate the transfer of fluctuations through the stages it would be necessary to know all the statistical distributions which are involved.

Photographic Amplifying Factors

A direct way of arriving at an overall equation such as (47) for the photographic process has been described by Shaw[51]. This is based on the simple two-parameter (T,m) model for the sensitivity distribution, with DQE then as defined by equation (19) of Chapter 4:

$$\varepsilon_{T,m}(mq) = \frac{1}{m} \times \varepsilon_Q(q). \tag{53}$$

This equation relates the DQE for the T,m model at the exposure level of mq absorbed quanta per grain to the DQE for constant quantum requirements, Q, at the exposure level of q absorbed quanta per grain. The random-loss parameter, m, thus acts as a filter in the sense of equation (47), and from equation (53) it follows that

$$\sigma_{OUT}^2 = m \times \frac{1}{DQE_Q} \sigma_p^2. \tag{54}$$

This equation relates only to those quanta absorbed by grains. If only a random fraction, $\frac{1}{n}$ of the total incident quanta are absorbed, then equation (54) can be redefined to include this fraction:

$$\sigma_{OUT}^2 = m \times n \times \frac{1}{DQE_Q} \sigma_p^2. \tag{55}$$

This now relates to all incident quanta, and defines two photon noise amplifying k-factors greater than unity, as in equation (47), and both of which are independent of exposure level.

It is possible to express the amplifying factor $1/DQE_Q$, which is a function of exposure level, as the product of three separate amplifying factors which relate to a grain threshold defect, a grain saturation defect, and a random-position grain defect. If these are denoted by TD, SD and RD respectively, then equation (55) becomes[51]:

$$\sigma_{OUT}^2 = m \times n \times TD \times SD \times RD\ \sigma_p^2. \tag{56}$$

5. DETECTIVE QUANTUM EFFICIENCY

The separation of the random grain defect follows directly from equation (12) of Chapter 3, relating the DQE for checkerboard and random grain models when the quantum requirements, Q, are constant:

$$\text{RD} = \frac{1}{e^{-q} \sum_{0}^{Q-1} \frac{q^r}{r!}}. \qquad (57)$$

The separation of the DQE of the checkerboard model for single-level grains needing Q quanta into DQEs relating to threshold and saturation

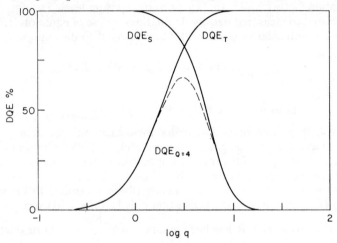

FIG. 21. DQE for a checkerboard model relating to multilevel grains with threshold and saturation defects based on 4 quanta, and the product (dashed curve) which yields the mono-sensitivity DQE curve for Q = 4.

defects is not so straightforward. However, as indicated by Shaw[51], we can write

$$\text{DQE}_Q = \text{DQE}_T \times \text{DQE}_S, \qquad (58)$$

where DQE_T denotes the DQE associated with a threshold defect, based on a multilevel imaging process with grains which could record all quanta absorbed at and above a threshold of Q quanta per grain; and DQE_S denotes the DQE associated with a saturation defect, based on a multilevel imaging process with grains which could record all quanta absorbed up to and including Q quanta per grain.

A general proof of this result is difficult, but follows from the definition of DQE for a multilevel model as in equation (17) of Chapter 1, where the functions f_1, f_2 and f_3 are defined by equations (20) to (22) of that chapter. For DQE_T it is necessary to substitute $T = Q$, $S \to \infty$; and for DQE_S it is necessary to substitute $T = 0$, $S = Q$. The product of the two DQEs at any given exposure level then gives the checkerboard DQE for grains with the appropriate Q-value, at this same exposure level. Figure 21 shows an example of this product according to equation (58) for the special case of $Q = 4$.

The analogy of spectrally-selective filters is evident for these defects which are functions of exposure level: the combination of two broad-band filters providing a narrow-band DQE curve. For this $Q = 4$ example the checkerboard and random grain DQE curves will be related by the random grain defect, $\dfrac{1}{e^{-q}\sum_0^3 q^r/r!}$, as shown in Fig. 3 of Chapter 3.

Equation (56) now expresses the output noise in terms of a five-stage amplification of the photon noise in the exposure, with three of these stages having amplification factors which are functions of exposure level. The output noise can be related to measured density fluctuations by use of equation (15). If the σ-values on both sides of equation (56) correspond to unit image area, then

$$\frac{G}{(\gamma \log_{10} e)^2} = m \times n \times \text{TD} \times \text{SD} \times \text{RD}\, \sigma_p^2,$$

and hence

$$G = m \times n \times \text{TD} \times \text{SD} \times \text{RD}\, (\gamma \log_{10} e)^2 \sigma_p^2. \tag{59}$$

Equation (59) demonstrates that under these assumptions, what is usually referred to as photographic granularity is merely amplified photon noise.

Numerical examples of the relative importance of the five defects are shown in the following table[51]. These are based on the T,m sensitivity model with $T = 4$, $m = 5$ (1 in 5 absorbed quanta contribute at random to the required number, 4). A value of $n = 2$ signifies that only half the total exposure quanta are absorbed by grains. It has been assumed that $\dfrac{Na}{A} = 2\cdot30$ in relating the absolute exposure per unit area to the exposure per grain. The exposure is also shown after passing through the n and m "filters", and hence this value can be compared with the requirements of $Q = 4$ quanta.

Average exposure quanta:		Constant defects	"Internal" exposure	Defects dependent on exposure level			Product of five defects	DQE %	NEQ per unit area
per unit area	per grain								
E_q	q	n	m	q/nm	TD	SD	RD	x	$\dfrac{100}{x}$	$\dfrac{E_q}{x}$
23	10	2	5	1	5·00	1·00	1·02	51·0	1·97	0·45
55	24	2	5	2·4	1·54	1·11	1·25	21·4	4·70	2·57
115	50	2	5	5	1·05	1·82	3·57	68·0	1·47	1·69

The three levels of exposure which are shown in this table are for that corresponding to maximum DQE, and a higher and lower level. In each case the product $nm = 10$ plays a dominant role in determining DQE. Of the three exposure-dependent defects, at low exposures the threshold defect dominates, while at high exposures the saturation defect and random grain defect become more important. Traditionally, the randomness of the photographic grain pattern has often been assumed to be the most important noise factor, but it is interesting to note that the maximum DQE value of 4·7% would only be increased to 5·9% if the random grain defect of 1·25 at this level were eliminated.

Factors due to the grain size distribution, the exposure distribution through the layer, etc., have not been included in this analysis. In the more general case it may be difficult to separate out the defects as a series of filters, and it is easier to express the photon noise amplification by

$$\sigma_{OUT}^2 = \frac{1}{DQE}\sigma_p^2. \qquad (60)$$

Then, as reasoned by Fellgett[47], the overall DQE is regarded as a measure of the overall photon noise amplification.

The presence of fog grains has been analysed[22] as acting in two principal ways. Exposure quanta absorbed by such grains are wasted, and in this sense fog grains act as a further amplifying mechanism on the photon noise. Also, fog grains appear in the image as an additional noise source. The mean-square density fluctuation due to fog grains and photon-produced grains will be directly additive at all exposure levels for mono-sized grains for which density is proportional to the number of image grains. Fog grains thus contribute in both a multiplicative and additive sense, and we can write

$$\sigma_{OUT}^2 = k\sigma_p^2 + \sigma_{fog}^2; \quad DQE = \frac{1}{k + \frac{\sigma_{fog}^2}{\sigma_p^2}}. \qquad (61)$$

The total amplifying factor k now includes a factor due to fog grains. Fog grains may cause a significant DQE reduction at low density levels[22].

References

1. Marchant, J. C. (1964). Exposure criteria for photographic detection of threshold signals. *J. Opt. Soc. Amer.*, **54**, 798.
2. Marchant, J. C. and Millikan, A. G. (1965). Photographic detection of faint stellar objects. *J. Opt. Soc. Amer.*, **55**, 907.
3. Wolfe, W. L. (1973). Photon number D^* figure of merit. *Appl. Opt.*, **12**, 619.
4. Shaw, R. (1972). The photographic process as a photon counting device. *J. Photogr. Sci.*, **20**, 174.
5. Shaw, R. (1965). The equivalent quantum efficiency of aerial films. *J. Photogr. Sci.*, **13**, 308.
6. Shaw, R. (1973). Some detector characteristics of the photographic process. *Optica Acta*, **20**, 749.
7. Jones, R. C. (1958). On the minimum energy detectable by photographic materials. *Photogr. Sci. Eng.*, **2**, 191; 198.
8. Jones, R. C. (1959). Quantum efficiency of detectors for visible and infrared radiation. *Advances in Electronics and Electron Physics*, **11**, 87. Academic Press, New York and London.

9. Zweig, H. J., Higgins, G. C. and MacAdam, D. L. (1958). On the information-detecting capacity of photographic emulsions. *J. Opt. Soc. Amer.*, **48**, 926.
10. Zweig, H. J. (1961). The relation of quantum efficiency to energy- and contrast-detectivity for photographic materials. *Photogr. Sci. Eng.*, **5**, 142.
11. Baum, W. A. (1962). The detection and measurement of faint astronomical sources. (Chapter 1), *In* "Astronomical Techniques", 2nd Ed., (ed. W. A. Hiltner). University of Chicago Press.
12. Hoag, A. A. and Miller, W. C. (1969). Application of photographic materials in astronomy. *Appl. Opt.*, **8**, 2417.
13. Fellgett, P. B. (1970). Design of astronomical telescope systems. *In* "Optical Instruments and Techniques", (ed. J. Home Dickson). Oriel Press, Newcastle.
14. Fellgett, P. B. (1961). On the necessary measurements for the characterization and optimum use of photographic materials for scientific purposes. *J. Photogr. Sci.*, **9**, 201.
15. Bird, G. R., Jones, R. C. and Ames, A. E. (1969). The efficiency of radiation detection by photographic films: state-of-the-art and methods of improvement. *Appl. Opt.*, **8**, 2389.
16. Miller, W. C. (1964). The application of pre-exposure to astronomical photography. *Publ. Astr. Soc. Pacific*, **76**, 328.
17. Guttman, A. (1968). Background-light sensitization of a fast emulsion. *Photogr. Sci. Eng.*, **12**, 146.
18. Porteous, R. L. and Shaw, R. (1967). Results presented at the Royal Astronomical Society Conference on "Astronomical Optics", London.
19. Shaw, R. and Shipman, A. (1969). Practical factors influencing the signal-to-noise ratio of photographic images. *J. Photogr. Sci.*, **17**, 205.
20. Smith, A. G., Schrader, H. W. and Richardson, W. W. (1971). Response of type IIIa-J Kodak spectroscopic plates to baking in various controlled atmospheres. *Appl. Opt.*, **10**, 1597. Scott, R. L. and Smith, A. G. (1974). Hypersensitization of Kodak 103a-0 plates by nitrogen baking. *Astron. J.*, **79**, 656.
21. Kodak advertisement *In*: *Scientific Amer.*, **228**, No. 3: 45 (1973).
22. Shaw, R. (1973). The influence of fog grains on detective quantum efficiency. *Photogr. Sci. Eng.*, **17**, 495.
23. Courtney-Pratt, J. S. (1961). A note on the possibility of photographing a satellite near the moon. *J. Photogr. Sci.*, **9**, 36.
24. Amoss, J. and Davidson, F. (1972). Detection of weak optical images with photon counting techniques. *Appl. Opt.*, **11**, 1793.
25. Falconer, D. G. (1970). Image enhancement and film-grain noise. *Optica Acta*, **17**, 693; (1971). Enhancement of photographically recorded imagery. *Procs. of the Soc. Phot. Opt. Instr. Engrs.* Page 191. Seminar on "Developments in Holography", Boston Mass.
26. Weinstein, F. S. (1969). RMS error in radiometric measurements when a photographic emulsion is used as a detector. *J. Opt. Soc. Amer.*, **59**, 108.
27. Shepp, A. and Kammerer, W. (1970). Increased detectivity by low gamma processing. *Photogr. Sci. Eng.*, **14**, 363.
28. Yu, F. T. S. (1972). Film-grain noise and signal-to-noise ratio. *Optik*, **36**, 434.
29. Barnard, T. W. (1972). Image evaluation by means of target recognition. *Photogr. Sci. Eng.*, **16**, 144.
30. Pryor, P. L. (1972). The three primary parameters of imaging systems and *DQE*. *Procs. of the Conference Electro-Optics* 1972, Page 189. Brighton.

31. Chambers, R. P. and Courtney-Pratt, J. S. (1969). Experiments on the detection of visual signals in noise using computer-generated signals. *Photogr. Sci. Eng.*, **13**, 286.
32. Hayen, L. and Verbrugghe, R. (1972). Film versus Plumbicon camera: a critical comparison between both systems in terms of sensitivity and signal-to-noise ratio. *Brit. Kinematogr. Sound and Telev.*, **54**, 324.
33. Proceedings of the *Soc. Photogr. Opt. Instr. Engrs.* Seminars on: "Computerized Imaging Techniques", Washington D.C., June 1967. "Image Information Recovery", Philadelphia, October 1968.
34. Kohler, R. J. and Howell, H. K. (1963). Photographic image enhancement by superimposition of multiple images. *Photogr. Sci. Eng.*, **7**, 241.
35. Zweig, H. J. (1965). Detective quantum efficiency of photodetectors with some amplification mechanisms. *J. Opt. Soc. Amer.*, **55**, 525.
36. Wilcock, W. L. and Baum, W. A. (1962). Astronomical tests of an imaging photomultiplier. *Advances in Electronics and Electron Physics*, **16**, 383. Academic Press, London and New York.
37. Catchpole, C. E. (1962). X-ray image intensification using multi-stage image intensifiers. *Advances in Electronics and Electron Physics*, **16**, 567. Academic Press, London and New York.
38. Rossmann, K. (1962). Recording of X-ray quantum fluctuations in radiographs. *J. Opt. Soc. Amer.*, **52**, 1162.
39. de Belder, M. (1972). Quantum noise, granularity and detail perceptibility. *Rontgenblatter*, **25**, 322.
40. Cain, D. G. (1972) X-ray image enhancement by least-mean-square estimation. *Appl. Opt.*, **11**, 2949.
41. Nitka, H. F. (1963). Sensitivity and detail recognition in the recording of light images and electron beams. *Photogr. Sci. Eng.*, **7**, 188.
42. Valentine, R. C. and Wrigley, N. G. (1964). Graininess in the photographic recording of electron microscope images. *Nature*, **203**, 713.
43. Hamilton, J. F. and Marchant, J. C. (1967). Image recording in electron microscopy. *J. Opt. Soc. Amer.*, **57**, 232.
44. Valentine, R. C. (1966). The response of photographic emulsions to electrons. *Advances in Optical and Electron Microscopy*, **1**, 180. Academic Press, London and New York.
45. Bacik, H., Coleman, C. I., Cullum, M. J., Morgan, B. L., Ring, J. and Stephens, C. L. (1972). The analysis of direct Spectracon exposures obtained on the Isaac Newton telescope. *Advances in Electronics and Electron Physics*, **33B**, 747. Academic Press, London and New York.
46. Rosenblum, W. M. (1968). Effect of photon distributions on photographic grain. *J. Opt. Soc. Amer.*, **58**, 60.
47. Fellgett, P. B. (1963). On the relevance of photon noise and of informational assessment in scientific photography. *J. Photogr. Sci.*, **11**, 31.
48. Shaw, R. (1972). Photon fluctuations and photographic noise. *J. Photogr. Sci.*, **20**, 64.
49. Fried, D. L. (1965). Noise in photoemission current. *Appl. Opt.*, **4**, 79.
50. Schade, O. H. (1964). An evaluation of photographic image quality and resolving power. *J. Soc. Motion Pict. Tel. Engrs.*, **73**, 81.
51. Shaw, R. (1973). Photon noise amplification due to photographic defects. *Photogr. Sci. Eng.*, **17**, 491.
52. Levi, L. (1973). On noise analysis of electro-optical systems. *Opt. Commun.*, **9**, 325.

Exercises

(1) A photographic plate used in astronomy has a peak DQE-value when exposed for 45 mins to night sky light at the prime focus of a 150 inch telescope, corresponding to an average exposure of 40 photons per μm^2 at the surface of the plate. The detection criterion is a S/N ratio of 4 (rms) in the image. If the incoming S/N ratio of a faint star is 32, will it be detected? (At this exposure level the photographic plate has a characteristic curve slope = 0·8 and rms density fluctuation = 0·008 as measured with a 50×50 μm scanning aperture.) It should be assumed that the incoming noise is dominated by quantum fluctuations.

If the properties of the plate are measured for normal exposure times, but reciprocity law failure reduces the DQE to one-half this value when the exposure time is 45 mins, according to what rms criterion could the star then said to be detected?

(2) Verify the DQE improvement made possible by the use of a filter when there is detector "overloading" (i.e., in terms of Fig. 10 show that DQE_f is replaced by DQE_g).

(3) Explain qualitatively why the maximum values of DQE and NEQ occur at different exposure levels for conventional photographic grains, and discuss the practical significance of this. The maximum DQE of a film is ε and occurs at exposure level E. Show that the effective DQE of the composite image from n exposures at exposure levels E/n must be less than ε, but that the effective DQE of the composite image from n exposures at exposure level E remains as ε.

(4) Give an account of the general principles to be observed in photographic detection problems and discuss the sense in which the input and output statistics should be matched.

Explain with simple examples why the principle of image intensification is to amplify without introducing fluctuations.

(5) A photographic process is exposed to $q = 2$ photons per μm^2 on average. At this level the slope of the D–q curve is 0·05 and the image noise has rms fluctuation of $\sigma = 0.04$ density units when measured with a 25×25 μm scanning aperture. What image area will be necessary to yield 100 noise-equivalent quanta?

(6) Describe the conditions under which the output image noise may be represented as an n-stage amplification of the input photon noise. Are there any reasons to suppose that the photographic process introduces additive noise during imaging?

Discuss the amplifying role of the random nature of the grain pattern, and explain why for mono-sized grains the size does not act as a photon noise amplifier, though acting as an amplifier in the speed sense.

(7) A photographic plate has a DQE which may be expressed as a function of the average quantum exposure level, q, per unit area by $\varepsilon = \dfrac{a}{e^q - 2q}$, where a is a constant. The plate is to be incorporated in a system for detecting low-contrast quantum-noise limited signals. Show that about 20% more exposure will be necessary to give the

same S/N ratio for an experiment designed for optimum NEQ per unit image area rather than for optimum DQE.

(8) From the NEQ plots for the six films shown in Fig. 3, derive the plots of log DQE–log$E$. Draw in the tangents of slope $+1$, and hence estimate in each case the potential pre-exposure advantages in DQE terms when the exposure is only one-fifth that for maximum DQE.

(9) Explain in general terms how the shape of the NEQ–Density curve is influenced by the size and sensitivity of the photographic grains.
Discuss the features of an unconventional process which would give a linear increase in NEQ with decrease in image density level.

(10) Why may it be more important in low-light-level detection problems to have a high degree of fog suppression than an increase in quantum sensitivity?
Show how NEQ with fog grains present can be expressed as a fraction of the value which would be relevant without fog grains, in a way which defines the relative importance of reductions due to additive and multiplicative changes.

6. Fourier Transforms, and the Analysis of Image Resolution and Noise

6.1 Fourier Transforms

Introduction

In previous chapters the image resolution and noise have been represented mainly by a single parameter: the former by the area—based on the spread function—corresponding to a single resolution "unit", and the latter by the product $G = A\sigma_A^2$. Although these parameters are adequate in many aspects of image analysis, it is often necessary to investigate the spatial image structure in more detail. For example, the photographic scientist may require to investigate the discrepancy between the resolution as based theoretically on grain dimensions and as measured experimentally in terms of the spread function, and in scientific photography it may be necessary to know the role of image noise on the profile of a recorded signal with intricate spatial structure. We may think of the resolution unit and the noise parameter, G, as the linking parameters between the macro scale (the characteristic curve) and the micro image scale, and it is to the latter which we now turn our attention.

In the analysis of spatial image detail we shall see that Fourier transform theory and the spatial frequency analysis of images can be greatly beneficial. For example, the modulation transfer function (MTF)—which is the Fourier transform (FT) of the spread function—may be a more useful quantity in practice. Spatial frequency sometimes proves a conceptual stumbling-block, and the analogy with temporal frequency in the frequency analysis of electrical communication systems is helpful in this respect.

The MTF describes the extent to which spatially sinusoidal exposure inputs are modulated in the image. From the viewpoint of desirable test objects for photographic image evaluation, Frieser[1], and later Selwyn[2], pointed out the interesting result that the image of a sinusoidal object is also sinusoidal but with reduced modulation depending only on the spatial frequency. However the main stimulus for applying Fourier methods to optical imaging systems came from Duffieux[3] and Luneberg[4] in the mid 1940s, who pointed out the full implications of replacing the convolution of spread functions by the product of their Fourier transforms. These concepts were also discussed in the optical context by Elias and colleagues[5,6], and Schade[7].

The extension of Fourier theory to the analysis of image noise followed as a natural conclusion. Wiener[8,9] pioneered methods of analysing random processes, and these methods had been widely adopted in the measurement

6. FOURIER TRANSFORMS

and specification of noise in electrical communication systems. The concept of the Wiener spectrum of the noise fluctuations was applied to the photographic case by Fellgett[10], Jones[11], and Zweig[12] in the mid 1950s. In broad terms the Wiener spectrum describes the mean-square noise fluctuation as a function of spatial frequency in the image.

Since the Fourier transform is an essential mathematical tool for understanding the MTF and Wiener spectrum, a brief introduction will be given in this section: more general and detailed texts on Fourier theory are listed as general reading at the end of the chapter. Practical details involved in numerical calculation of Fourier transforms will be stressed, since this is an important aspect often neglected in more theoretical texts. Fuller photographic details concerning the MTF and the Wiener spectrum will be the main topics of Chapters 7 and 8.

Definitions

The Fourier transform $F(u)$ of a function $f(x)$ is defined as:

$$F(u) = \int_{-\infty}^{+\infty} f(x)e^{-2\pi i u x} dx, \qquad (1)$$

and the inverse relationship is:

$$f(x) = \int_{-\infty}^{+\infty} F(u)e^{+2\pi i u x} du. \qquad (2)$$

Upper and lower case versions of the same letter will be used to denote Fourier transform pairs, except where this clashes with commonly accepted photographic notation. It should be noted that definitions can be found in the literature which involve a factor 2π in the equivalent equations to (1) and (2), but this is of no importance if the notation remains consistent.

The Fourier transform of a function only exists if the function obeys certain mathematical conditions; for example the function must be finite in the sense of the inequality:

$$\int_{-\infty}^{+\infty} |f(x)|^2 dx < \infty \qquad (3)$$

However we can summarize all these conditions by saying that any physically-realizable (or "well-behaved") function has a Fourier transform.

The units of the variable u are the reciprocal of those of x. In the optical and photographic case $f(x)$ is usually a function of distance, and therefore $F(u)$ is a function of spatial frequency, which is usually defined in terms of cycles/mm. If $u = 0$ in equation (1), then

$$F(0) = \int_{-\infty}^{+\infty} f(x) dx; \qquad (4)$$

and similarly from equation (2),

$$f(0) = \int_{-\infty}^{+\infty} F(u) du. \qquad (5)$$

Equations (4) and (5) state that the area under a function equals the value of the central ordinate of its transform. Some examples of Fourier transform pairs are shown in Fig. 1, and inspection of these reveals that all functions on the left hand side have area a, and all functions on the right have area of unity. The functions and their transforms shown in Fig. 1 are real and even. In general a complex function has a complex transform, a real function has a transform whose real part is even and imaginary part is odd (called a hermitian function), and an even function has an even transform.

The transforms shown in Fig. 1 may be derived by direct application of equation (1). For the first example:

$$f(x) = \mathrm{rect}\left(\frac{x}{a}\right) = 1, |x| < \frac{a}{2};$$

$$= 0 \text{ elsewhere.}$$

$$F(u) = \int_{-\infty}^{+\infty} f(x) e^{-2\pi i u x} dx = \int_{-a/2}^{+a/2} e^{-2\pi i u x} dx = \frac{\sin \pi a u}{\pi u}. \quad (6)$$

The function $\dfrac{\sin \pi x}{\pi x}$ occurs often, and is denoted by sinc (x), and so equation (6) can be written as

$$F(u) = a \,\mathrm{sinc}\,(au).$$

Although a function must satisfy the condition of equation (3) if its Fourier transform is to exist, there are a number of commonly-used functions (such as cosine, or a constant) which do not satisfy this condition, but for which it would be convenient to define an effective transform. Such transforms may be defined in terms of the Dirac delta function, $\delta(u)$:

$$\delta(u) = 0, u \neq 0; \quad \int_{-\infty}^{+\infty} \delta(u) \, du = 1. \quad (7)$$

This function is essentially an infinitely narrow spike of infinite height, and is conveniently drawn as a vertical arrow of unit height, as in Fig. 2, which shows some transform pairs involving the delta function.

Delta functions have a number of interesting properties, one of which we shall use shortly in connection with the sampling theorem is the so-called sifting property:

$$\int_{-\infty}^{+\infty} \delta(u) f(a-u) \, du = \int_{-\infty}^{+\infty} \delta(a-u) f(u) \, du = f(a). \quad (8)$$

The transform of the constant function $f(x) = 1$ follows from the transform pair

$$\mathrm{rect}\left(\frac{x}{a}\right) \stackrel{\mathrm{FT}}{\leftrightarrows} a \,\mathrm{sinc}\,(au)$$

shown in Fig. 1. As the width a of the rectangle function increases, the transform becomes taller and narrower. In the limit, as $a \to \infty$, the rectangle

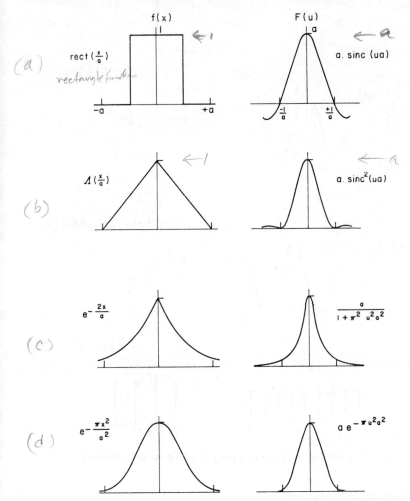

FIG. 1. Some Fourier transform pairs.

function becomes the constant function and its transform becomes the delta function:

$$1 \underset{\text{FT}}{\rightleftharpoons} \delta(u)$$

A more general form of this transform pair may be found using the shift theorem of Table I:

$$e^{-2\pi i u a} \underset{\text{FT}}{\rightleftharpoons} \delta(x-a).$$

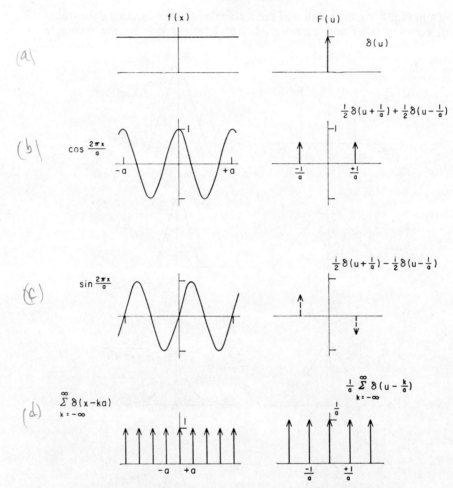

FIG. 2. Some Fourier transform pairs involving the delta function.

Two Dimensional Transforms

The two dimensional Fourier transform $F(u,v)$ of the function $f(x,y)$ is defined by:

$$F(u,v) = \iint_{-\infty}^{+\infty} f(x,y) \, e^{-2\pi i(ux+vy)} \, dx \, dy; \tag{9}$$

and the inverse relationship is now

$$f(x,y) = \iint_{-\infty}^{+\infty} F(u,v) \, e^{+2\pi i(ux+vy)} \, du \, dv. \tag{10}$$

Some examples of two dimensional transform pairs are shown in Fig. 3.

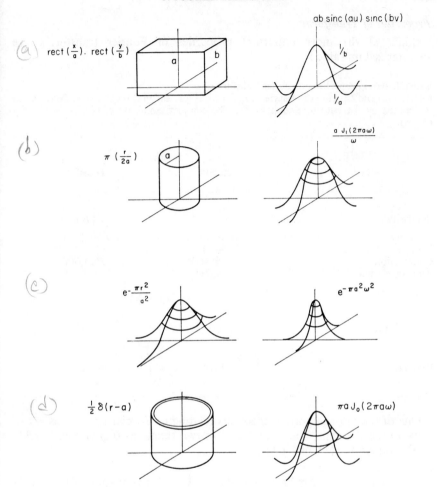

FIG. 3. Some two-dimensional Fourier transform pairs ($r^2 = x^2 + y^2$; $\omega^2 = u^2 + v^2$).

If a two dimensional function is rotationally symmetric, then its Fourier transform may be simplified to a transform of one variable called the Hankel transform:

$$F(\omega) = 2\pi \int_0^\infty f(r) J_0(2\pi r\omega)\, r dr, \qquad (11)$$

where $\omega^2 = u^2 + v^2$, $r^2 = x^2 + y^2$, and $J_0(\ )$ denotes the zero order Bessel function of the first kind[13]. For example, if $f(r)$ is the circular "top hat" function of unit height and radius a, as shown in Fig. 3, then

$$F(\omega) = 2\pi \int_0^a J_0(2\pi r\omega)\, r dr = \frac{a}{\omega} J_1(2\pi a\omega),$$

where $J_1(\ )$ is the first order Bessel function[13].

Properties of Transforms

Some of the more important properties of Fourier transforms are summarized in Table I.

TABLE I. Basic properties of Fourier transforms.
(Note: $F(u)$ and $G(u)$ are the Fourier transforms of $f(x)$ and $g(x)$ respectively. Convolution is denoted by ⊛; autocorrelation by ◇. The complex conjugate of $f(x)$ is denoted by $f^*(x)$.).

Name	Fourier Pairs Function	Transform
Similarity	$f(ax)$	$\left\|\frac{1}{a}\right\| F\left(\frac{u}{a}\right)$
Linearity	$c_1 f(x) + c_2 g(x)$	$c_1 F(u) + c_2 G(u)$
Shift	$f(x - a)$	$F(u) e^{-2\pi i u a}$
Convolution	$f(x) \circledast g(x)$	$F(u) G(u)$
Autocorrelation	$f(x) \diamond f(x)$	$\|F(u)\|^2$
	Equalities	
Multiplication	$\int_{-\infty}^{+\infty} f^*(x) g(x)\, dx =$	$\int_{-\infty}^{+\infty} F^*(u) G(u)\, du$
Parseval	$\int_{-\infty}^{+\infty} \|f(x)\|^2\, dx \quad =$	$\int_{-\infty}^{+\infty} \|F(u)\|^2\, du$

One of the most important theorems relating to Fourier transforms is the convolution theorem. The convolution of two functions $f(x)$ and $g(x)$ yields a third function, $h(x)$, defined by:

$$h(x) = \int_{-\infty}^{+\infty} f(x_1) g(x - x_1) dx_1 = \int_{-\infty}^{+\infty} f(x - x_1) g(x_1) dx_1. \quad (12)$$

We denote this by

$$h(x) = f(x) \circledast g(x).$$

The convolution theorem (see Table I) states that the transform of a convolution of two functions is equal to the product of the transforms of each function:

$$\int_{-\infty}^{+\infty} \left\{ \int_{-\infty}^{+\infty} f(x_1) g(x - x_1)\, dx_1 \right\} e^{-2\pi i u x} dx = F(u)\, G(u). \quad (13)$$

The autocorrelation of a function is essentially a self-convolution:

$$h(x) = \int_{-\infty}^{+\infty} f^*(x_1) f(x + x_1)\, dx_1,$$

6. FOURIER TRANSFORMS

which is denoted by

$$h(x) = f(x) \diamondsuit f(x).$$

The Sampling Theorem

It may not be possible to represent practical functions such as spread functions by simple analytical expressions, and the numerical calculation of their Fourier transforms (for example using a computer) may then be necessary.

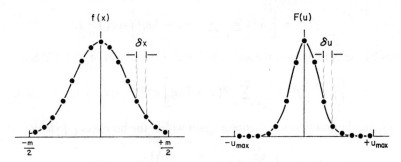

FIG. 4. Illustration of the sampling theorem. If $f(x)$ is defined within the range m, then $F(u)$ is fully described by points $\delta u = \dfrac{1}{m}$ apart. Conversely, if the range of interest of $F(u)$ is $2u_{\max}$, then $f(x)$ may be sampled at intervals not greater than $\delta x = \dfrac{1}{2u_{\max}}$.

In this case some decision has to be made on how to sample the function. This situation is illustrated in Fig. 4, where a function $f(x)$ is sampled at intervals δx over a range m. At what intervals δu, and up to what frequency $u_{\max}$, can the transform be calculated? The answers to these questions turn out to be:

$$\delta u = \frac{1}{m}; \qquad (14)$$

$$u_{\max} = \frac{1}{2\delta x}, \qquad (15)$$

and they follow from the sampling theorem[14,15]:—

If a function is known within a range X, its transform is completely specified by points $\dfrac{1}{X}$ apart.

Applying the sampling theorem to $f(x)$ we set $X = m$, and hence obtain equation (14); applying it to $F(u)$ we set $X = 2u_{\max}$, and hence obtain equation (15). The maximum frequency, $u_{\max}$, is known as the Nyquist or folding frequency.

The sampling theorem may be proved as follows. For simplicity we take $f(x)$ as a real and even function defined within the range $-\dfrac{m}{2}$ to $+\dfrac{m}{2}$. If $f(x)$ is convoluted with a "comb" of delta functions of period m (see Fig. 28), the convolution is also periodic, and defined by:

$$f(x) \circledast \sum_{k=-\infty}^{+\infty} \delta(x - km) \tag{16}$$

The function $f(x)$ can be recovered from the convolution if the convolution is multiplied by a rectangle function of width m:

$$f(x) = \left[f(x) \circledast \sum_{k=-\infty}^{\infty} \delta(x - km) \right] \text{rect}\left(\frac{x}{m}\right).$$

Thus by the convolution theorem the Fourier transform of $f(x)$ will be:

$$F(u) = \left\{ \int_{-\infty}^{+\infty} \left[f(x) \circledast \sum_{k=-\infty}^{\infty} \delta(x - km) \right] e^{-2\pi i u x} \, dx \right\} \circledast m \, \text{sinc}\,(mu).$$

Applying the convolution theorem again within the brackets { } yields:

$$F(u) = \left\{ \sum_{k=-\infty}^{+\infty} F(u) \, \delta\left(u - \frac{k}{m}\right) \right\} \circledast \text{sinc}\,(mu)$$

$$= \left\{ \sum_{k=-\infty}^{+\infty} F\left(\frac{k}{m}\right) \right\} \circledast \text{sinc}\,(mu). \tag{17}$$

Thus $F(u)$ is completely specified by the values $F\left(\dfrac{k}{m}\right)$ which are $\dfrac{1}{m}$ apart. If values of $F(u)$ are required at intermediate points then we may interpolate using sinc (mu).

As an example, suppose that a function of distance is sampled at 100 points which are 1 μm apart; i.e., $m = 100 \, \mu\text{m} = 0{\cdot}1$ mm:

$$\delta u = \frac{1}{m} = 10 \text{ cycles/mm}; \quad u_{\max} = \frac{1}{2\delta x} = 500 \text{ cycles/mm}.$$

The Fourier transform of this function is completely specified by values 10 cycles/mm apart, up to a frequency of 500 cycles/mm.

Computational Methods

When calculating Fourier transforms on a digital computer, the integral expression as in equation (1) is replaced by a summation. The continuous function $f(x)$ is replaced by a step-wise function f_j, as in Fig. 5, where

$$f_j = f(j\delta x),$$

and where the index j runs from (approximately) $-\dfrac{N}{2}$ to $+\dfrac{N}{2}$, covering N points in total. In practice it may be more convenient if the index runs from

0 to N − 1, and in this case the Fourier integral expression may be written as the summation

$$F_k = \delta x \underbrace{\exp -\left(\frac{2\pi i \left(\frac{N}{2} + \frac{1}{2}\right) k}{N}\right)}_{\substack{\text{scale} \\ \text{factor}}} \underbrace{\sum_{j=0}^{N-1} f_j \exp -\left(\frac{2\pi i j k}{N}\right)}_{\substack{\text{discrete} \\ \text{Fourier} \\ \text{transform}}}; \quad (18)$$

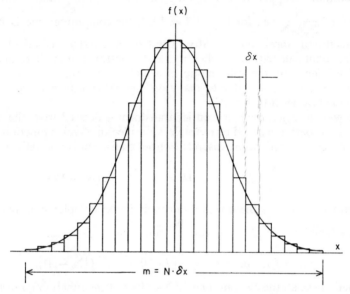

Fig. 5. The approximation of a continuous function by a stepwise function for the purposes of numerical calculation.

where $j = 0, 1, 2, \ldots, (N-1)$;

$$k = -\frac{N}{2}, \ldots, 0, \ldots, +\frac{N}{2} \text{ (approximately)}.$$

The values F_k given by equation (18) are the values of the transform at the frequencies $k\delta u$. By the sampling theorem $\delta u = \frac{1}{N\delta x}$, and the frequency range extends from $-u_{\max}$ to $+u_{\max}$ where $u_{\max} = \frac{1}{2\delta x}$. The right hand side of equation (18) may be split into three parts as shown; a scale factor, a phase factor, and a discrete Fourier transform. The phase factor arises because a discrete form of the shift theorem has been applied to enable the index j to run from 0 to N − 1, and the $\frac{1}{2}$ arises because sample points are taken at the centre of each step, as indicated in Fig. 5.

As a result of the sampling thorem N complex values are required to specify the transform of a function which itself has N complex values. However only N/2 complex values are required to specify the transform of a real function, since the transform is then hermitian. Calculation of the discrete Fourier transform normally involves N^2 complex multiplications and additions, and for a large number of values the computing time can be excessively long. An algorithm due to Cooley and Tukey[16], usually referred to as a "fast" Fourier transform, reduces the number of complex multiplications and additions to approximately $2N \log_2 N$. The computing time is reduced by a factor of $\dfrac{N}{2\log_2 N}$ i.e., for $N = 2^{10} = 1024$, the computing time is reduced by a factor of approximately fifty. The Cooley-Turkey method eliminates the redundant calculations implicit in the straightforward method, and relies for its efficiency on the splitting-up of a large transform into a number of smaller ones. It is applicable to many orthogonal transformations, and not just to the Fourier transform[17].

The way in which the fast Fourier transform is derived from the discrete transform is best illustrated by example. Consider an N-value function (where N is an even number) whose discrete Fourier transform is normally written:

$$F_k = \sum_{j=0}^{N-1} f_j e^{-\frac{2\pi i j k}{N}}; \quad (k = 0, 1, \ldots (N-1)). \tag{19}$$

To evaluate all the F_k there would normally be N^2 complex multiplications. Let the function be split into two $\dfrac{N}{2}$ value functions, a_j and b_j, such that

$$a_j = f_{2j}; \quad b_j = f_{2j+1}; \quad (j = 0, 1, \ldots (\tfrac{1}{2}N - 1)).$$

In other words all the even-numbered N-values of the function go to form a, and all the odd-numbered values go to form b. The discrete Fourier transforms of a_j and b_j will then be:

$$A_k = \sum_{j=0}^{\tfrac{1}{2}N-1} a_j e^{-\frac{4\pi i j k}{N}};$$

$$B_k = \sum_{j=0}^{\tfrac{1}{2}N-1} b_j e^{-\frac{4\pi i j k}{N}}. \qquad (k = 0, 1, \ldots (\tfrac{1}{2}N - 1))$$

The required transform F_k will be defined as

$$F_k = A_k + B_k e^{-\frac{2\pi i k}{N}} \quad (k = 0, 1, \ldots (\tfrac{1}{2}N - 1)).$$

It is noted that for the values of k from $\dfrac{N}{2}$ to $(N-1)$ the expressions for A_k and B_k repeat themselves, and it follows that

$$F_{k+\tfrac{1}{2}N} = A_k - B_k e^{-\frac{2\pi i k}{N}} \quad (k = 0, 1, \ldots (\tfrac{1}{2}N - 1)).$$

Thus in computing F_k as the sum of the discrete transforms of the two sequences we require

$$2(\tfrac{1}{2}N)^2 + N \simeq \frac{N^2}{2}$$

complex multiplications and additions. If $\frac{N}{2}$ is also an even number, each sequence can be split into two further subsequences, and so on. The method is therefore most efficient for $N = 2^n$, and in this case the total number of complex operations required is approximately $2N \log_2 N$. Further details of this algorithm may be found in the literature[17,18], and fast Fourier transform subroutines are now standard software on most computers[19].

If the number of values to be transformed is small, then a straightforward method of calculating the discrete Fourier transform may be more convenient, and here a simple cyclic method is outlined. The discrete Fourier transform may be written as:

$$F_k = \sum_{j=0}^{N-1} f_j \left(e^{-\frac{2\pi i k}{N}}\right)^j \quad (k = 0, 1, \ldots (N-1)).$$

If we set $\alpha = e^{-\frac{2\pi j k}{N}} = \cos\left(\frac{2\pi k}{N}\right) - i \sin\left(\frac{2\pi k}{N}\right)$,

then the transform may be expressed as:

$$F_k = \{[(f_{N-1}\alpha + f_{N-2})\alpha + f_{N-3}]\alpha + \ldots\} + f_0.$$

This repetitive operation is well-suited to computer calculation using complex arithmetic. If real arithmetic is used throughout, the cumulative sums of the real and imaginary parts can be expressed by recurrence relations in the form:

$$R_{j+1} = R_j \cos\frac{2\pi k}{N} + I_j \sin\frac{2\pi k}{N} + \text{Real}(f_{j+1});$$

$$I_{j+1} = R_j \cos\frac{2\pi k}{N} - I_j \sin\frac{2\pi k}{N} + \text{Imag}(f_{j+1});$$

where R and I are the real and imaginary parts of the cumulative sum, and Real and Imag denote the real and imaginary parts of the function f.

Computational Examples

To illustrate some of the factors that are important when calculating discrete Fourier transforms, we take a function whose exact Fourier transform is known:

$$e^{-\frac{|x|}{a}} \rightleftharpoons \frac{2a}{1 + 4\pi^2 a^2 u^2}, \quad \text{and we define } a = 5 \ \mu m.$$

This function is fairly similar in overall shape to the line spread function of a typical silver halide emulsion, and is shown on the left hand side of Fig. 6(a); the FT is not unlike the MTF of a typical silver halide film.

The crosses in Fig. 6(a) are the sample points used to calculate the discrete Fourier transform from equation (18). By taking N = 201 and $\delta x = 0.5$ μm, the computed transform values (+) are seen to give a good fit to the true transform. The values of the transform occur at intervals of approximately 10 cycles/mm, up to a maximum of ± 1000 cycles/mm (although the results are only plotted in the range 0–120 cycles/mm), and are symmetrical about

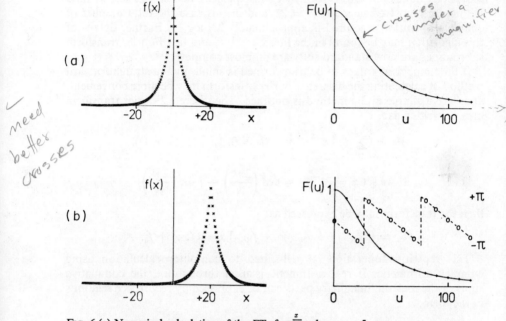

FIG. 6 (a) Numerical calculation of the *FT* of $e^{-\frac{x}{a}}$, where $a = 5$ μm:

——— true *FT*;
× × × sample points;
+ + + computed points.

(b) A shifted version of the function: symbols as above, but:
+ + + computed modulus;
○ ○ ○ computed phase.

the origin. If the function is shifted, as in Fig. 6(b), then the modulus is unchanged but there is a periodic phase component as predicted by the shift theorem of Table I.

Many of the values of $e^{-\frac{|x|}{5}}$ plotted in Fig. 6 are very small, and it is interesting to investigate the influence of truncating the function by setting it to zero outside a certain range, since this is often necessary in practice. The effect is shown in Fig. 7, where the function is set to zero outside the ranges of:

(a) -40 to $+40$ μm; (b) -20 to $+20$ μm; (c) -10 to $+10$ μm; (d) -5 to $+5$ μm.

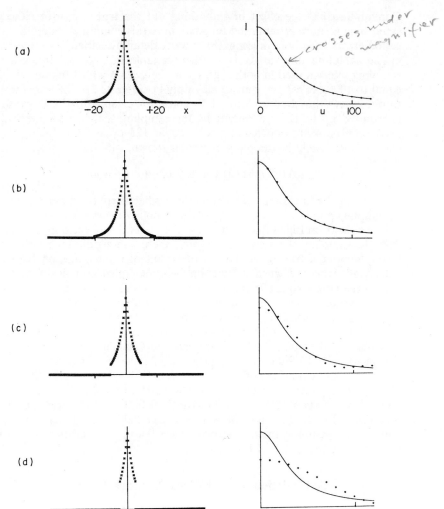

FIG. 7. The effect of truncation on computed FTs (see text for details).

The total range of -50 to $+50$ μm (N = 201, $\delta x = 0.5$ μm) has been maintained in each case, so the values of the Fourier transform are still spaced at 10 cycles/mm.

The error introduced by truncation is most pronounced at low spatial frequencies. The reason for this can be understood if we consider the truncated function to be the product of the original function $f(x)$ and a rectangle function of width m, $\text{rect}\left(\dfrac{x}{m}\right)$. By the convolution theorem the FT of this product will be

$$F(u) \circledast m\,\text{sinc}\,(mu)$$

Convolution has the effect of smoothing out the true transform, and this smoothing is more pronounced for small m in this computed example. Truncation errors have been investigated in more detail by Tatian[20].

The sampling interval δx is another variable of practical interest. The sampling interval used in both Figs. 6 and 7 gives a Nyquist frequency, u_{max}, equal to 1000 cycles/mm, but we may only be interested in the transform up to one or two hundred cycles/mm. Suppose for example that the transform is required up to 125 cycles/mm: by the sampling theorem the interval δx need only be 4 μm = 0·004 mm for u_{max} = 125 cycles/mm. In Fig. 8 the effect of increasing the sampling interval is shown, with intervals of:

(a) 0·5 μm; (b) 1 μm; (c) 2 μm; (d) 4 μm.

In each case the total range of the function has been kept the same, the number of sampling points thus decreasing as the sampling interval increases.

It is clear from Fig. 8 that if the sampling interval is selected in accordance with the sampling theorem ($\delta x = 4\ \mu$m), there is a significant positive bias in the computed transform. This is the result of a phenomenon known as "aliasing", whereby when a function being sampled contains frequencies greater than the Nyquist frequency—which is the case here—these frequencies can contribute to the frequencies below the Nyquist frequency. Thus ideally a value should be selected for δx corresponding to a frequency above which there is only a small amplitude rather than the maximum frequency of interest.

Aliasing will be discussed in more detail in Chapter 8 in connection with the measurement of the Wiener spectrum of noise. For typical line spread functions there is no firm rule, but as a guide it has been suggested that the absolute error in the computed MTF will be less than 0·005 if the sampling interval is less than 25% of the half-width at half-height of the line spread function[21]. In our example the half-width at half-height is approximately 4 μm, so the sampling interval should be less than 1 μm, and this seems to be confirmed by the results of Fig. 8.

6.2 Input-output Relationships for Linear Stationary Systems

Basic Theory

A system may be defined as that which produces a set of output functions from a set of input functions. Physically the system may be an electrical circuit, in which case the input and output are time-varying voltages or currents; it may be an incoherent optical system with spatially-varying intensities as input and output; or it may be a photographic process with varying exposure as the input and, say, density as the output. Provided that certain assumptions are made about each of these systems, their input-output relationships can be analysed in a similar manner. The fact that the inputs and outputs are one-dimensional for electrical systems (functions of time) and two-dimentional for optical and photographic systems (functions of space) is of little importance. The method of analysis of electronic and optical systems are in fact so similar that Elias[6] described how he was looking forward to the

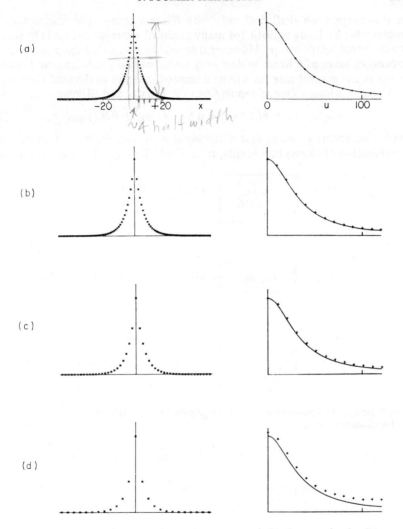

FIG. 8. The effect of sampling interval on computed FTs (see text for details).

first design for a camera lens featuring independent bass and treble control knobs!

A system can be represented by an operator S{ }, which acts on input functions to produce output functions. If $f(x,y)$ represents the input and $g(x,y)$ represents the output, then

$$g(x,y) = S\{f(x,y)\}.$$

This relationship between input and output is shown schematically in Fig. 9(a). To develop the relationship further we must restrict the nature of the system.

In this chapter we shall deal only with *linear* systems. The assumption of linearity is physically realistic for many practical systems, and leads to simple input-output relationships. However it should be noted that the photographic process behaves as a linear system only under special conditions, and is non-linear in the general case, as will be discussed in detail in the next chapter.

A system is linear if for all inputs $f_1(x,y)$ and $f_2(x,y)$ and all constants a and b:

$$S\{af_1(x,y) + bf_2(x,y)\} = a\,S\{f_1(x,y)\} + b\,S\{f_2(x,y)\}. \tag{20}$$

Basically, linearity means that if the input is broken down into an additive combination of elementary inputs, each of which gives a known output, then

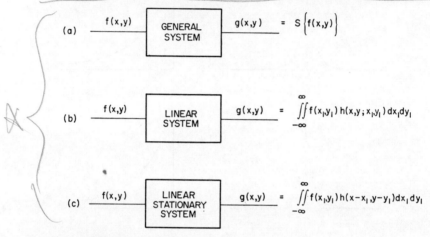

FIG. 9. Schematic representations of: (a) a general system; (b) a linear system; (c) a linear and stationary system.

the total output is found simply by adding the weighted values of the known outputs.

Using the sifting property of the delta function, as in equation (8), any input may be considered to be a linear combination of weighted and displaced delta functions:

$$f(x,y) = \iint_{-\infty}^{+\infty} f(x_1, y_1)\,\delta(x - x_1)\,\delta(y - y_1)\,dx_1 dy_1$$

The output of the system is defined by

$$g(x,y) = S\left\{\iint_{-\infty}^{+\infty} f(x_1, y_1)\,\delta(x - x_1)\,\delta(y - y_1)\,dx_1 dy_1\right\}.$$

If $f(x_1, y_1)$ is regarded as a weighting function applied to the delta functions, we may use the linearity property to bring the operator within the integral:

$$g(x,y) = \iint_{-\infty}^{+\infty} f(x_1, y_1)\,S\{\delta(x - x_1)\delta(y - y_1)\}\,dx_1\,dy_1.$$

6. FOURIER TRANSFORMS

If we let $h(x, y; x_1, y_1)$ be the response of the system at output coordinates (x, y) due to a delta function input at (x_1, y_1), then

$$h(x, y; x_1, y_1) = S\{\delta(x - x_1)\,\delta(y - y_1)\}.$$

The input and output of the system are now related by the equation:

$$g(x, y) = \iint_{-\infty}^{+\infty} f(x_1, y_1)\, h(x, y; x_1, y_1)\, dx_1\, dy_1. \qquad (21)$$

This relationship between the input and output of a linear system is shown schematically in Fig. 9(b). It is clear from equation (21) that a linear system is completely specified by its response to a delta function input. This response is usually called the impulse response function for electrical systems, and for optical and photographic systems it is called the point spread function.

A second assumption, particularly appropriate in the photographic case, reduces the input-output relationship to a simple convolution formula. If the point spread function depends only on the differences $(x - x_1, y - y_1)$ and not on each variable separately, i.e.,

$$h(x - x_1, y - y_1) \equiv h(x, y; x_1, y_1),$$

then the system may be said to be stationary. Electrical networks containing only fixed resistors, capacitors and inductances are stationary, since their characteristics do not vary with time. Optical systems are stationary only if the point spread function maintains a constant shape over all regions of the image. For optical systems this is not generally the case, since the aberrations of lenses leads to variations over the image area. However, the photographic process usually satisfies this stationarity condition and the shape of the photographic point spread function will then be constant over the entire image area.

It must be remembered that the photographic spread function is of course a limiting statistical function, and cannot describe any individual quantum/grain interaction, applying only as a representation for very large numbers of such interactions. The relationship between individual events and the overall spread function will be discussed in Chapter 7 in terms of Monte Carlo modelling. Similarly, from our previous considerations of the quantum processes involved in latent image formation, it is clear that the individual quantum/grain interaction is completely non-linear.

For a system that is both linear and stationary, the input-output relationship becomes:

$$g(x, y) = \iint_{-\infty}^{+\infty} f(x_1, y_1)\, h(x - x_1, y - y_1)\, dx_1\, dy_1$$

$$= \iint_{-\infty}^{+\infty} f(x - x_1, y - y_1)\, h(x_1, y_1)\, dx_1\, dy_1. \qquad (22)$$

In other words the output is equal to the input convoluted with the point spread function, and this is represented schematically in Fig. 9(c). Due to the physical implications of convolution this is often referred to as a "smearing", "blurring" or "smoothing" of one function by another.

When the photographic process is exposed to an exposure distribution $f(x_1, y_1)$, the above convolution relationship can be deduced using a simple physical argument. In Fig. 10 the exposure distribution $f(x_1, y_1)$ is to be imaged onto a photographic material which lies in the (x_1, y_1)-plane. Assuming

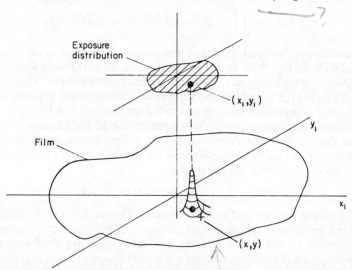

FIG. 10. Imaging an exposure distribution, in terms of the spread function (see text)

stationarity, the output density at a point (x, y) due to a single point (x_1, y_1) in the input exposure will be simply

$$f(x_1, y_1) \, h(x - x_1, y - y_1).$$

Assuming linearity, the total density at (x, y) denoted by $g(x, y)$ due to all points in the input exposure will be:

$$g(x, y) = \iint_{-\infty}^{+\infty} f(x_1, y_1) \, h(x - x_1, y - y_1) \, dx_1 \, dy_1.$$

Applying similar physical reasoning to electronic and optical systems also leads to the convolution relationship between input and output.

The convolution relationship is not only simple, but is also ideally suited to the practical application of Fourier transform techniques. Suppose the functions $F(u, v)$, $G(u, v)$ and $H(u, v)$ are the Fourier transforms of $f(x, y)$, $g(x, y)$ and $h(x, y)$ respectively. The functions $F(u, v)$ and $G(u, v)$ describe the contribution due to each spatial frequency which is present in the input and output, while the function $H(u, v)$ describes the fraction of each spatial frequency contribution that is transmitted by the system, and is thus called the transfer function of the system. By the convolution theorem, as in Table I;

$$G(u, v) = F(u, v) H(u, v). \tag{23}$$

Equation (23) states that at any spatial frequency (u, v) the amplitude and phase of the output are equal to those of the input times the system transfer function.

The main advantages of the spatial frequency (u, v) description, as in equation (23), over the distance (x, y) description, as in equation (22), is that the complicated process of convolution is replaced by straightforward multiplication. As an example, we consider a common problem in image recording. A photographic process is to be used to measure an intensity distribution, $f(x, y)$; the point spread function is known, and the output density distribution can be measured on a microdensitometer. Given $h(x, y)$ and $g(x, y)$, the problem is to calculate $f(x, y)$. Inspection of the convolution formula of equation (22) shows that in general this would be difficult. However, in terms of spatial frequency the solution is found by simple division:

$$F(u, v) = \frac{H(u, v)}{G(u, v)}.$$

The original input can then in principle be deduced by taking the inverse Fourier transform of $F(u, v)$. Thus given the output, the easiest way of finding the input is to work in terms of spatial frequency rather than space. On the other hand, it may be convenient in some simple cases when the input is given to use the convolution relationship, and we shall use this relationship to calculate the images of lines, edges and one-dimensional sine waves.

The Line Spread Function

The line spread function of an imaging system is defined as the response of the system to a line input. A line input may be represented by a single delta function, $\delta(x_1)$, which lies along the y_1 axis as shown in Fig. 11. The line spread function depends only on the x-variable, and using the convolution relationship between input and output, will be defined as

$$l(x) = \int\int_{-\infty}^{+\infty} \delta(x - x_1) h(x_1, y_1) \, dx_1 \, dy_1.$$

By the sifting property of the delta function;

$$l(x) = \int_{-\infty}^{\infty} h(x, y_1) \, dy_1. \tag{24}$$

Thus the line spread function is obtained from the point spread function by integrating over one variable.

Point spread functions of photographic processes are usually rotationally symmetrical, i.e.,

$$h(r) \equiv h(x, y), \text{ where } r^2 = x^2 + y^2.$$

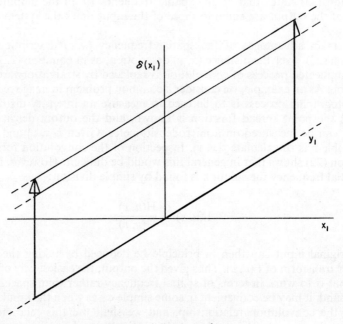

FIG. 11. A line input $\delta(x_1)$.

In this case the point spread function and line spread function are related as Abel transform pairs[22]:

$$l(x) = 2 \int_x^\infty h(r)(r^2 - x^2)^{-\frac{1}{2}} r \, dr$$

$$h(r) = -\frac{1}{\pi} \frac{d}{dr} \left(\int_r^\infty \frac{l(x) r \, dx}{x(x^2 - r^2)^{\frac{1}{2}}} \right). \tag{25}$$

Although calculation of the line spread function from the point spread function is quite straightforward, the reverse operation is seen to be rather complicated, and this has been discussed in some detail by Marchand both for rotationally symmetrical[23] and non-symmetrical[24] point spread functions.

If the photographic point spread function is rotationally symmetrical it will be completely specified by a section $h(r)$, or by the one-dimensional line spread function, $l(x)$. In practice the line spread function is often the more useful of the two functions, and it may also be easier to measure experimentally. Suppose the input is an opaque edge lying along the y-axis such that

$$f(x) = 0 \ (x < 0); f(x) = 1 \ (x \geqslant 0).$$

Applying the convolution relationship, the edge spread function, $e(x)$, is given by

$$e(x) = \int\int_{-\infty}^{+\infty} f(x - x_1) h(x_1, y_1) \, dx_1 \, dy_1$$

$$= \int_{-\infty}^{\infty} l(x_1) \, dx_1. \tag{26}$$

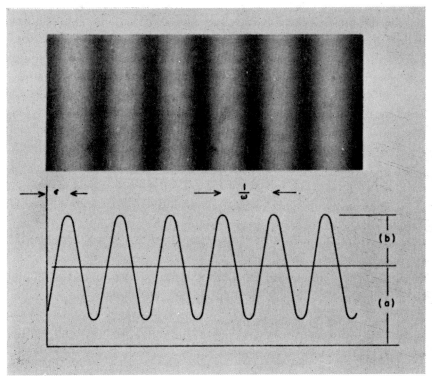

FIG. 12. A one-dimensional sine wave input, with spatial frequency ω, modulation $\frac{b}{a}$, and phase ε.

The edge spread function is an integration of the line spread function, as was illustrated in Fig. 20 of Chapter 2. From equation (26) it follows that the line spread function is given by the differential of the edge spread function:

$$l(x) = \frac{d}{dx}(e(x)). \tag{27}$$

The Modulation Transfer Function

It is now supposed that the input exposure is a one-dimensional sinusoidal distribution of the form shown in Fig. 12. This may be defined as:

$$f(x) = a + b\cos(2\pi\omega x + \varepsilon),$$

where ω is the one-dimensional spatial frequency (or line frequency), and ε is a measure of the phase. It will be shown that the output, or image, is also sinusoidal with the same spatial frequency ω as the input, but with a change of amplitude, or modulation. The ratio of the output modulation to the input modulation depends on the spatial frequency, and turns out to be equal to the modulus of the Fourier transform of the spread function. It is this ratio of output to input modulation that is called the *modulation transfer function*, or MTF. Although this is now common terminology[25], in older literature it may be described as the sine-wave response, frequency response, contrast transfer, etc.

The input modulation is defined by:

$$M_{in} = \frac{f_{max} - f_{min}}{f_{max} + f_{min}} = \frac{b}{a}.$$

From the convolution relationship, the output is

$$g(x) = \iint_{-\infty}^{+\infty} f(x - x_1, y - y_1) h(x_1, y_1) \, dx_1 \, dy_1$$

$$= \iint_{-\infty}^{+\infty} (a + b \cos(2\pi\omega(x - x_1) + \varepsilon)) h(x_1, y_1) \, dx_1 \, dy_1.$$

Integrating with respect to y_1:

$$g(x) = \int_{-\infty}^{+\infty} l(x_1)(a + b \cos(2\pi\omega(x - x_1) + \varepsilon)) dx_1, \qquad (28)$$

where $l(x_1)$ is the line spread function. Using the expansion of $\cos(A - B)$ and normalizing the spread function such that its area is unity, i.e.,

$$\int_{-\infty}^{+\infty} l(x_1) \, dx_1 = 1,$$

equation (28) may then be written as

$$g(x) = a + b \cos(2\pi\omega x + \varepsilon) \int_{-\infty}^{+\infty} l(x_1) \cos(2\pi\omega x_1) \, dx_1$$

$$+ b \sin(2\pi\omega x + \varepsilon) \int_{-\infty}^{+\infty} l(x_1) \sin(2\pi\omega x_1) \, dx_1;$$

or

$$g(x) = a + b \cos(2\pi\omega x + \varepsilon) C(\omega) + b \sin(2\pi\omega x + \varepsilon) S(\omega); \qquad (29)$$

where

$$C(\omega) - i\, S(\omega) = T(\omega) = \int_{-\infty}^{+\infty} l(x_1) e^{-2\pi i \omega x_1} \, dx_1.$$

The function $T(\omega)$ is the optical transfer function, and $C(\omega)$ and $-S(\omega)$ are its real and imaginary parts. The optical transfer function is the Fourier transform of the line spread function.

If $M(\omega)$ and $\phi(\omega)$ are the modulus and phase of the optical transfer function, defined by

$$M(\omega) = \sqrt{C^2(\omega) + S^2(\omega)}; \quad \phi = \tan^{-1}\left(\frac{-S(\omega)}{C(\omega)}\right);$$

then $C(\omega) = M(\omega) \cos \phi(\omega)$; $S(\omega) = -M(\omega) \sin \phi(\omega)$.
Equation (29) now reduces to

$$g(x) = a + M(\omega) \, b \cos(2\pi\omega x + \varepsilon + \phi(\omega)). \tag{30}$$

Equation (30) shows that the output $g(x)$ is also sinusoidal, and has the same frequency as the input. The output modulation will be defined by

$$M_{out} = \frac{g_{max} - g_{min}}{g_{max} + g_{min}} = M(\omega)\frac{b}{a}.$$

Thus the ratio of the output modulation to the input modulation is simply equal to $M(\omega)$—the modulus of the Fourier transform of the line spread function.

Since the area under the spread function has been defined as unity, the MTF will be normalized to unity at zero spatial frequency:

$$M(0) = \left| \int_{-\infty}^{+\infty} l(x_1) \, dx_1 \right| = 1.$$

If the line spread function is both real and even, the imaginary part and the phase of the transfer function will be zero. In this case the real part of the transfer function, the optical transfer function and the modulation transfer function, will all be identical.

Symmetry Properties

Imaging in either optics or photography is essentially a two-dimensional process, and consequently the basic functions that describe the performance of these systems are functions of two variables. Whereas two-dimensional space variables are commonplace, the significance of two-dimensional spatial frequency is not so immediately obvious, and in Fig. 13 two examples are given. A two-dimensional spatial frequency is a more general case of the one-dimensional form considered earlier. For each pattern as in Fig. 13 we may assign a line frequency ω and an angle θ, such that

$$\omega^2 = \sqrt{u^2 + v^2}; \quad \theta = \tan^{-1}\frac{v}{u}.$$

The two-dimensional transfer function is related to the one-dimensional transfer function in a simple way. The one-dimensional transfer function (i.e., the Fourier transform of the line spread function) describes the way in which a two-dimensional sine-wave with angle θ is modulated by the system. Thus the shape of the one-dimensional optical transfer function is equal to the shape of a *section* of the two-dimensional one, the section being taken through the origin at angle θ. By varying the angle from 0 to π the complete two-dimensional transfer function is generated. When the point spread function is rotationally symmetrical a single one-dimensional transfer function completely defines the two-dimensional transfer function.

Since the point spread function and line spread function are Abel transform pairs when there is rotational symmetry, and since the line spread function and the one-dimensional OTF are Fourier transform pairs, it follows that the

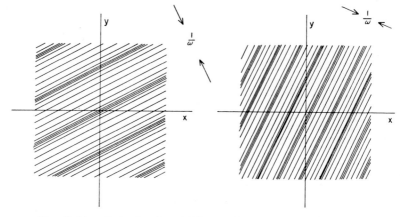

FIG. 13. Two-dimensional spatial frequencies: ω is the line frequency.

point spread functional and OTF are Hankel transform pairs[22]. The transform relationships are as below:

ABEL

$$l(x) = 2 \int_x^\infty h(r)(r^2 - x^2)^{-\frac{1}{2}} r \, dr$$

$$h(r) = -\frac{1}{\pi} \frac{d}{dr} \left(\int_r^\infty \frac{l(x) r \, dx}{x(x^2 - r^2)^{\frac{1}{2}}} \right); \tag{31}$$

HANKEL

$$T(\omega) = 2\pi \int_0^\infty h(r) J_0(2\pi\omega r) r \, dr$$

$$h(r) = 2\pi \int_0^\infty T(\omega) J_0(2\pi\omega r) \omega \, d\omega; \tag{32}$$

FOURIER

$$T(\omega) = \int_{-\infty}^{+\infty} l(x) e^{-2\pi i \omega x} \, dx$$

$$l(x) = \int_{-\infty}^{+\infty} T(\omega) e^{+2\pi i \omega x} \, d\omega. \tag{33}$$

J_0 again denotes the zero-order Bessel function. We shall see that a similar set of relationships apply to the various one-dimensional functions that describe the properties of image noise.

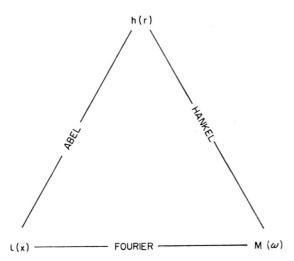

FIG. 14. The relationship between the point spread function, $h(r)$, the line spread function, $l(x)$, and the transfer function, $M(\omega)$.

6.3 Noise Analysis

Random Processes

In our analysis of linear stationary systems in the previous section, it was assumed that both the input and output could be specified exactly. However, as already observed, at the quantum/grain level of interaction a statistical approach is necessary. In previous chapters the image noise has been specified by the large-scale parameter G, and we shall now turn our attention to more complete descriptions of the fluctuations in the output.

The methods used to analyse the noise in a photographic image are much the same as those used to analyse any other type of random process. Here the discussion will be confined to the general methods of description, and their photographic application will be given more detailed treatment in Chapter 8. The general literature on random processes is usually concerned with time series. However, as with linear systems analysis, these methods can be readily adapted to two-dimensional functions of image area.

Suppose that the density distribution of a uniformly exposed and developed area of film is denoted by $D(x, y; 1)$, and that a second sample which has been exposed and developed in an identical way is denoted by $D(x, y; 2)$. By repeating this procedure many times we can generate an ensemble of density distributions $D(x, y; i)$. This ensemble defines a random process, and each member of the ensemble is called a realization of the process. Figure 15 shows some realizations of a one-dimensional process, which might be the image density of a sample of uniformly exposed and developed film as scanned in a microdensitometer. Although in the general case all members of the ensemble are required to describe a random process, it is intuitively obvious in the photographic case that a single realization is sufficient provided that this is large enough to provide a satisfactory statistical sample. If all the statistics of a random process can be determined from a single realization, the process is called *ergodic*, and it may be assumed that photographic noise has this property. As a result of ergodicity, averages that would normally be taken

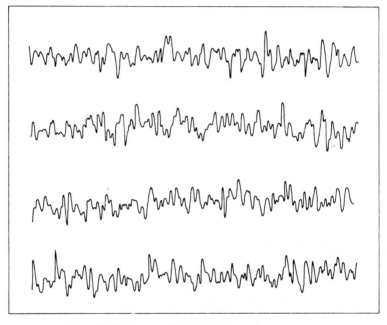

FIG. 15. Different realizations of a random process.

over the ensemble may usually be replaced by space averages, although occasionally—as we shall see for the Wiener spectrum—both space and ensemble averages should be considered.

Another feature of the density distribution of a sample of uniformly exposed and developed film is that it is *statistically stationary*; i.e., the statistics are the same over all areas and are not influenced by a shift of origin (this is analogous to the definition of a stationary system earlier in the chapter). Whilst statistical stationarity is usually assumed for photographic noise, it must be emphasized that this holds only for uniform exposure and development. This may be a limiting practical factor in experimental measurement, since it can be quite difficult to produce large image areas satisfying these conditions.

First Order Statistics

The first order probability distribution function $P(D_1)$ of a stationary ergodic process gives the probability that the random process D is less than or equal to some value D_1, as a function of D_1; i.e.,

$$P(D_1) = \text{prob}\,[D \leqslant D_1].$$

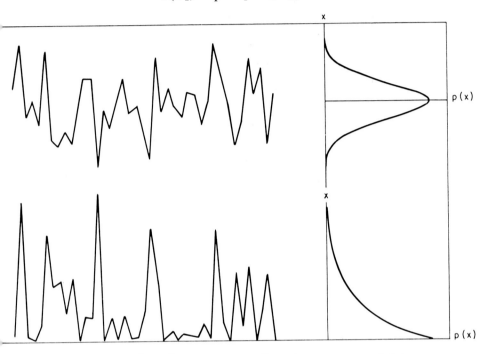

FIG. 16. Realizations of two (computer-generated) random processes, and their corresponding probability-density functions.

The corresponding first order probability density function is obtained by differentiation:

$$p(D_1) = \frac{d}{dD_1} P(D_1).$$

The term "probability density" is somewhat unfortunate in this context where we are concerned with photographic density, but it is a standard term and no real confusion need arise. The probability density function may be interpreted as the probability that any value of the process selected at random lies in the range D_1 to $D_1 + dD_1$:

$$p(D_1)\, dD_1 = \text{prob}[D_1 < D \leqslant D_1 + dD_1]. \tag{34}$$

Realizations of two random processes are shown in Fig. 16, together with their probability density functions. It can be seen that the upper process is symmetrical about the most probable value, whereas the second process is highly asymmetrical.

To distinguish in a quantitative manner between the two processes shown in Fig. 16, it is necessary to specify their probability density functions. However the main difference between two random processes is often only in the magnitude of the fluctuations, as in Fig. 17. In this case it is adequate to compare the moments of the probability distribution. The mean value, D, and the variance, σ^2, of a random process are equal to the first and second central

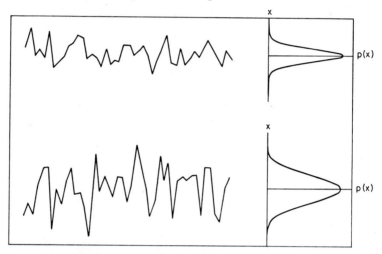

FIG. 17. Realizations of two (computer-generated) Gaussian random processes, and their corresponding probability-density functions.

moments respectively, and for a stationary ergodic process may be expressed as:

$$D = \lim_{X, Y \to \infty} \frac{1}{2X} \frac{1}{2Y} \int_{-X}^{+X} \int_{-Y}^{+Y} D(x, y) \, dx \, dy; \quad (35)$$

$$\sigma^2 = \lim_{X, Y \to \infty} \frac{1}{2X} \frac{1}{2Y} \int_{-X}^{+X} \int_{-Y}^{+Y} (D(x, y) - D)^2 \, dx \, dy$$

$$= \left\{ \lim_{X, Y \to \infty} \frac{1}{2X} \frac{1}{2Y} \int_{-X}^{+X} \int_{-Y}^{+Y} D(x, y)^2 \, dx \, dy \right\} - D^2. \quad (36)$$

The variance is equal to the mean of the square minus the square of the mean of the random process.

Using a model to represent the photographic image fluctuations, Selwyn[26] demonstrated that the density D_A as measured with an aperture of area A should follow a Gaussian distribution:

$$p(D_A) = \frac{1}{\sigma_A \sqrt{2\pi}} e^{-\frac{(D_A - D)^2}{2\sigma_A^2}}. \quad (37)$$

The main assumption in arriving at this result is that the aperture area A over which the density is measured at any single aperture position is large compared with a grain, and so many grains lie within the aperture. Whether or not a developed grain is present at a particular point in the image may be taken to be a random variable with some arbitrary probability density function. If a large number of random variables are added together, the probability density function of the sum tends to a Gaussian distribution irrespective of the distribution of the component random variables. This result is known as the central limit theorem[27]. Consequently the total number of grains lying

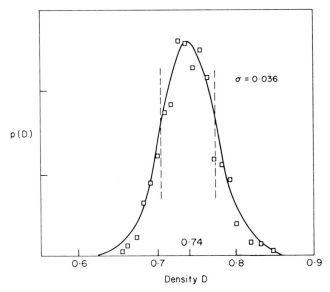

FIG. 18. An example of the probability-density function of image noise fluctuations, compared with the Gaussian distribution having the same mean and standard deviation (solid curve)[28].

within the aperture will tend to follow a Gaussian distribution, as will the resulting image density if the Nutting formula holds. The measured density is found experimentally to give a good fit to a Gaussian distribution in most practical cases[28], and some typical experimental results are shown in Fig. 18.

Although the value of σ_A is a useful measure of the extent of the density fluctuations, it will depend on the aperture A with which the density is measured. As A increases the average number of developed grains lying within the aperture will increase, and if the area is increased by a factor k, then the relative fluctuation in the number of grains lying within the aperture decreases by a factor $\sqrt{k}$, and thus

$$\sigma_A \propto \frac{1}{\sqrt{A}}.$$

Reasoning along these lines, Selwyn introduced a granularity coefficient S, that is independent of the scanning aperture provided the area is large:

$$S = \sigma_A \sqrt{2A}. \tag{38}$$

This independence of the aperture area was deduced in Chapter 1 during the more general analysis of image density fluctuations. Since we have used the parameter $G = A\sigma_A^2$, the relationship between S and G will be $S^2 = 2G$. Some experimental results showing that S and G are essentially independent of A were given in Fig. 22 of Chapter 2.

Since the density fluctuation is typically Gaussian, the first order statistics of the image noise are completely specified by one of the parameters σ_A, S or G. Numerically these quantities are found from the density fluctuation $\Delta D(x, y) = D(x, y) - D$, using a simplified version of equation (36) such as

$$\sigma_A^2 = \frac{1}{N} \sum_{i=1}^{N} (\Delta D_i)^2. \tag{39}$$

The Autocorrelation Function

In order to specify the spatial structure of image noise it is necessary to investigate higher-order probability density functions. The second order (or joint) probability density function of a stationary ergodic process gives the probability that at (x, y) the process lies in the range D_1 to $D_1 + dD_1$, and at $(x + \xi, y + \eta)$ the process lies in the range D_2 to $D_2 + dD_2$:

$$p(D_1, D_2)\, dD_1\, dD_2 = \text{prob}\, [D_1 < D(x, y) \leqslant D_1 + dD_1$$
$$\text{and } D_2 < D(x + \xi, y + \eta) \leqslant D_2 + dD_2].$$

Although this second order probability function is in general quite complicated, fortunately for Gaussian processes it is completely specified by its first joint moment—just as the first order statistics are completely specified by D and σ_A.

This first joint moment defines the autocorrelation function, which for a stationary ergodic process may be written as

$$C(\xi, \eta) = \lim_{X, Y \to \infty} \frac{1}{2X} \frac{1}{2Y} \int_{-X}^{+X} \int_{-Y}^{+Y} D(x, y)\, D(x + \xi, y + \eta)\, dx\, dy. \tag{40}$$

In practice it is customary to work in terms of the autocorrelation function of the fluctuations:

$$C_{\Delta D}(\xi, \eta) = \lim_{X, Y \to \infty} \frac{1}{2X} \frac{1}{2Y} \int_{-X}^{+X} \int_{-Y}^{+Y} \Delta D(x, y)\, \Delta D(x + \xi, y + \eta)\, dx\, dy. \tag{41}$$

It is noted that the scale value is simply equal to the measured variance:

$$C_{\Delta D}(0, 0) = \lim_{X, Y \to \infty} \frac{1}{2X} \frac{1}{2Y} \int_{-X}^{+X} \int_{-Y}^{+Y} \Delta D(x, y)^2\, dx\, dy = \sigma_A^2.$$

The suffix ΔD will henceforth be omitted, and it will be assumed that the autocorrelation function relates to the density fluctuation.

Since the autocorrelation function includes the variance, it completely specifies the first and second order statistics of the measured density fluctuations, provided the process is Gaussian. Furthermore, it can be shown that for a Gaussian process all the higher order probability density functions and their moments can be specified in terms of the autocorrelation function, and so this function defines the whole random process. One of the disadvantages of the autocorrelation function is that its shape depends critically on the aperture used in measurement, as shown for the example in Figure 19.

The effect of the measuring system on the autocorrelation function may be analysed in the following way. Suppose that the density fluctuation is represented by $\Delta D(x, y)$ as would be measured with a very small aperture and a

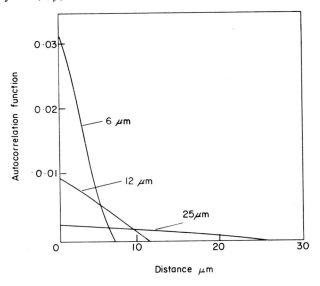

FIG. 19. The effect of the diameter of the scanning aperture on the measured autocorrelation function[12].

perfect optical system. If the aperture and optical system of a microdensitometer used to measure $\Delta D(x, y)$ have a combined spread function $h(x, y)$, this will constitute the effective measuring aperture. If the density fluctuation is small, then the microdensitometer system acts as a linear stationary system, with the "actual" fluctuation $\Delta D(x, y)$ as the input and the measured fluctuation $\Delta D'(x, y)$ as the output. By the convolution relationship of equation (22):

$$\Delta D'(x, y) = \int\int_{-\infty}^{+\infty} h(x_1, y_1)\, \Delta D(x - x_1, y - y_1)\, dx_1\, dy_1.$$

Similarly,

$$\Delta D'(x + \xi, y + \eta) = \int\int_{-\infty}^{+\infty} h(x_2, y_2) \Delta D(x - x_2 + \xi, y - y_2 + \eta)\, dx_2\, dy_2.$$

Hence the measured autocorrelation function $C'(\xi, \eta)$ will be as defined according to equation (41):

$$C'(\xi, \eta) = \underset{X, Y \to \infty}{\text{limit}} \frac{1}{2X} \frac{1}{2Y} \int_{-X}^{+X} \int_{-Y}^{+Y} \Delta D'(x, y) \Delta D'(x + \xi, y + \eta) \, dx \, dy.$$

If we substitute the above integral expressions for $\Delta D'(x, y)$ and $\Delta D'(x + \xi, y + \eta)$ we obtain an overall expression for $C'(\xi, \eta)$ which involves integration over $dx \, dy \, dx_1 \, dy_1 \, dx_2 \, dy_2$. However inspection of the expressions readily shows that in convolution notation we may write

$$C'(\xi, \eta) = C(\xi, \eta) \circledast h(-x, -y) \circledast h(x, y). \tag{42}$$

Thus the measured autocorrelation function is equal to the actual autocorrelation function convoluted twice with the point spread function of the measuring system. To extract the actual autocorrelation function it again becomes simpler to transcribe equation (42) into the spatial frequency domain.

The Wiener Spectrum

The Wiener (or noise-power) spectrum of the fluctuations of a stationary ergodic process is defined by:

$$W(u, v) = \underset{X, Y \to \infty}{\text{limit}} \left\langle \frac{1}{2X} \frac{1}{2Y} \left| \int_{-X}^{+X} \int_{-Y}^{+Y} \Delta D(, x\, y) \, e^{-2\pi i (ux + vy)} \, dx \, dy \right|^2 \right\rangle, \tag{43}$$

where the symbol $\langle \rangle$ denotes the ensemble average. The Wiener–Khintchin theorem then states that the Wiener spectrum and the autocorrelation function are Fourier transform pairs:

$$W(u, v) = \int \int_{-\infty}^{+\infty} C(\xi, \eta) \, e^{-2\pi i (u\xi + v\eta)} \, d\xi \, d\eta; \tag{44}$$

$$C(\xi, \eta) = \int \int_{-\infty}^{+\infty} W(u, v) \, e^{+2\pi i (u\xi + v\eta)} \, du \, dv. \tag{45}$$

It follows that the variance of the measured density fluctuation is the volume under the two-dimensional Wiener spectrum:

$$\sigma_A^2 = C(0, 0) = \int \int_{-\infty}^{+\infty} W(u, v) \, du \, dv. \tag{46}$$

The autocorrelation function and Wiener spectrum are essentially equivalent measures of image noise in much the same way as the point spread function and the MTF are equivalent ways of describing image spread or resolution. For a Gaussian process such as photographic image noise, either function gives a complete statistical description of the random process. In Chapter 8 it will be shown that the parameter G used in previous chapters is simply equal to the scale value of the Wiener spectrum:

$$G = A\sigma_A^2 = \frac{S^2}{2} = W(0, 0). \tag{47}$$

If the Fourier transform is taken of both sides of equation (42), then

$$W'(u, v) = W(u, v)|T(u, v)|^2, \qquad (48),$$

where W'(u, v) is the measured Wiener spectrum,
W(u, v) is the actual Wiener spectrum,
T(u, v) is the optical transfer function of the measuring system.

Extracting the actual Wiener spectrum from the measured spectrum is thus simply a matter of dividing the measured function by the square of the MTF of the measuring system, as opposed to a de-convolution in terms of the autocorrelation function.

Symmetry Properties

When the two-dimensional image noise is isotropic, (i.e., its statistical properties are independent of direction in the image) it can be represented by a single variable, just as the point spread function can be so represented when there is rotational symmetry. For example the functions of one variable can be written in terms of polar coordinates:

$$\omega = \sqrt{u^2 + v^2}; \qquad \zeta = \sqrt{\xi^2 + \eta^2}.$$

In this case

$$W(\omega) = 2\pi \int_0^\infty C(\zeta) J_0(2\pi\omega\zeta) \, \zeta \, d\zeta$$

$$C(\zeta) = 2\pi \int_0^\infty W(\omega) J_0(2\pi\omega\zeta) \, \omega \, d\omega. \qquad (49)$$

Again $J_0(\ )$ denotes the zero-order Bessel function. The functions $W(\omega)$ and $C(\zeta)$ are sections through the origins of the corresponding two-dimensional functions $W(u, v)$ and $C(\xi, \eta)$ and are Hankel transform pairs.

We can also express the functions in the integral form:

$$W_1(u) = \int_{-\infty}^{+\infty} W(u, v) \, dv$$

$$C_1(\xi) = \int_{-\infty}^{+\infty} C(\xi, \eta) \, d\eta. \qquad (50)$$

Making the substitution $\zeta = \sqrt{\xi^2 + \eta^2}$ in equation (50) yields

$$C_1(\xi) = 2 \int_\xi^\infty C(\zeta)(\zeta^2 - \xi^2)^{-\frac{1}{2}} \, \zeta \, d\zeta. \qquad (51)$$

Thus $C_1(\xi)$ and $C(\zeta)$ are Abel transform pairs, and a similar relationship exists between $W_1(u)$ and $W(\omega)$. The complete inter-relationship[29] is summarized in Figure 20.

If the Wiener spectrum of a two-dimensional isotropic random process is measured using a long and narrow slit and scanning at right angles to the length of the slit, then the one-dimensional Wiener spectrum that is obtained is

$W(\omega)$, i.e., a section of the two-dimensional spectrum. It can be seen however from the relationships of Fig. 20 that the Fourier transform of this function is an integral form of the two-dimensional autocorrelation function rather than a section of it. On the other hand, if a very small circular aperture is used to scan the image in one direction, the Wiener spectrum obtained is in an integral form and the autocorrelation function is now a section of the two-dimensional

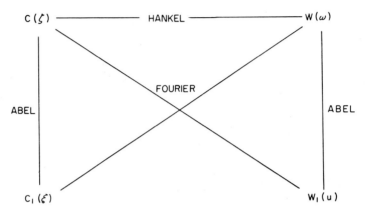

FIG. 20. Relationship between the various one-dimensional functions that describe isotropic noise.

function. The implications of the ways in which the one and two dimensional functions are related will be discussed in more detail in Chapter 8.

Computational Examples

The expressions for the autocorrelation function and Wiener spectrum, as in equations (41) and (43), imply that large noise samples must be used in measurement. For practical numerical calculations it is necessary to take a large number of sample points in order to obtain good estimates to these functions. The computations involved can be time-consuming and costly even on modern high-speed computers, and so its is important to know how the calculated functions converge to the functions required, and the relative errors which are involved. Fuller practical details will be discussed in Chapter 8, but here some simple examples are considered as an introduction to problems of convergence and accuracy.

Sequences of Gaussian random variables of zero mean and unit variance were generated using a standard computer subroutine[30]. Part of these sequences are illustrated in Fig. 21, and the first 1024 values in each sequence are tabulated in Tables 1 and 2 of Appendix I. Process A consists of an uncorrelated set of random variables whose autocorrelation function should converge to a delta function and whose Wiener spectrum should be independent of frequency. Process B is a correlated set of random variables with an autocorrelation function nominally equal to the triangle function $\Delta(\xi/10)$, and Wiener spectrum defined by $\text{sinc}^2(10u)$.

6. FOURIER TRANSFORMS

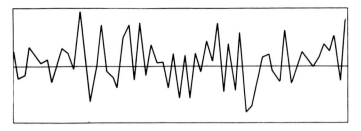

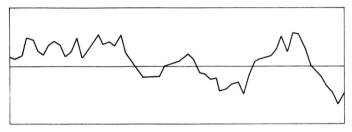

FIG. 21. Two computer-generated random processes, with zero mean and unit variance. Upper, Process A; Lower, Process B.

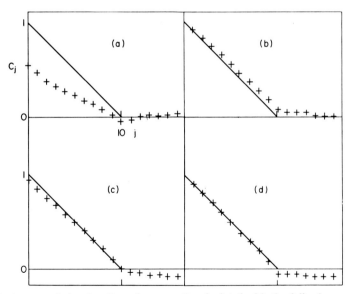

FIG. 22. Autocorrelation function of process B, calculated with the following number of data values: (a) 250; (b) 500; (c) 1,000; (d) 2,000.

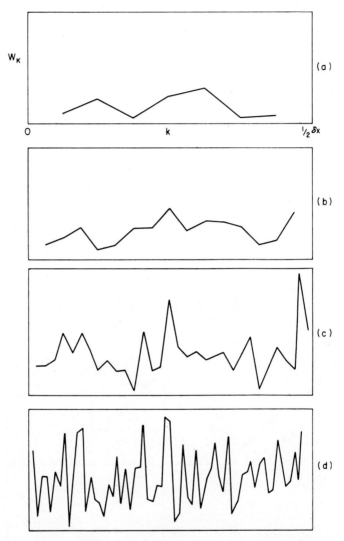

FIG. 23. Wiener spectra of process A, calculated with the following number of data values: (a) 16; (b) 32; (c) 64; (d) 128.

Figure 22 shows the autocorrelation function of process B as calculated from the formula

$$C_j = \frac{1}{N} \sum_{i=1}^{N} \Delta D_i \Delta D_{i+j} \quad (j = 0, 1, 2, \ldots)$$

for values of N equal to 250, 500, 1000, and 2000. The convergence of the calculated function to the limiting function shows clearly from this plot.

The Wiener spectrum is shown for process A in Fig. 23, where the formula:

$$W_k = \frac{1}{N} \left| \sum_{j=1}^{N} \Delta D_j \, e^{-\frac{2\pi i k j}{N}} \right|^2$$

has been used to calculated the spectrum. Calculations are shown for values of N equal to 16, 32, 64 and 128 (all numbers being powers of two since the fast Fourier transform was used). In accordance with the sampling theorem the values of the Wiener spectrum appear at closer intervals ($\delta u = 1/N\delta x$, where the data points are separated by $\delta x = 1$ in each case) as more data values

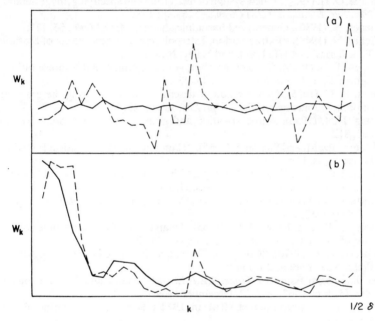

FIG. 24. Wiener spectra of processes A and B, using: – – – – – – 1 × 64; and ———— 32 × 64 data values.

are used in the computation. However, there is no evidence of a decrease in error, and the measured function does not appear to converge as N increases. This is why the proper definition of the Wiener spectrum, as in equation (43), includes the ensemble average, and for the spectrum to converge it is necessary to take both space and ensemble averages.

Figure 24 shows the computed Wiener spectra for processes A and B for 1 × 64 and 32 × 64 (= 2 048) data values. A more satisfactory convergence to the limiting Wiener spectra is now obtained.

References

1. Frieser, H. (1935). Concerning the resolution of photographic layers. *Kinotechnik*, **17**, 167.

2. Selwyn E. W. H. (1948). The photographic and visual resolving power of lenses. *Photogr. J.*, **88B**, 6, 46.
3. Duffieux, P. M. (1970). "The Fourier Integral and its Application in Optics". (2nd Ed.), Masson and Cie, Paris.
4. Luneberg, R. K. (1964). "The Mathematical Theory of Optics". University of California Press. (This book is based on lectures at Brown University USA in 1944.)
5. Elias, P., Grey, D. S. and Robinson, D. Z. (1952). Fourier treatment of optical processes. *J. Opt. Soc. Amer.*, **42**, 127.
6. Elias, P. (1953). Optics and communication theory. *J. Opt. Soc. Amer.*, **43**, 229.
7. Schade, O. H. (1954). A new system of measuring and specifying image definition. *Nat. Bur. Standards (US) Circular* 526, April 29.
8. Wiener, N. (1930). Generalized harmonic analysis. *Acta Math.*, **55**, 117.
9. Wiener, N. (1949). "Extrapolation, Interpolation, and Smoothing of Stationary Time Series". M.I.T. Press, and Wiley, New York.
10. Fellgett, P. B. (1953). Concerning photographic grain, signal-to-noise ratio, and information. *J. Opt. Soc. Amer.*, **43**, 271.
11. Jones, R. C. (1955). New method of describing and measuring the granularity of photographic materials. *J. Opt. Soc. Amer.*, **45**, 799.
12. Zweig, H. J. (1956). Autocorrelation and granularity. *J. Opt. Soc. Amer.*, **46**, 805, 812.
13. Abramowitz, M. and Stegun, I. (1965). "Handbook of Mathematical Functions". Dover Press, New York.
14. Shannon, C. E. and Weaver, W. (1949). "The Mathematical Theory of Communication". University of Illinois Press, Urbana.
15. Whittaker, E. T. (1915). Expansions of the interpolation theory. *Proc. Roy. Soc. Edinburgh*, **35**, 181.
16. Cooley, J. W. and Tukey, J. W. (1965). An algorithm for the machine computation of Fourier series. *Maths. of Comput.*, **19**, 297.
17. Andrews, H. C. (1970). "Computer Techniques in Image Processing". Academic Press, New York and London.
18. See the special issues of *Trans. I.E.E.E.* on: "Audio and Electroacoustics". *AU–15*, June 1967; *AU–17*, December 1969.
19. For example, subroutine HARM in IBM's Scientific Subroutine Package, Version 3.
20. Tatian, B. (1971). Asymptotic expansions for correcting truncation error in transfer function calculations. *J. Opt. Soc. Amer.*, **61**, 1214.
21. Metz, C. E., Strubler, K. A. and Rossmann, K. (1972). Choice of line spread function sampling distance for computing the MTF of radiographic screen-film systems. *Phys. Med. Biol.*, **17**, 638.
22. Jones, R. C. (1958). On the point and line spread functions of photographic images. *J. Opt. Soc. Amer.*, **48**, 934.
23. Marchand, E. W. (1964). Derivation of the point spread function from the line spread function. *J. Opt. Soc. Amer.*, **54**, 915.
24. Marchand, E. W. (1965). From line to point spread function: the general case. *J. Opt. Soc. Amer.*, **55**, 352.
25. Ingelstam, E. (1961). Nomenclature for Fourier transforms of spread functions. *Photogr. Sci. Eng.*, **5**, 282.
26. Selwyn, E. W. H. (1935). A theory of graininess. *Photogr. J.*, **75**, 571.
27. Papoulis, A. (1965). "Probability, Random Variables and Stochastic Processes". McGraw-Hill, New York.

28. Mees, C. E. K. and James, T. H. (1966). "The Theory of the Photographic Process". (3rd Ed.), (Pages 525–526 give a bibliography on this experimental work). MacMillan, New York.
29. Baudry, P. Desprez, R. and Preteseille, D. (1968). Recent contribution to the study of photographic granularity. *J. Photogr. Sci.*, **16**, 132.
30. For example, subroutine GAUSS in IBM's Scientific Subroutine Package, Version 3.

Recommended Reading

FOURIER TRANSFORMS

Bracewell, R. (1965). "The Fourier Transform and its Applications". McGraw-Hill, New York.
Champeney, D. C. (1973). "Fourier Transforms and their Physical Applications". Academic Press, London and New York.
Jennison, R. C. (1961). "Fourier Transforms and Convolutions". Pergamon Press, Oxford.
Lighthill, M. J. (1964). "Fourier Analysis and Generalized Functions". Cambridge University Press, Cambridge.

FOURIER THEORY OF IMAGE FORMATION

Born, M. and Wolf, E. (1970). "Principles of Optics". (4th Ed.), Pergamon Press, Oxford.
Goodman, J. W. (1965). "Introduction to Fourier Optics". McGraw-Hill, New York.
Linfoot, E. H. (1964). "Fourier Methods in Optical Image Evaluation". Focal Press, London.

RANDOM PROCESSES

Davenport, W. B. and Root, W. L. (1958). "An Introduction to the Theory of Random Signals and Noise". McGraw-Hill, New York.
See also Reference 27.

Exercises

(1) Write a computer programme to calculate the real and imaginary parts, modulus and phase, of the Fourier transform of N points spaced δx apart. Check the proprogramme by calculating the transform of the function (x/a), with $a = 10$ μm.

(2) Show that the Fourier transform of $e^{-\frac{|x|}{a}}$ is $\frac{2a}{1 + 4\pi^2 a^2 u^2}$, and use a Fourier transform programme to verify this result. With $a = 5$ μm, investigate the effect of truncation of the function on its transform, as in Figure 7.

(3) A line spread function is zero outside a total range of 50 μm, and the MTF is to be calculated up to 200 cycles/mm. It is suspected that no spatial frequencies greater than 400 cycles/mm are present in the measured line spread function. What values

would you assign to the sampling interval δx, and at what interval $\delta \omega$ will values appear in the MTF?

(4) In a given camera system the line spread function of the lens and film may be represented by Gaussian functions of the form:

$$l_{\text{lens}}(x) = e^{-a^2 x^2}; \quad l_{\text{film}}(x) = e^{-b^2 x^2}$$

Derive an expression for the overall MTF of the system. If $a = 300$ mm^{-1} and $b = 100$ mm^{-1}, at what spatial frequency will the system MTF fall to 0·20?

(5) Show that for a two-dimensional imaging system which is linear and stationary,

$$L(x) = \int_{-\infty}^{+\infty} P(x, y) \, dy,$$

where $P(x, y)$ is the point spread function, and $L(x)$ is the line spread function.

An aerial camera system records images that are degraded by image motion during exposure. Suppose that the image moves a distance a at constant velocity during exposure. Derive an expression for the modulation transfer function due to image motion, and sketch its form for $a = 25$ μm.

(6) Explain why the autocorrelation function is often used in the evaluation of photographic granularity. How is the autocorrelation function related to the rms density fluctuation and to what extent is it influenced by the measuring aperture?

The granularity of a uniformly exposed and processed sample of film is measured with a scanning aperture of 20 μm diameter. The mean-square density fluctuation is found to be 0·0035. Assuming the Selwyn law to hold, what would the mean-square fluctuation be if measured with apertures of 5 μm and 80 μm in diameter?

Can you suggest reasons why the Selwyn law might not hold for apertures of these sizes?

(7) The line spread function of a film, which is measured under low contrast exposure conditions and is symmetrical about the origin, is as tabulated:

distance μm	normalized spread function
0	1·00
2	0·84
4	0·53
6	0·34
8	0·23
10	0·16
12	0·11
14	0·07
16	0·04
18	0·02
20	0·00

6. FOURIER TRANSFORMS

A slit of variable width is contact-printed onto the film, the exposure being in the linear part of the macroscopic density-exposure curve. As the slit width is decreased the resultant density at the centre of the image of the slit, $D(0)$, also decreases if the overall exposure remains the same.

(a) Obtain an expression for the central density, $D(0)$, in terms of the slit width and the line spread function.

(b) If the value of the central density is 0·30 for a very wide slit, find by a graphical method or otherwise the central densities for slit widths of 20, 10 and 5 μm.

(c) Derive an expression between $D(0)$ and the film MTF.

(d) Explain briefly how you would calculate the MTF given the density distribution $D(x)$ of the image of a slit of known width.

(8) Calculate the probability density function, the autocorrelation function, and the Wiener spectrum of each of the two processes tabulated in Appendix I, indicating the errors involved in each case.

(9) Draw sketch diagrams to indicate the nature of the convolutions of the following line functions:
    (a) a rectangle function with an identical rectangle;
    (b) a triangle function with an identical triangle;
    (c) a triangle function with a rectangle function of half the total width of the triangle.

Explain why the autocorrelation function can be thought of in physical terms as the self-scanning of a function.

A fine-grained film whose Wiener spectrum is known to be constant up to 400 cycles/mm is scanned using a microdensitometer whose line spread function is a rectangle function of width 4 μm. Sketch the form of the measured autocorrelation function and the measured Wiener spectrum.

7. The Modulation Transfer Function

7.1 Linearity of the Photographic Process
Input and Output Parameters

The establishment of the Fourier approach to optical and photographic systems was followed by the practical application to photographic image evaluation by Schade[1] and others[2] in the 1950s. Schade [3] reviewed much of his earlier work in a comprehensive publication in 1964, and Perrin[4] has given a good historical survey and bibliography.

At the onset of these practical applications, which were rapidly followed by many others, it might have seemed that a complete specification of photographic imaging properties would be possible, with a unique transfer function linking input and output spatial frequencies. That this has not proved the case is due to the high degree of non-linearity of conventional photographic processes under normal conditions of use.

The non-linearity of the macroscopic characteristic curve can be allowed for by expressing the output in terms of "effective exposure" rather than density or transmittance, but dealing with non-linearities resulting from development adjacency effects is more difficult. The presence of this second form of non-linearity results in there being no unique transfer function for the photographic process. However for the special case in which the input results in a low contrast output, and adjacency effects are absent or very small, a unique transfer function may prove a satisfactory representation, and the existence of a transfer function under these conditions is one of the main justifications for the application of Fourier methods. Although many practical applications of photography may not meet these requirements, they may be realistic in applications where the photographic process is used in the recording and detection of low contrast signals, and especially at high spatial frequencies.

In applying the techniques of Chapter 6 to practical photographic processes it is necessary to know the relevant non-linearities. Kelly[5] described a three-stage model which might be used to specify the input-output relationships in an adequate manner, and we shall go through the assumptions leading up to this model, starting from a simple one-stage model.

Photographic images are formed by incoherent light in the majority of cases, i.e., with optical systems which are linear in intensity. If the two elements, optical and photographic, are to be cascaded as a system, then the input of the photographic element must be proportional to the signal intensity, or the two will interact non-linearly. Exposure ($E = $ I.t) is proportional to the input intensity, and all the models we shall consider here have exposure as the input.

In those cases where the photographic input is coherent or partially-coherent light, then exposure will no longer be the relevant input variable. It is still true of course that the silver halide grains respond to the energy of the radiation, but the light scattering process which precedes quantum image formation will be governed by the coherence of the radiation. Thus for example, the input parameter of interest for the photographic process in holographic applications will be some quantity which is proportional to the complex amplitude of the radiation[6].

The quantity chosen to describe the photographic output will depend on whether the film is to be studied as a recording medium or as a light scattering medium. When the photographic process is used as a recording medium, then quantities such as density, opacity, transmittance, or even the number of image grains, may be taken as output, and these may be expressed in terms of the "effective exposure" if required.

When the photographic element is used as part of a communication system, then the output quantity which is appropriate will be goverened by the system used to extract the recorded information. Quantitative information is often extracted using an incoherent optical system, and in this case intensity transmittance is the relevant parameter. However, if the information is to be extracted using coherent light, as with laser flying-spot scanners or in halographic reconstruction, then amplitude transmittance is the relevant output quantity. For visual assessment density is usually considered the most practical output measure, although the visual system is by no means linear in its response to density (or log intensity).

A One-Stage Model

The simplest model representing photographic image formation is that of a single linear filter, as illustrated in Fig. 1, and this model is identical to the

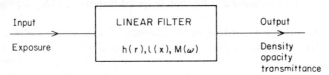

FIG. 1. A simple one-stage model of photographic image formation.

linear system whose properties were discussed in Chapter 6. If the photographic process is isotropic, the imaging properties may then be specified by either the point spread function $h(r)$ or the line spread function $l(x)$, or the modulation transfer function $M(\omega)$ in the spatial frequency domain. If this simple model is to provide an adequate representation of the photographic process it will be required that:

(i) the macroscopic input-output curve is linear;
(ii) no spurious development effects occur.

Practical processes will satisfy both these requirements if the input exposure has very low contrast, and for such low contrasts it may be possible to take either density, opacity or transmittance as the output parameter, since they will all be related in an approximately linear manner.

As the contrast of the input exposure is increased, either condition (i) or (ii) may be the first to be seriously violated. If development is carried out by a developer designed to encourage adjacency effects, then condition (ii) will not be satisfied and the model will not be valid. On the other hand, by suitable choice of development conditions it is possible to eliminate virtually all these effects, and condition (i) will then be the first to be violated. There are no well-defined rules concerning the maximum permissible input modulation to give

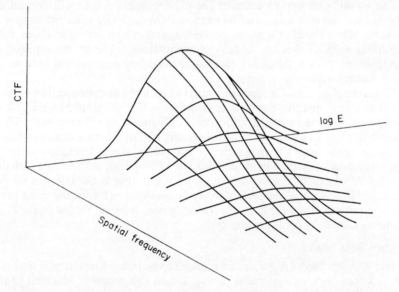

FIG. 2. Contrast transfer as a function of spatial frequency and log exposure.

approximately linear recording in the absence of adjacency effects, since this will depend on the particular output parameter involved, on the working region of the macroscopic input-output curve, and of course on the degree of non-linearity that can be tolerated. As a general guide however it is often assumed that if the image density differences are below 0·4 or 0·5, and if the working region is in the straight-line region of the characteristic D-log$E$ curve, then the amplitude of the harmonics generated by sine-wave inputs will be of an acceptably low magnitude in the absence of adjacency effects.

An exceptional case of good linearity between transmittance and exposure occurs when a reversal film is processed to a gamma of -1. From the general expression $T \propto E^{-\gamma}$ it follows that transmittance T is directly proportional to exposure E when $\gamma = -1$. By suitable processing and pre-exposure techniques it is thus possible to achive linear recording in this input-output sense for fairly high input contrasts when adjacency effects are small. However the general rule is that a simple one-stage linear filter is only a valid description of the imaging properties when the input contrast is low.

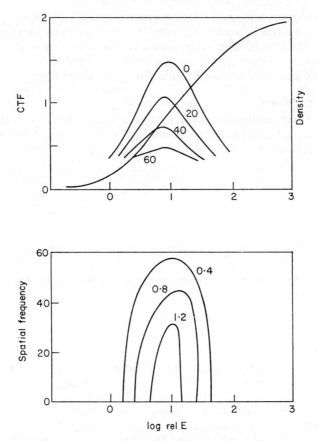

FIG. 3. Upper: contrast transfer related to the characteristic curve, showing contours of constant spatial frequency. Lower: contours of equal contrast shown on a spatial frequency —log exposure plot.

In low contrast recording applications it is often convenient to multiply the modulation transfer function by the slope, gamma, of the D-log$E$ curve, the combined function then being called the contrast transfer function (CTF), where

$$M_c(\omega) = \gamma M(\omega), \tag{1}$$

and $M_c(\omega)$ denotes the CTF. Gamma itself is of course a function of the exposure level, and the complete CTF characteristics can be represented on a three-dimensional plot, as in Fig. 2. It is sometimes more convenient to plot two-dimensional sections of such a three-dimensional representation, and examples are shown in Fig. 3. The upper plot shows the CTF as a function of log exposure for contours of fixed spatial frequency, in relation to the macroscopic D-log$E$ curve. The lower plot of Fig. 3 shows contours of fixed contrast plotted on a spatial frequency—log exposure diagram. This is a useful mode of expression when there are practical criteria for the lowest acceptable contrast in the image. The area inside the appropriate contour then defines the permitted spatial frequency—log exposure working region. Examples of contrast transfer functions of a number of recording films are given in reference 7.

A Two-Stage Model

This model takes into account the macroscopic non-linearity of the photographic process, but still assumes that adjacency effects due to development are negligible. It is based on the concept of "effective exposure", first suggested by

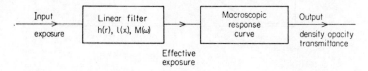

FIG. 4. A two-stage model of photographic image formation.

Frieser[8]. The formation of the photographic image is considered in two stages, as illustrated in Fig. 4:

(a) light scattering inside the photographic layer, linear in intensity, or exposure;

(b) latent image formation, and development.

The first stage may be divided into two further parts. The effect of the finite thickness of the layer combined with the angle of the image-forming rays is to act as a low pass filter, as will be discussed later in the chapter. The main degradation at this stage however is due to be scattering of light within the photographic layer. In general this process is not linear[9], but the deviations from linearity appear to be so small that they may have negligible practical significance.

The second stage is not linear because of the statistical nature of latent image formation and development. However this stage is assumed to be a point-to-point transfer of effective exposure to the relevant output parameter. In other words the resultant density at a point is assumed to be governed by the effective exposure at that point, and is not affected by the effective exposure at neighbouring points.

The MTF of a photographic process which is represented by the two-stage model refers in fact only to the transfer characteristics of the first stage. This effective exposure MTF is found in practice by comparing the incident exposure modulation with the apparent modulation inside the photographic layer derived from the transmittance or density distribution in the image and using the macroscopic characteristic curve. This is illustrated in Fig. 5. The transfer function derived in this way is identical to the transfer function measured when using a low contrast input and defining density or transmittance as the output. The concept of effective exposure is a useful one since it enables an MTF to be defined that is unaffected by macroscopic non-linearities. This transfer function is unique in that it does not depend on the exposure configuration used in its measurement. For example, effective exposure transfer functions determined

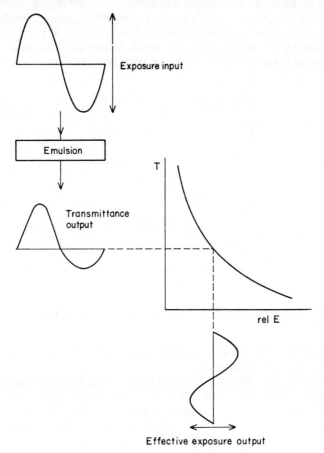

FIG. 5. The effective exposure MTF: the ratio of the effective exposure modulation to the input exposure modulation is the modulation transfer factor.

directly by observing the decrease in modulation for a sinusoidal input, and indirectly via the line spread function, will be identical.

The effective exposure MTF describes the scattering properties of the undeveloped photographic layer, but does not directly describe the properties of the recorded exposure in terms of density or transmittance unless the recorded exposure is of low contrast.

Adjacency Effects

Both models considered so far assume that the density has a point-to-point relationship with the effective exposure. Although this state of affairs may be approached using suitable processing methods, the assumption is not generally valid for most film-development combinations. As discussed in Chapter 2, substantial adjacency effects may occur during development and are the result of the spatial diffusion of developer and development-inhibiting compounds[10]. Of the various manifestations of adjacency effects the Eberhard effect occurs in the image of a bright line or star, and the density of a smaller image is found to be greater than that of a larger or wider image with the same exposure. Over a limited range of image sizes a decrease in size results in an increase in

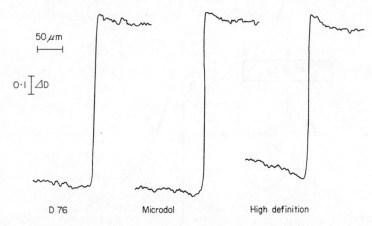

FIG. 6. Step-function responses to an input contrast of 25:1, for three developers[5].

density, this increase more than compensating for the usual reduction of density caused by light scattering.

The so-called edge effect is illustrated in Fig. 6 for three different developer types. Active developer diffuses from the low density side of the edge to increase the density on the high density side, while bromide ions and other development-inhibiting products diffuse from the high-density side to reduce the density on the low density side.

The magnitude of adjacency effects depends on the developer composition, development time, temperature and agitation, the composition of the photographic layer itself, and on the exposure level and contrast. A so-called vigorous developer such as Kodak D-19 tends to give small adjacency effects, whereas developers of low concentration and high pH tend to enhance such effects. A long development time tends to give small adjacency effects compared with a shorter time, and the addition of a silver halide solvent may increase the adjacency effects due to physical development[11]. Enhanced adjacency effects are produced by viscous developers[12]. They can also be obtained by the inclusion of development-inhibiting couplers in the layer[13], as used for masking in colour processes.

Adjacency effects may result in images which are subjectively "sharper" than that of corresponding images produced in the absence of these effects, and consequently are often considered to be a desirable feature to be deliberately built in to the photographic process. On the other hand their presence complicates the input-output relationship to such an extent that they may be undesirable when, for example, extracting data from the image by automatic means.

As pointed out by Schade[3], adjacency effects constitute an effective negative feedback component during image formation, the magnitude of which depends on both the exposure level and contrast, as well as the spatial frequency. The effect on the line spread function and the MTF is shown in Fig. 7.

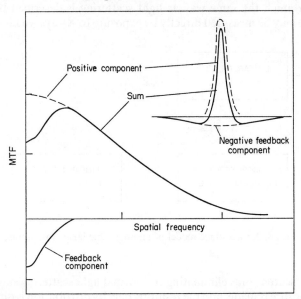

FIG. 7. The effect of a negative feedback component due to adjacency effects on the line spread function and the MTF[3].

The negative feedback component is sometimes called the "chemical" transfer function. It is interesting to note that similar effects occur in different types of imaging device (for example, image orthicons and the human eye).

When adjacency effects are negligible the effective exposure model will allow Fourier analysis to be applied to the light scattering properties of photographic processes, and a single transfer function independent of exposure and development conditions[14] will suffice as a complete description. Transfer functions found by different methods will then be identical[15,16]. However in the presence of adjacency effects the measured transfer function will vary with the exposure configuration (lines, edges, sine waves etc.), and also with all development parameters[17].

Experimental results confirm that when chemical diffusion effects are present the difference between the effective exposure "transfer functions" measured using sine wave and line test objects can be large. Consequently in the presence of adjacency effects there is no unique transfer function for the photographic process. Measured MTF curves are still useful for comparing the properties of different processes exposed and developed in the same way, but they are no longer modulation transfer functions as such. They are not unique, and they do not describe how input spatial frequencies are modulated during recording under any general conditions.

A Three-Stage Model

Adjacency effects can be described analytically by a "chemical" spread function in much the same way as light scattering is described by a spread function. It may be measured directly by exposure to X-rays which undergo no

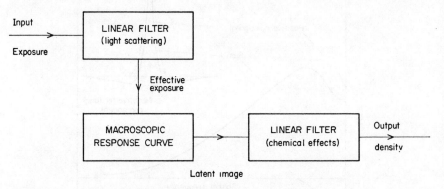

FIG. 8. A three-stage model of photographic image formation.

appreciable scatter, thus eliminating the normal light scattering stage. Usually an edge exposure is made, and the resulting density distribution is differentiated and normalized to give the chemical line spread function[18].

Kelly[5] has used the concept of the chemical transfer function in a three-stage model of photographic image formation, and this is represented in Fig. 8. The first stage is optical diffusion, resulting in an effective exposure distribution inside the layer. Using a macroscopic density-exposure curve, this effective exposure distribution is converted on a point-to-point basis to a latent image distribution in the second stage. In the third (chemical diffusion) stage the latent image distribution is converted to a density distribution. Only the second stage is non-linear. This model assumes that the magnitude of the chemical diffusion is linearly related to the density of the developed image. In reality it is more likely to be linearly related to the mass of developed silver, M, and the relationship between M and D is not usually linear, but of the form

$$M = PD^n,$$

where P is the photometric equivalent, and the constant n may lie in the range 0·5–1·0. More elaborate forms of this model have been developed[18,19] to take account of this remaining non-linearity, and Higgins[20] has given a general survey of methods of dealing with non-linearities.

To summarize, in scientific applications of photography when the photographic process is used to record quantitative data about an intensity or exposure distribution, there are two possibilities. Either the linearity conditions are satisfied and a linear filter can be used to describe the imaging properties, or the conditions are violated and a more complex model is used. In our treatment here we generally assume that the former applies, since many scientific applications involve low contrast signals, and also because the advantages of the simplicity of this model outweigh the disadvantages of the possible introduction of error.

7.2 Measurement of the MTF

Sine Wave Methods

Methods based on the use of spatial sine waves are commonly used in the practical measurement of the MTF. Other types of method involve Fourier transformation of the measured line spread function, and the use of coherent optical systems, and will be discussed shortly. General descriptions and extensive bibliographies can be found in reviews by Ooue[21] and Dainty[22].

Suppose that a film is exposed to a one-dimensional exposure distribution, $E(x)$, of the form:

$$E(x) = a + b\cos(2\pi\omega x), \qquad (2)$$

where ω denotes the spatial frequency and b/a denotes the modulation. After development the density or transmittance distribution of the image is scanned by a microdensitometer system. For a low contrast exposure the ratio of the output modulation to the input modulation gives the modulation transfer factor for the spatial frequency ω. In order to determine the effective exposure MTF the measured density or transmittance should be transferred through a microcharacteristic curve for each spatial frequency[23], but in practice the macroscopic curve is often used for all spatial frequencies.

The main problem associated with this method lies in the production of a spatially-sinusoidal exposure of known modulation. A relatively straightforward method is to photograph a variable area test chart of the type illustrated in Fig. 9, using a cylindrical lens[24,25], a slit pupil[26], or other "smearing" techniques[27]. A scanning system coupled to a time-varying modulation of a slit exposure can also be used to achieve the required exposure distribution[28,29], and one such system is illustrated in Fig. 10. A long narrow slit is imaged onto the film, and rotating polarisers with a quarter wave plate between them provide a time-varying sinusoidal intensity of known and adjustable modulation. In the arrangement shown in Fig. 10 the imaged slit is scanned across the film using a mirror system, and this particular method has the disadvantage that the lens is used at a range of angles. The relative merits of various types of scanning systems have been discussed by Langner and Müller[30], and an improved version of the apparatus shown in Fig. 10 has been described[31]. Defocused images of Moire patterns have also been used to produce an exposure distribution which is approximately sinusoidal[32].

The above methods of producing sinusoidal distributions are often used for

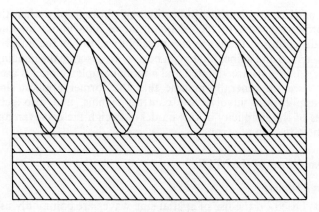

FIG. 9. A variable-area sinusoidal test-chart.

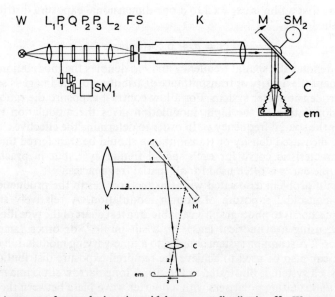

FIG. 10. Apparatus for producing sinusoidal exposure distributions[28]: W, tungsten lamp; L_1 and L_2 lenses; P_1 rotating polariser driven by synchronous motor SM_1; Q quarter wave plate; P_2 and P_3 polarisers with planes of polarization fixed at angles α and β to Q; S, entrance slit of collimator K; M, mirror rotated slowly by synchronous motor SM_2; C, camera lens; em, photographic emulsion.

the preparation of a large scale master chart, which can then be photographed, typically at a reduction of 25:1, when measuring MTFs.

For methods that involve imaging of variable-area or variable-transmittance chart, the measured MTF is that of the imaging lens—film—microdensitometer combination. Assuming that these three components combine linearly:

$$M_{film}(\omega) = \frac{M_{meas}(\omega)}{M_{lens}(\omega)M_{\mu dens}(\omega)}. \quad (3)$$

A correction based on equation (3) is only approximate for a number of reasons. The microdensitometer correction is valid only under restricted conditions as will be investigated in Chapter 9. The OTF of the imaging lens may not be unique, and will for example depend on the field angle. The transfer function of any imaging system depends on the plane of focus, and consequently for thick photographic layers the measured MTF depends on the numerical aperture of the imaging lens[31], and examples will be shown later in this section.

Intereference fringes have also been used to produce sinusoidal exposure distribution[33,34]. However it should be noted that if coherent light is used at the exposure stage, then the scattering which occurs within the layer will be linear in the complex amplitude of the radiation, and not in intensity, and in general the "transfer function" obtained using a coherent exposure is different from the incoherent MTF.

Due to the inconvenience of producing sinusoidal exposures, a square-wave chart of the type designed by Sayce is sometimes used, as shown in Fig. 11. A square-wave "transfer function" $M_1(\omega)$ is measured using the square-wave exposure, where ω represents the fundamental frequency. A calculation due to Coltman[35] based on the Fourier series of a square wave then gives an approximation to the true MTF:

$$M(\omega) = \frac{\pi}{4}\left(M_1(\omega) + \frac{1}{3}M_1(3\omega) - \frac{1}{5}M_1(5\omega)\ldots\right). \quad (4)$$

A simple laboratory method of measuring the MTF may use a square-wave chart. The chart is imaged onto the film using a high quality lens with known OTF under the experimental conditions, and the resulting image is scanned in a microdensitometer, typically using a 2 × 500 μm slit and an objective of numerical aperture 0·25. A microdensitometer trace such as shown in Fig. 11 is obtained, and the data is transferred through the macroscopic response curve to give the effective exposure square-wave transfer function (although of course this is not a real transfer function). The sine-wave response is calculated from equation (4), and this response is divided by the transfer functions of the imaging lens and microdensitometer to give an estimate of the transfer function of the film alone. Further details of a typical square-wave method are given in reference 36.

Equipment for routine MTF measurement based on sine-wave test objects is described in the literature[37,38].

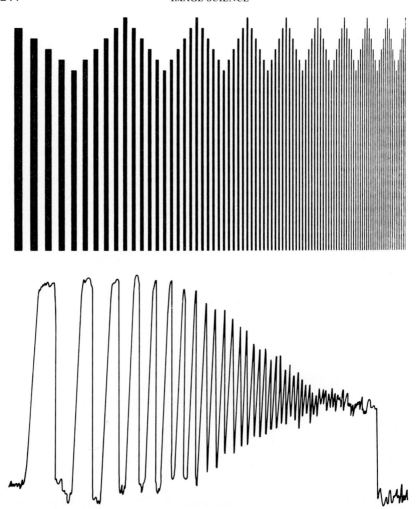

FIG. 11. Upper: a modified Sayce chart. Lower: a microdensitometer trace of an image of a Sayce chart recorded on a medium-speed film.

Spread Function Methods

For a linear photographic system the modulation transfer function is equal to the modulus of the Fourier transform of the line spread function:

$$M(\omega) = \left| \int_{-\infty}^{+\infty} l(x)\, e^{-2\pi i \omega x}\, dx \right|. \tag{5}$$

Due to the inherent non-linearity of the photographic process the MTF deduced from spread function measurements will not generally be the same as that obtained from sine-wave methods.

exposure, and it is simple to test for linearity in a given exposure since the diffraction pattern of the developed image should contain only the odd harmonics, and the presence of even harmonics will indicate a departure from linearity. The main disadvantage of this method is that the diffraction orders for the original grating decrease in intensity as the square of the order number, i.e., as 1, 1/9, 1/25, 1/49, ... This decrease is of course even faster for the photographic reproduction, and usually it is possible to measure only the first few orders.

For the measurement of MTFs of holographic and other high resolution films it is customary to use two-beam interference to produce a sinusoidal input[46,47]. The light diffracted by the film is proportional to the square of the MTF, but for holographic purposes the square root is not usually taken, and it is more useful to plot the diffraction efficiency as a function of offset angle (spatial frequency).

A further coherent light processing method makes use of the "white" noise of a laser-produced "speckle" pattern[48]. For a linear photographic process the Wiener spectrum of the output, $W_{out}(\omega)$, is related to the Wiener spectrum of the input, $W_{in}(\omega)$, by

$$W_{out}(\omega) = W_{in}(\omega)(M(\omega))^2, \qquad (9)$$

where $M(\omega)$ is the MTF of the photographic process of interest. If $W_{in}(\omega)$ is constant over the required spatial frequency range, then the output Wiener spectrum will be simply the square of the MTF. The Weiner spectrum of the complex amplitude fluctuations of the film may be conveniently measured in a coherent optical system such as that shown in Fig. 14. The method is claimed to be rapid and simple, although precautions must be taken to ensure linear recording. A similar method involving incoherent light exposures can also be used[77].

Another way of analysing photographic images using coherent light has recently been developed, called the multiple-sine-slit microdensitometer[49]. In this instrument sinusoidal fringes produced by two-beam interference are projected onto the developed image, and the integrated transmitted light flux is measured. The advantage of the large sample area of the diffraction method is retained, but the effect of phase variations is eliminated. It can also be shown that the transfer function of such a microdensitometer is independent of spatial frequency, and so it can be used to examine very high spatial frequencies in the image.

Finally, it is possible to use coherent light techniques to examine the scattering properties of undeveloped layers, but the relationship between this MTF and conventionally-measured MTFs is not obvious[50].

Practical Results

Some typical examples of measured photographic MTFs are shown in Fig. 16 and 17 on arithmetic and logrithmic scales respectively[51,52]. It can be seen that the three general purpose pictorial films have fairly similar MTFs despite a range of about forty in sensitometric speeds between the fastest and the slowest of these. The Lippmann-type emulsion has exceptionally good response over

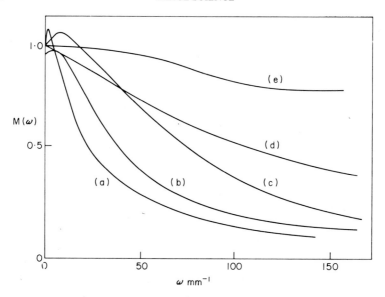

FIG. 16. Some representative MTFs, plotted on arithmetic scales: (a) fast panchromatic emulsion; (b) portrait film; (c) fine grain film; (d) Kalvar type 10; (e) Lippmann type emulsion. (From references 47, 51 and 52).

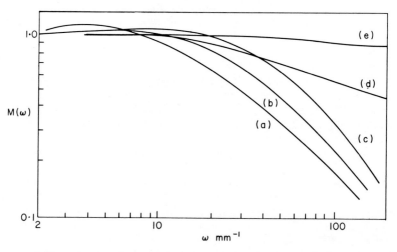

FIG. 17. Representative MTFs, plotted on a logarithmic scale (key as for Figure 16).

this spatial frequency range, and the Kalvar film designed for microfilm applications also has a relatively good response.

The scattering of light within the photographic layer is wavelength dependent, and consequently we would expect the MTF to depend on the wavelength distribution of the exposing light source. The MTFs of two films exposed to three different colours of illuminant are shown in Fig. 18. It can be seen that the variation is different for the two films, and in general there is little systematic

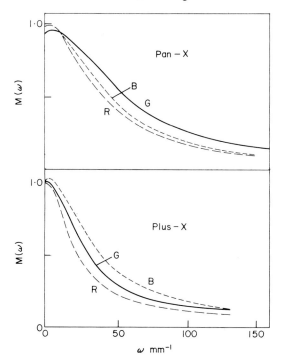

FIG. 18. The effect of exposure wavelength on the MTF of two films[52]. Exposures to tungsten lamp (2854°K) plus: R, W29 filter; G, W61 filter; B, W98 filter.

variation. For panchromatic films the results given with exposures in the green region usually differ little from those obtained using "white" light.

On theoretical grounds the variations of the MTF with wavelength might be expected to follow a similar trend as the variation of absorption with wavelength: the greater the absorption, the less the light scatter and the greater the transfer factor at any particular frequency. It is more difficult to theorise on any change in MTF when the spectral absorption stays constant but the grain size changes. If the grain diameter was very much smaller than the wavelength of light, Rayleigh scattering would occur, and the scattered intensity would be proportional to λ^{-4}, and blue light would always be scattered most strongly. However, for conventional silver halide layers in which the grain diameter is of the same order as the wavelength of light, the scattered light is more directional and the problem becomes extremely complicated. The fact that multiple scattering occurs and that there is a range of grain sizes excludes the straightforward application of the classical Mie treatment. Experimental results on the influence of grain size on scattering have been reported by Klein[75].

Although the effective exposure MTF would be expected to be independent of exposure (or density) level in the absence of adjacency effects, as shown in Fig. 19, in some cases it is found to vary to a small extent. This is partly due to the depth-wise build-up of the image through the layer such that the effective thickness of the image increases with exposure. For thick layers it is found that

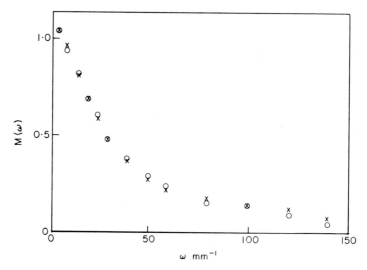

FIG. 19. The effect of average density level on the MTF of a medium-speed film developed in Kodak D-19[14]; ○○○, $\bar{D} = 0.35$; × × ×, $\bar{D} = 1.70$.

on this account the MTF decreases with increasing exposure[53], but on the other hand for thin layers the MTF may increase with exposure level due to adjacency effects. Likewise the exposure variable of input contrast will influence the MTF due to these effects. Of course the CTF also varies with exposure conditions in the way that gamma varies[7,54].

For a given developer the evidence is that the effective exposure MTF is independent of development time, as demonstrated in Fig. 20. Different developers vary considerably in their ability to produce adjacency effects, and the MTF may depend critically on developer composition. The MTFs of a film developed in two different developers are shown in Fig. 21. At this stage it should again be emphasised that these curves are not transfer functions as such in the strict sense. The difference between these two curves is due entirely to adjacency effects, and the magnitude of this difference will be a function of contrast and spatial configuration of the input exposure. Different methods of agitation during development can produce substantial differences in the degree

7. THE MODULATION TRANSFER FUNCTION

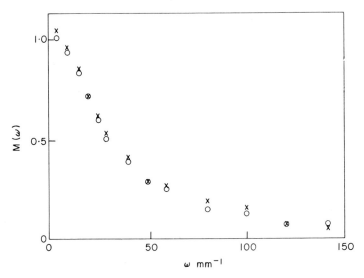

FIG. 20. The effect of development time on the MTF of a medium speed film developed in Kodak D-19 for: ○○○, 2 mins; × × ×, 10 mins[14].

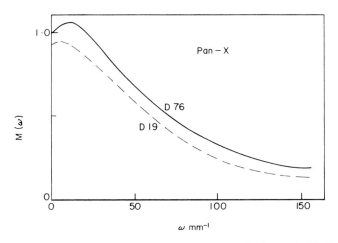

FIG. 21. The effect of developer type on the MTF of a fine grain film[51].

of adjacency effects, as shown in Fig. 22. It has also been demonstrated[76] that when high-resolution silver halide emulsions are weakly developed to a very

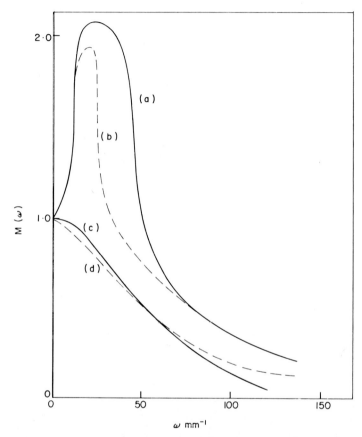

FIG. 22. The effect of different methods of development for a slow panchromatic film exposed to a sinusoidal target of modulation 0·8 and developed for 3 mins in Kodak D-158[17]: (a) spiral development, uniform rotation of spiral; (b) spiral development, irregular rotation of spiral; (c) brush development, uniform motion; (d) brush development, irregular motion.

low γ, adjacency effects may influence the MTF even at spatial frequencies in excess of 1000 cycles/mm.

Many of the methods used to measure the MTF involve the imaging of a test object using a well-corrected lens system. It is generally assumed that if the film and incoherent optical system interact linearly, then the imaging system can be allowed for by dividing the measured MTF by that of the optical system. However, since most photographic layers have thicknesses of 5 to 20 μm, the effective transfer function of the optical system is not the "in focus" one, but rather some average function over the depth of the layer. For a given layer thickness, the defocusing effect will be most severe for optical systems of high numerical aperture, and Fig. 23 shows the MTFs (corrected by the in-focus transfer factors of the imaging lens) for six spatial frequencies, plotted as a function of the aperture of the imaging system[31,55]. The measured MTF factors are reduced at small f-numbers (high numerical apertures) because the in-focus transfer factors used to correct the measured values are too high. It should also be noted that since the image thickness tends to increase with exposure level, the problem of MTF variation with numerical aperture will be

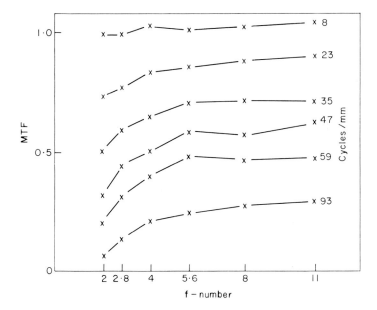

FIG. 23. The influence of the aperture of the exposing lens on the MTF of a fine grain film[55].

closely linked to the three-dimensional distribution of exposure in the photographic layer[56,57]. A similar problem arises in microdensitometry, and will be discussed further in Chapter 9.

Influence of Image Noise

Noise may enter into the measurement of photographic MTFs in a number of ways. The main contribution is usually from the image noise, but for example electronic and mechanical noise may be added during microdensitometry. The effect of noise is to introduce a random error on the optical transfer function which gives both a random error and a positive bias to the modulation transfer function. This can be illustrated by considering the effect of noise on the edge trace method of measurement[58,59].

Suppose that the density distribution, $e(x)$, over a low contrast image of an edge is written as the sum of an average density distribution, $\langle e(x) \rangle$, that would be obtained in the absence of noise, and a fluctuating noise component, $n(x)$:

$$e(x) = \langle e(x) \rangle + n(x). \tag{10}$$

Since the image noise will be a function of image density level, this simple additive equation is valid only for low contrast images, where the noise amplitudes on either side of the edge will be nearly identical in the statistical sense.

The relationship between the measured spread function and that which would be obtained in the absence of noise will be defined by

$$l(x) = \langle l(x) \rangle + \frac{d}{dx} n(x).$$

Since the OTF is equal to the Fourier transform of the line spread function, it follows that

$$T(\omega) = \langle T(\omega) \rangle + FT\left[\frac{d}{dx} n(x)\right].$$

If the FT of the noise sample $n(x)$ is defined as $N(\omega)$, where the average value of $|N(\omega)|^2$ defines the Wiener spectrum of the noise, then

$$n(x) = \int N(\omega) e^{2\pi i \omega x} d\omega.$$

Hence:

$$\frac{d}{dx} n(x) = \int 2\pi i \omega \, N(\omega) e^{2\pi i \omega x} d\omega,$$

and it follows that

$$FT\left[\frac{d}{dx} n(x)\right] = 2\pi i \omega \, N(\omega).$$

The relationship between OTFs may thus be written as

$$T(\omega) = \langle T(\omega) \rangle + 2\pi i \omega N(\omega). \tag{11}$$

Due to the factor ω in equation (11), the effect of noise is an increase in error in the measured OTF with increase in spatial frequency. This is illustrated in Fig. 24, which shows a vector plot of the spread due to noise in terms of the real and imaginary parts of the OTF, as a function of spatial frequency.

The MTF (which it is remembered is the modulus, $\sqrt{R^2 + I^2}$, of the OTF), will be influenced by this spread due to noise as illustrated in the vector diagram of Fig. 25. The average OTF, denoted by $\langle T(\omega) \rangle$, is drawn at zero phase, and the length of this vector is equal to the MTF that would be obtained in the absence of noise. The noise component vector, denoted by N, adds to this average OTF each particular noise sample having a phase value δ. The measured MTF, denoted by $M(\omega)$, will thus have a random error, and moreover, if all values of δ are equiprobable, then the average MTF, $\langle M(\omega) \rangle$, will always be greater than the value obtained in the absence of noise.

7. THE MODULATION TRANSFER FUNCTION

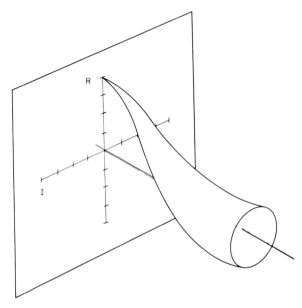

FIG. 24. A vector plot of the possible spread of OTF values due to noise[59].

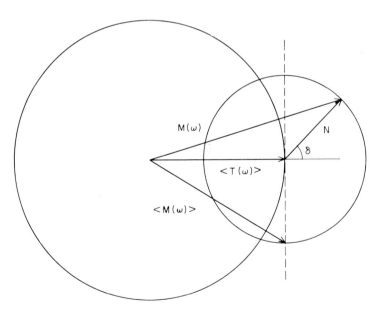

FIG. 25. Signal and noise interaction in MTF measurement. If all values of δ are equiprobable the average MTF (denoted by $< M(\omega) >$) is greater than the value obtained in the absence of noise.

Due to the positive bias introduced by noise, to improve the accuracy in MTF measurement it is necessary either to increase the S/N ratio of the measurement system (as for example by increasing the scanning slit length and thereby decreasing the influence of image noise), or to apply a correction function to the measured MTF. One such correction function is based on the fact that the phase component of the Fourier transform of the noise sample varies rapidly as a function of spatial frequency, compared with the slowly-varying or constant phase of the transfer function in the absence of noise. The phase transfer function is in fact a useful quantity for estimating the errors present in a single MTF measurement[59].

7.3 Analysis and Application of the MTF

Empirical MTF Models

Various models have been proposed to interpret the observed nature of the photographic MTF, in attempts to specify the experimental curve by a small number of characteristic parameters. Table I shows four sets of expressions[60,61] which correspond to single-parameter MTF curves. The expressions are shown for the point and line spread functions, in addition to those for the MTF, and the three functions will be related by Abel, Hankel and Fourier transforms, as discussed in Chapter 6.

TABLE I. Empirical expressions for the point spread function, $h(r)$, the line spread function, $l(x)$, and the MTF, $M(\omega)$. K_0 and K_1 denote Bessel functions of the third kind[65].

	$h(r)$	$l(x)$	$M(\omega)$
(a)	$\dfrac{1}{2\pi k_1} K_0\left(\dfrac{r}{k_1}\right)$	$\dfrac{1}{2k_1} e^{-\frac{|x|}{k_1}}$	$\dfrac{1}{1 + (2\pi k_1 \omega)^2}$
(b)	$\dfrac{1}{2\pi k_2{}^2} e^{-\frac{r}{k_2}}$	$\dfrac{|x|}{\pi k_2{}^2} K_1\left(\dfrac{|x|}{k_2}\right)$	$\dfrac{1}{[1 + (2\pi k_2 \omega)^2]^{\frac{3}{2}}}$
(c)	$\dfrac{1}{2\pi k_3} \dfrac{1}{\left[1 + \left(\frac{r}{k_3}\right)^2\right]^{\frac{3}{2}}}$	$\dfrac{1}{\pi k_3} \dfrac{1}{1 + \left(\frac{x}{k_3}\right)^2}$	$e^{-2\pi k_3 |\omega|}$
(d)	$\dfrac{1}{4\pi k_4{}^2} e^{-\left(\frac{r}{2k_4}\right)^2}$	$\dfrac{1}{2k_4\sqrt{\pi}} e^{-\left(\frac{x}{2k_4}\right)^2}$	$e^{-(2\pi k_4 \omega)^2}$

Plots of the empirical expressions are shown in Fig. 26, with normalization such that

$$2\pi \int_0^\infty h(r) r \, dr = \int_{-\infty}^{+\infty} l(x) \, dx = M(0) = 1.$$

Various two-parameter formulae have also been suggested, and attempts have been made to fit MTFs in the presence of adjacency effects[15].

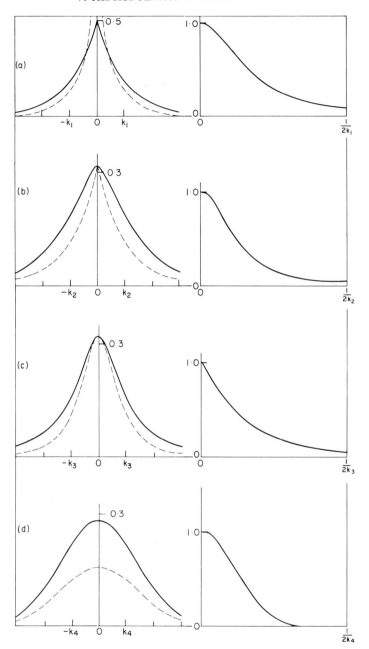

FIG. 26. Empirical functions for the point and line spread functions and the MTF: – – – – point spread function, ——— line spread function. The vertical scale for the spread functions refers to the values of the line spread function.

The expressions shown in Table I are of an empirical nature, with the exception of the first set, which were suggested by Frieser[8,62,63] as being a valid description of the effects of light scatter within the photographic layer. According to this model the line spread function has a negative exponential shape defined by

$$l(x) = \frac{1}{2k_1} e^{-\frac{|x|}{k_1}}. \tag{12}$$

The reasoning behind this expression is based on the fact that scattered light due to absorption and boundary losses will decrease as the distance from the exposure point increases: for a narrow-slit exposure this decrease will proceed in an essentially exponential manner, since the decrease of intensity through an element dx at distance x from the slit will be proportional to the prevailing intensity. A steeper decrease would be expected from a point exposure, since the exponential decrease is augmented by a decrease due to radial expansion. According to this logic model, (b) of Table I will not be valid.

A more general expression of which equation (12) is a special case has been derived by Gilmore[64], using a relatively simple diffusion calculation.

As pointed out by Frieser[63], equation (12) does not allow for light that is only very weakly scattered, for which an additional term is necessary, giving a corrected equation:

$$l(x) = \frac{\rho}{2k_1} e^{-\frac{|x|}{k_1}} + \frac{1-\rho}{2c_1} e^{-\frac{|x|}{c_1}}. \tag{13}$$

In this equation ρ denotes the fraction of light that is strongly scattered, and hence $1 - \rho$ is the fraction that is weakly scattered. If c_1 is very small, equation (13) reduces to

$$l(x) = \frac{\rho}{2k_1} e^{-\frac{|x|}{k_1}} + (1-\rho)\,\delta(x), \tag{14}$$

where δ denotes the delta function. It can be shown that the MTF corresponding to this corrected line spread function is defined by

$$M(\omega) = \frac{\rho}{1 + (2\pi k_1 \omega)^2} + (1-\rho). \tag{15}$$

The form of the spread function and MTF based on equation (14) and equation (15) is shown in Fig. 27. Frieser[62] and Haase and Muller[66] have investigated the relationship between the constants k_1 and ρ and the optical properties (specular and diffuse transmittances) and other properties (grain size and shape, coating weight, layer thickness).

Monte Carlo MTF Models

Monte Carlo methods of simulation are particularly suited to practical problems where it is required to build up a limiting statistical picture of a process involving too large a number of conditional probabilities to allow an analytical solution. As was pointed out in previous chapters, such modelling

7. THE MODULATION TRANSFER FUNCTION 261

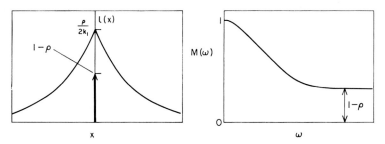

FIG. 27. The forms of the line spread function and MTF suggested by Frieser[63] (see equations (14) and (15)).

has provided one of the most useful studies of the implications of the "internal" factors operating during latent image formation. Here we are essentially concerned with the "external" events prior to the absorption of quanta by grains. From studies described in the literature it is apparent that Monte Carlo

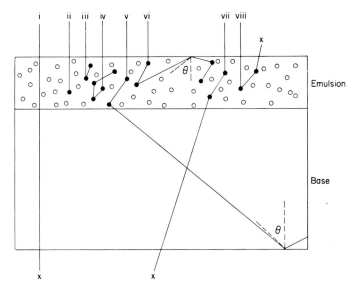

FIG. 28. Some possible paths of a photon after entering the photographic layer[68].

methods are well-suited for studying the scattering properties of photographic layers. De Belder et al.[67] gave the first detailed account of such an approach, in 1965, and more recent investigations are due to Wolfe et al.[68] and DePalma and Gasper[69].

Figure 28 illustrates some possible "paths" of a photon entering a photographic layer. A photon may pass straight through the layer, as shown by path (i): it may be absorbed at the first collision (ii): after one collision and scattering (iii): after colliding and being scattered many times (iv). In (v) there is back surface halation; in (vi) front surface halation; and in (vii) and (viii) the photon leaves the layer without being absorbed. If a large number of photons (typically 2000) are considered to enter the layer at one point, and the number of absorptions at a radius r is calculated on the basis of a random walk, then a plot of this number as a function of the radius will approximate to the effective exposure point spread function. The MTF can then be found by Hankel

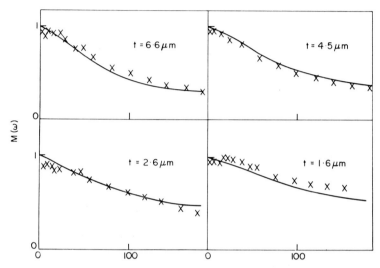

FIG. 29. Comparison of the experimental (———), and (× ×) calculated MTFs for layers of different thickness[69].

transformation of the point spread function. Realistic simulation is possible if six parameters are known:

(1) the thickness of the layer;
(2) the refractive indices of the gelatin and the base;
(3) the probability p that a photon which collides with a grain will be absorbed;
(4) the probability distribution, $p(\theta)$, for angular deviation which a photon undergoes when it collides with a grain but is not absorbed;
(5) the probability distribution function, $p(x)$, for the distance travelled by a photon between collisions with grains.

If all these parameters are available it is straightforward to compute a statistical approximation to the point spread function. The main practical difficulties are in the experimental measurement of the probability distributions. To measure $p(\theta)$ a diluted emulsion sample is placed in a gonioradiometer and illuminated either normally or at $\theta/2$. The apparent radiance of the sample is measured using a photomultiplier, and assuming that there is no shadowing of the grains this measurement yields $p(\theta)$. The function $p(x)$ can be obtained from the slope of the curve of specular density against layer thickness and finally, if it is assumed that p is independent of the direction of the photons it is possible to calculate it from absorption and scattering coefficients[68,69].

Figure 29 illustrates the agreement that is obtained between MTFs computed in this way and those measured experimentally in the absence of adjacency effects. This method is particularly powerful for use as an aid to emulsion manufacture, since a large number of emulsion variables can be tested for their influence on light scattering in the layer, and hence layers (and multilayers) can be designed for minimum scatter and optimum MTFs. Also the influence of the introduction to the layer of, for example, screening dyes, can be predicted in advance.

Practical Resolution Criteria

Prior to the introduction of image evaluation by the MTF, more pragmatic methods of evaluation were used, and some of these are still commonly applied

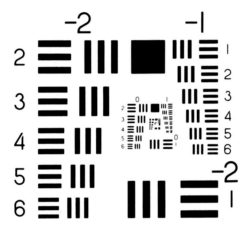

FIG. 30. The *USAF* resolution test chart.

where less sophisticated criteria are adequate. Thus it is instructive to consider their relationship with the MTF.

Figure 30 shows a type of target which is often used to measure the resolving power of a photographic system, and there are many variations on this type. The size of the elements in each successive group of bars usually diminishes in geometric progression, the nominal frequency being taken as the reciprocal of the bar-plus-space distance. The actual design of the chart can influence the estimated resolving power in a critical manner, the chart contrast and the bar aspect ratio being two of the more important features. Photographic negatives exposed to such charts are subsequently examined—usually visually with a micrscope—and the spatial frequency of the "just resolvable" bars is then defined as the resolving power of the system.

From this brief description it is clear that resolving power is not merely a measure of the ability of the photographic layer to record fine detail: rather, it is a measure of the complete lens/photographic/microscope/visual system, and the overall ability to detect special types of signal in image noise. Any part of the overall system can influence the value of resolving power that is obtained, but if the imaging properties of one element in the system are substantially worse than those of all the other elements, then it is possible to arrive at a value of resolving power which is sensibly attributable to that element.

Resolving power is essentially a statistical quantity. In simple terms, two bars will be resolved if the dip in central image density is greater than some

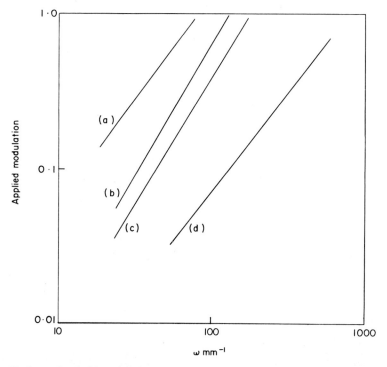

FIG. 31. Some threshold modulation curves for a three-bar target[70], ranging from a high-speed film (a), to a fine grain film (d).

threshold quantity determined by the total noise, both photographic and visual. Thus resolving power is not simply a measure of the light scattering properties of the photographic layer, but also depends on photographic gamma and noise, as well as all the other elements in the overall system.

For all these reasons resolving power is useful only in practical applications of photography with well-defined aims, as in aerial reconnaissance[70], where it can be measured using simple equipment and can be given a fairly realistic interpretation. It has the disadvantage that the elements of a system cannot be cascaded to obtain the overall resolving power, as is the case when using transfer functions. However a technique has been developed whereby the resolving power of, say, a lens/photographic system can be determined using the MTF of the lens and a threshold photographic modulation curve. In this way transfer functions are used to describe all elements prior to the photographic stage, and these may be cascaded. The resolving power of the complete system is then calculated, and specified by a single figure-or-merit.

The photographic threshold modulation curve[71] is found by measuring over the spatial frequency range of interest, the applied modulation necessary in the exposure to give a just detectable photographic image, using a specified test target. Some typical threshold modulation curves are shown in Fig. 31.

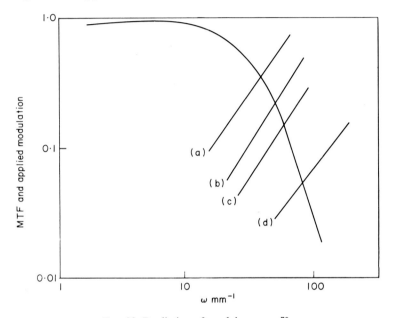

FIG. 32. Prediction of resolving power[70].

The applied modulation necessary to produce a just detectable image is found in practice to increase in a linear manner with spatial frequency when both are plotted logarithmically, as shown. If the MTF of the lens and the threshold photographic modulation curve are plotted on the same graph, the point of intersection indicates the resolving power of the lens/photographic combination, as illustrated in Fig. 32.

Sometimes photographic recording is used for resolving power tests of lenses, and it is easily understood why anomalies may occur. This is demonstrated by the example of Fig. 33: if a fast film is used lens (a) will give a higher value of resolving power than (b), but if a slower film is used the situation is reversed. These and other aspects of resolving power as a practical criterion are discussed in the literature[14,70,72].

Another common practical criterion is that of single bar contrast. Suppose that a long slit or bar image of width a, whose exposure distribution is described by the function $\text{rect}(x/a)$, is imaged photographically in the linear region of the density-exposure curve. The image density which results in excess of the background density will be given by the convolution formula:

$$D(x) = k \int_{-\infty}^{+\infty} \text{rect}\left(\frac{x'}{a}\right) l(x - x') \, dx',$$

where k is a constant and $l(x)$ denotes the photographic line spread function. The central image density in excess of the background is thus defined by:

$$D(0) = k \int_{-\frac{1}{2}a}^{+\frac{1}{2}a} l(x') \, dx'. \tag{16}$$

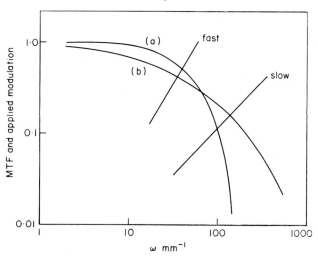

FIG. 33. Illustration of the importance of using the correct film when using resolving power tests as a measure of lens performance.

If the line spread function can be taken as zero outside a range x_1, then the central density of the bar image will be the same for all bar widths greater than x_1, and the integration limits of equation (16) can be set to infinity. However for bar widths less than x_1 the central density will decrease as the width a decreases. A useful measure of the single bar contrast is thus defined by the ratio

$$C(a) = \frac{\int_{-\frac{1}{2}a}^{+\frac{1}{2}a} l(x') \, dx'}{\int_{-\infty}^{+\infty} l(x') \, dx'}. \tag{17}$$

The equivalent expression for $C(a)$ in terms of the MTF may readily be shown to be

$$C(a) = \frac{\int_{-\infty}^{+\infty} \text{sinc}(a\omega) M(\omega) \, d\omega}{\int_{-\infty}^{+\infty} \text{sinc}(a\omega) \, d\omega}. \tag{18}$$

From equation (18) it is apparent that the MTF acts as a weighting function for any particular bar width. For criteria based on single bars the bar width should thus be chosen to give emphasis to the spatial frequency range which is of practical relevance. This will depend on the viewing magnification, the visual system, and other factors which are appropriate to the overall experimental system.

The main advantage of the single bar contrast over other single number parameters is that its measurement is objective, and is not dependent on the image noise since it is a function only of the bar width and the photographic MTF. The interaction of multiple bar targets and the MTF has also been investigated[73].

Systems Cascade of MTFs

In order to benefit from the advantages of linear analysis, the photographic process is often studied in terms of effective exposure. However, as is commonly the practical case, when a print is made from a photographic negative the linearity of the system generally breaks down, since the transmittance-exposure curve is not linear. However, it has been shown both theoretically and experimentally[74] that the transfer functions may often still be cascaded with little error, because, in the absence of adjacency effects, the harmonics generated by the non-linearities tend to cancel each other out.

Over the straight line region of the D-log$E$ curve, the relationship between the transmittance T and the exposure E will be defined as

$$T = kE^{-\gamma}, \qquad (19)$$

where gamma is as usual the slope of the D-log$E$ curve, and k is a constant. When the exposure consists of a sinusoidal exposure distribution, the distribution of effective exposure in the photographic layer is of the form

$$E(x) = 1 - A \cos x, \qquad (20)$$

where A denotes the modulation of the combination of test object, lens and film. From equations (19) and (20) the transmittance of the negative will be given by:

$$\begin{aligned} T(x) &= k(1 - A \cos x)^\gamma \\ &= k(1 + \gamma A \cos x + \frac{\gamma(\gamma - 1)}{2!} A^2 \cos^2 x + \ldots) \\ &= k(Q_0 + Q_1 \cos x + Q_2 \cos 2x + \ldots), \end{aligned} \qquad (21)$$

where the Q-values may be evaluated using the multiple-angle equivalents for the powers of angles. The harmonic content of the transmittance distribution will increase as both γ and A increase.

When a transmittance distribution as expressed by equation (21) is printed photographically, the relative exposure distribution in the print can be expressed as

$$E'(x) = k(Q_0 M_0 + Q_1 M_1 \cos x + Q_2 M_2 \cos 2x + \ldots),$$

where the values M_i represent the combined MTF of the printing system and print material.

It is useful to define the modulation as the peak-to-peak amplitude of the variations of the effective exposure, divided by the average level. The effect of even harmonics on the modulation is shown in the upper plot of Fig. 34. These harmonics do not change the amplitude of the pattern, but change the mean

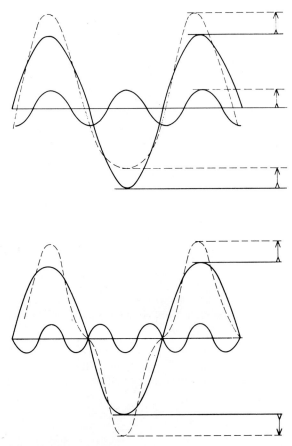

FIG. 34. Upper: combination of fundamental and second harmonic; Lower: combination of fundamental and third harmonic[74].

level by an amount equal to the amplitude of the harmonics. On the other hand the odd harmonics—as shown in the lower diagram—leave the mean level unchanged but increase the amplitude by an amount equal to the amplitude of the harmonics. Thus the odd harmonics add to the numerator of the modulation ratio and the even harmonics add to the denominator, and so the print modulation, M_p, may be expressed as

$$M_p = \frac{M_1Q_1 + M_3Q_3 + M_5Q_5 + \ldots}{M_0Q_0 + M_2Q_2 + M_4Q_4 + \ldots}. \tag{22}$$

From equation (22) it is evident that the effects of the even and odd harmonics will have a tendency to cancel each other out. If the modulation of the input pattern, and the MTFs of the negative and positive and printing system are known, then it is possible to calculate the error introduced by the harmonics. It is found that typically the error is less than 10% for an exposure modulation of 70% when $\gamma = 0.7$, and decreases rapidly as the modulation decreases.

Figure 35 shows the experimental and predicted values of the MTF through successive printing stages. The predicted values were based on a straightforward

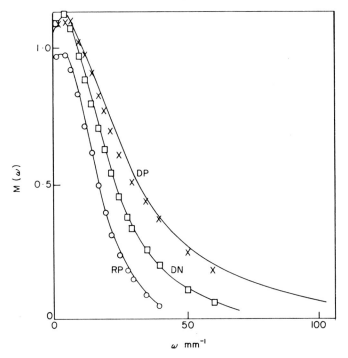

FIG. 35. The MTF at successive printing stages[74]: DP—stage 1, positive from camera negative; DN—stage 2, negative from positive of stage 1; RP—stage 3, positive from negative of stage 2.

MTF cascade, and the agreement is seen to be good in spite of the obvious presence of adjacency effects. Thus cascading can still often be used for photographic systems analysis without introducing serious error due to the inherent non-linearities of the photographic process. The concept of MTF is a useful one even for non-linear photographic processes, but care must be taken in applying it in the general case.

References

1. Schade, O. H. Image gradation, graininess, and sharpness in television and motion picture systems; I. Image structure and transfer characteristics. *J. Soc. Motion*

Pict. Tel. Engrs., **56**, 136, (1951); IV. Image analysis in photographic and television systems. *J. Soc. Motion Pict. Tel. Engrs.*, **64**, 593, (1955).
2. Ingelstam, E., Djurle, E. and Sjögren, B. (1956). Contrast-transmission functions determined experimentally for asymmetrical images and for the combination of lens and photographic emulsion. *J. Opt. Soc. Amer.*, **46**, 707.
3. Schade, O. H. (1964). An evaluation of photographic image quality and resolving power. *J. Soc. Motion Pict. Tel. Engrs.*, **73**, 81.
4. Perrin, F. H. Methods of appraising photographic systems. I. Historical review. *J. Soc. Motion Pict. Tel. Engrs.*, **69**, 151, (1960); II. Manipulation and significance of the sine-wave response function. *J. Soc. Motion Pict. Tel. Engrs.*, **69**, 239, (1960).
5. Kelly, D. H. Systems analysis of the photographic process. I. A three-stage model. *J. Opt. Soc. Amer.*, **50**, 269 (1960); II. Transfer function measurements. *J. Opt. Soc. Amer.*, **51**, 319 (1961).
6. Lorber, H. W. (1970). A theory of granularity and bleaching for holographic information recording. *IBM J. Res. Dev.*, **14**, 515.
7. "Contrast Transfer Data for Ilford Recording Films". A technical publication by Ilford Limited, 1963.
8. Frieser, H. (1935). Concerning the resolution of photographic layers. *Kinotechnik*, **17**, 167.
9. Frieser, H. and Metz, H. J. (1963). The relationship between density and exposure in a photographic layer. *Z. Agnew. Phys.*, **15**, 184.
10. Barrows, R. S. and Wolfe, R. N. (1971). A review of adjacency effects in silver photographic images. *Photogr. Sci. Eng.*, **15**, 472.
11. Barnes, J. C., Johnston, G. J. and Moretti, W. J. (1964). The chemistry of monobaths: effect of thiosulphate ion on image structure. *Photogr. Sci. Eng.*, **8**, 312.
12. Barnes, J. C., Bahler, W. W. and Johnston, G. J. (1965). Rapid processing of panchromatic negative film by the application of a viscous monobath. *J. Soc. Motion Pict. Tel. Engrs.*, **74**, 242.
13. Barr, G. R., Thirtle, J. R. and Vittum, P. W. (1969). Development-inhibiting-releasing (DIR) couplers in colour photography. *Photogr. Sci. Eng.*, **13**, 74.
14. Mees, C. E. K. and James, T. H. (1966). "The Theory of the Photographic Process". (3rd. Ed.) Chapter 23, MacMillan, New York.
15. Eyer, J. A. (1958). Spatial frequency response of certain photographic emulsions. *J. Opt. Soc. Amer.*, **48**, 938.
16. Powell, P. G. (1961). The frequency response of certain photographic reproductions in the presence and absence of diffusion effects during development. *J. Photogr. Sci.*, **9**, 312.
17. Hendeberg, L. O. (1960). The contrast transfer function of periodical structures in a photographic emulsion developed with adjacency effects. *Arkiv för Fysik*, **16**, 457.
18. Nelson, C. N. (1971). Prediction of densities in fine detail in photographic images. *Photogr. Sci. Eng.*, **15**, 82.
19. Simonds, J. L. (1965). Analysis of nonlinear photographic systems. *Photogr. Sci. Eng.*, **9**, 294.
20. Higgins, G. C. (1971). Methods for analyying the photographic system including the effects of nonlinearity and spatial frequency response. *Photogr. Sci. Eng.*, **15**, 106.
21. Ooue, S. (1961). The photographic image. "Progress in Optics", (ed. E. Wolf.), VII, Chapter VI, North-Holland, Amsterdam.

22. Dainty, J. C. (1971). Methods of measuring the modulation transfer function of photographic emulsions. *Optica Acta*, **18**, 795.
23. De Belder, M., Jespers, J. and Verbrugghe, R. (1965). On the evaluation of the modulation transfer function of photographic materials. *Photogr. Sci. Eng.*, **9**, 314.
24. Lamberts, R. L. (1959). Measurement of the sine-wave response of a photographic emulsion. *J. Opt. Soc. Amer.*, **49**, 425.
25. Lamberts, R. L. (1963). The production and use of variable-transmittance sinusoidal test objects. *Appl. Opt.*, **2**, 273.
26. Desprez, R. and Pollet, J. (1964). Production and use of a sine-wave test object. *J. Photogr. Sci.*, **12**, 202.
27. Scott, F. (1965). The production of variable-transmission sinusoidal patterns and other images. *Photogr. Sci. Eng.*, **9**, 86.
28. Hendeberg, L. O. (1960). Contrast transfer function of the light diffusion in photographic emulsions. *Arkiv för Fysik*, **16**, 417.
29. Shaw, R. (1962). The application of Fourier techniques and information theory to the assessment of photographic image quality. *Photogr. Sci. Eng.*, **6**, 281.
30. Langner, G. and Müller, R. (1967). The evaluation of the modulation transfer function of photographic materials. *J. Photogr. Sci.*, **15**, 1.
31. Hendeberg, L. O. (1967). Photographic transfer functions experimentally related to illuminating apertures. *Arkiv för Fysik*, **33**, 481.
32. Edgar, R. F., Lawrenson, B. and Ring, J. (1967). An optical analogue Fourier transformer. *J. de Physique*, **28**, (C2), 73.
33. Murty, M. V. R. K. (1954). An interferometric method of producing variable-frequency sine-wave pattern. *J. Opt. Soc. Amer.*, **44**, 468.
34. Grum, F. (1963). Modification and use of a Michelson interferometer to produce variable-frequency sinusoidal patterns. *Photogr. Sci. Eng.*, **7**, 96.
35. Coltman, J. W. (1954). The specification of imaging properties by response to a sine wave input. *J. Opt. Soc. Amer.*, **44**, 468.
36. Pospíšil, J. and Bumba, V. (1971). Measurement of the modulation transfer function of negative black-and-white photographic materials by means of the method with rectangular parallel wave grating. *Optik*, **34**, 136.
37. Lamberts, R. L., Straub, C. M. and Garbe, W. F. (1965). Equipment for the routine evaluation of the modulation transfer function of photographic emulsions. I. The camera; II. The microdensitometer; III. Evaluating and plotting instrument. *Photogr. Sci. Eng.*, **9**, 331, 335, 340.
38. Goddard, M. C. and Gendron, R. G. (1969). An MTF meter for film. *Photogr. Sci. Eng.*, **13**, 150.
39. Scott, F., Scott, R. M. and Shack, R. V. (1963). The use of edge gradients in determining modulation-transfer functions. *Photogr. Sci. Eng.*, **7**, 345.
40. Jones, R. A. (1967). An automated technique for deriving MTFs from edge traces. *Photogr. Sci. Eng.*, **11**, 102.
41. Jones, R. A. and Yeadon, E. C. (1969). Determination of the spread function from noisy edge scans. *Photogr. Sci. Eng.*, **13**, 200.
42. Bracewell, R. (1965). "The Fourier Transform and its Applications". Page 117. McGraw-Hill, New York.
43. Thiry, H. (1960). Resolving power of photographic materials as determined from interferometer patterns. *Photogr. Sci. Eng.*, **4**, 19.
44. Thiry, H. (1963). Spectometric measurement of the acutance of photographic materials. *J. Photogr. Sci.*, **11**, 121.

45. Swing, R. E. and Shin, M. C. H. (1963). The determination of modulation-transfer characteristics of photographic emulsions in a coherent optical system. *Photogr. Sci. Eng.*, **7,** 350.
46. Friesem, A. A., Kozma, A. and Adams, G. F. (1967). Recording parameters of spatially modulated coherent wavefronts. *Appl. Opt.*, **6,** 851.
47. Vander Lugt, A. and Mitchel, R. H. (1967). Technique for measuring modulation transfer functions of recording media. *J. Opt. Soc. Amer.*, **57,** 372.
48. Hariharan, P. (1970). Evaluation of modulation transfer function of photographic materials using a laser speckle pattern. *Appl. Opt.*, **9,** 1482.
49. Biedermann, K. and Johansson, S. (1972). Evaluation of the modulation transfer function of photographic emulsions by means of a multiple-sine-slit microdensitometer. *Optik*, **35,** 391.
50. Thiry, H. (1972). A method of measurement of the optical transfer function of nondeveloped photographic emulsion layers. *Photogr. Sci. Eng.*, **16,** 430.
51. "Modulation Transfer Data for KODAK films." Eastman Kodak Sales Service Pamphlet P. 49, 2nd and 3rd Edits.
52. Rabedeau, M. E. (1965). The microimage characteristics of a Kalvar film. *Photogr. Sci. Eng.*, **9,** 58.
53. Langner, G. (1963). Investigations regarding the measurement of the modulation transfer function and possibilities for its designation by a numerical value. *J. Photogr. Sci.*, **11,** 150.
54. Scott, F. and Rosenau, M. D. (1961). Film response as a function of exposure. *Photogr. Sci. Eng.*, **5,** 266.
55. Hendeberg, L. O. (1963). Dependence of the transfer function of photographic emulsion on the numerical aperture of the optical system. *J. Opt. Soc. Amer.*, **53,** 1114.
56. Berg, W. F. (1969). The photographic emulsion layer as a three-dimensional recording medium. *Appl. Opt.*, **8,** 2407.
57. Strübin, H. (1968). The depth distribution of light in photographic emulsions. *Phot. Korres.*, **104,** 5, 26, 53.
58. Blackman, E. S. (1968). Effects of noise on the determination of photographic system modulation transfer functions. *Photogr. Sci. Eng.*, **12,** 244.
59. Yeadon, E. C., Jones, R. A. and Kelly, J. T. (1970). Confidence limits for individual modulation transfer function measurements based upon the phase transfer function. *Photogr. Sci. Eng.*, **14,** 153.
60. Paris, D. P. (1961). Approximation of the sine-wave response of photographic emulsions. *J. Opt. Soc. Amer.*, **51,** 988.
61. Jones, R. C. (1958). On the minimum energy detectable by photographic materials: III. Energy incident on a microscopic area of the film. *Photogr. Sci. Eng.*, **2,** 198.
62. Frieser, H. (1971). The dependence of the modulation transfer function on the properties of photographic layers. "The Photographic Image", (ed. S. Kikuchi). Page 141. Focal Press, London.
63. Frieser, H. (1960). Spread function and contrast transfer function of photographic layers. *Photogr. Sci. Eng.*, **4,** 324.
64. Gilmore, H. F. (1967). Models of the point spread function of photographic emulsions based on a simplified diffusion calculation. *J. Opt. Soc. Amer.*, **57,** 75.
65. Abramowitz, M. and Stegun, I. (1965). "Handbook of Mathematical Functions". Chapter 9. Dover Press, New York.

66. Haase, G. and Muller, H. (1960). Investigation of light diffusion in the photographic layer. *Optik*, **17**, 1.
67. De Belder, M., De Kerf, J., Jespers, J. and Verbrugghe, R. (1965). Light diffusion in photographic layers: its influence on sensitivity and modulation transfer. *J. Opt. Soc. Amer.*, **55**, 1261.
68. Wolfe, R. N., Marchand, E. W. and DePalma, J. J. (1968). Determination of the modulation transfer function of photographic emulsions from physical measurements. *J. Opt. Soc. Amer.*, **58**, 1245.
69. DePalma, J. J. and Gasper, J. (1972). Determining the optical properties of photographic emulsions by the Monte Carlo method. *Photogr. Sci. Eng.*, **16**, 181.
70. Brock, G. C. (1970). "Image Evaluation for Aerial Photography". Focal Press, London.
71. Lauroesch, T. J., Fulmer, G. G., Edinger, J. R., Keene, G. T. and Kerwick, T. F. (1970). Threshold modulation curves for photographic films. *Appl. Opt.*, **9**, 875.
72. Ballantyne, J. G. (1970). Cobb chart resolution as a tool in photographic system design. *J. Photogr. Sci.*, **18**, 117, 185.
73. Charman, W. N. (1964). Spatial frequency spectra and other properties of conventional resolution targets. *Photogr. Sci. Eng.*, **8**, 253.
74. Lamberts, R. L. (1961). Sine-wave response techniques in photographic printing. *J. Opt. Soc. Amer.*, **51**, 982.
75. Klein, E. (1957). Investigation of the scattering of photographic layers. *Phot. Korres.*, **93**, 51.
76. Biedermann, K. and Johansson, S. (1974). Development effects and the MTF of high resolution photographic materials for holography. *J. Opt. Soc. Amer.*, **64**, 862.
77. Buschmann, H. T. (1974). Fast method of determining the MTF of photographic materials. *Photogr. Sci. Eng.*, **18**, 29.

Exercises

(1) A single-bar test object in which the bar is twice as bright as the surround is exposed onto a film such that the geometrical bar width is 20 μm in the image. At the relevant exposure level $\gamma = 0\cdot7$ and the photographic MTF is defined by curve (b) of Figure 16. If the criterion for detectability of a bar is that the density at the centre must be greater than that of the background by 0·1, will the bar be detectable in this case?

(2) Describe with the help of sketch diagrams the influence of adjacency effects on the images of narrow lines, edges, sine-waves, and three-bar test targets.
   Explain why the presence of adjacency effects may be advantageous in some photographic applications, and disadvantageous in others.

(3) Discuss the assumptions that are made in the derivation of the expression for the line spread function:

$$l(x) = \frac{1}{2k} e^{-\frac{|x|}{k}} .$$

Explain why the line spread function of real films may not conform to this equation.

It is found for a particular film that the line spread function approximates to this expression if $k = 5$ μm. At what spatial frequency does the MTF fall to 0·20 in this case?

(4) Discuss the advantages of edge methods and sine-wave methods for measuring photographic MTFs, and describe in detail the essential features of an edge method. What are the factors which influence the choice of scanning slit dimensions for edge image analysis?

(5) "In recent years . . . considerable attention has . . . been given to coherent optical systems used in data processing, holography and related applications; such systems are linear in light amplitude. Photographic films often play a dual role in these systems: they can be used to record a light distribution or to modulate a light wave, or for both these functions. In a hybrid system, they can be used first to record an intensity function in a noncoherent system and then to modulate a light wave in a coherent one. Hence, the manner in which photographic film is said to be linear must be carefully defined". (Reference 47).

Make a detailed appraisal of this statement, with special emphasis on the applicability of MTF techniques.

(6) What features of Monte Carlo modelling make this a suitable technique for investigating the influence of the properties of the photographic layer on light scattering?

Write a flow diagram for a computer programme to evaluate the MTF using a Monte Carlo method. Indicate the input data which would be necessary, and describe how the programme could be used to investigate the influence of the thickness of the layer.

(7) A conventional medium-speed film is used to record high contrast interference fringes whose spatial frequency is approximately 30 cycles/mm. Explain in detail how you would determine the fringe contrast from microdensitometer scans of the recorded image.

Outline the advantages and disadvantages of interference methods for MTF measurement.

(8) Describe the conditions under which MTFs can be cascaded to describe the operation of printing from a negative. To what extent might it be appropriate to apply MTF analysis to xerographic processes?

A film of transfer function $T_f(\omega)$ is exposed to a random noise pattern with a white noise spectrum ($W(\omega)$ = const). The image is scanned with a microdensitometer having a slit of width a and an optical system whose transfer function approximates to a triangle function with cut-off frequency of $1/b$. Derive the appropriate expression by which the measured Wiener spectrum, $W'(\omega)$, can be used to calculate the film transfer function.

How will the constants a and b influence the accuracy of the calculation?

7. THE MODULATION TRANSFER FUNCTION

(9) Explain why the average measured photographic MTF is always greater than the true MTF unless the effects of image noise are allowed for. Show how this error can be expressed in terms of the experimental S/N ratio, and explain the significance of the phase transfer function in correcting for image noise.

(10) The square-wave response of a film is found by experiment to be as follows:

cycles/mm	square-wave response
0	1·000
10	·995
20	·975
30	·940
40	·880
50	·805
60	·710
80	·530
100	·400
120	·305
140	·240

Calculate the MTF of the film.

In practice, allowance must be made for the degrading effect of the camera lens used to photograph the square-wave chart, and the microdensitometer. How should their effect be allowed for, and should the correction be made before or after the conversion to the sine-wave response?

8. Image Noise Analysis and the Wiener Spectrum

8.1 Photographic Wiener Spectrum Relationships

Introduction

The majority of modern methods of image noise evaluation are based on the original analysis of Wiener[1] for stationary, ergodic, processes, and the subsequent application of Wiener spectrum techniques to photographic images, notably by Schade[2], Fellgett[3], Jones[4] and Zweig[5], during the 1950s.

In Chapter 6 we saw that the Wiener spectrum provides a complete description of the statistical properties of a Gaussian process. The Wiener spectrum of image density fluctuations is essentially a function of two variables, but due to the statistically isotropic nature of photographic imaging properties it can be specified completely as a function of a single spatial frequency variable. The analogy between one-dimensional time series analysis—for which Wiener spectrum techniques were first devised—and the analysis of two-dimensional space series, then becomes even closer. We shall also see that in some cases photographic image noise has a Wiener spectrum which is "white" in two dimensions. In these special cases the entire two-dimensional spectrum can be specified by a single value which is directly related to the noise parameter G, which has been used in earlier chapters.

Because the density fluctuations due to image noise are usually small, photographic nonlinearity does not pose a serious problem as in MTF analysis. Although strictly speaking the Wiener spectrum should be defined in terms of transmittance fluctuations when cascading through linear systems, in practice it is often more convenient to work in terms of density fluctuations, and we shall see that this may usually be done without the introduction of any serious error.

The autocorrelation function and the Wiener spectrum have been demonstrated to be Fourier transform pairs, and hence both functions contain the same fundamental information about a random process. However, for the same reason that the MTF is often a more convenient representation than the spread function, so the Wiener spectrum is usually a more convenient manner of representation than the autocorrelation function: convolution of functions of space is replaced by cascade of functions of spatial frequency. The Wiener spectrum is thus more simple to manipulate in systems analysis with noise included, and hence Wiener spectrum techniques are widely used, for example, in the analysis of film-television systems[2,6].

8. IMAGE NOISE ANALYSIS

In Chapter 6 we saw that both the autocorrelation function and the Wiener spectrum of random isotropic image noise may be fully described by a one-dimensional function that is either a section through or an integral over the two-dimensional function. The one-dimensional function may be measured directly by one-dimensional scanning of the noise pattern, provided that the scanning aperture is of appropriate shape. A section of the Wiener spectrum is obtained if a long narrow scanning slit is used, whereas an integral form results if a "point" scan is made in a single direction. These exact scanning relationships are of great practical importance, and will now be explored in more detail. We need also to investigate the exact relationship between the Wiener spectrum and the noise parameter $G = A\sigma_A^2$ which has been widely used in previous chapters.

One-Dimensional Scans

By equation (42) of Chapter 6 it was shown that the measured autocorrelation function, $C'(\xi, \eta)$, is equal to the actual function, $C(\xi, \eta)$, convoluted twice with the point spread function, $h(x, y)$ of the measuring system. For a conventional microdensitometer the point spread function of the total scanning system is itself a convolution of the point spread function of the imaging objective and the transmittance distribution of the scanning aperture. For a well-corrected objective the size of the point spread function is usually small compared with typical scanning aperture dimensions, and in this case the overall point spread function of the scanning system, $h(x, y)$, is due to the scanning aperture alone.

Suppose that a scan is made in the x-direction, and the autocorrelation function is calculated only as a function of ξ ($\eta = 0$). If the noise pattern is stationary, then the autocorrelation so obtained will be independent of the actual scan direction. Hence, as represented in Fig. 1, it will be independent of the value of y.

If we write the autocorrelation function in terms of the Wiener spectrum, then by equation (45) of Chapter 6:

$$C(\xi, \eta) = \int\int_{-\infty}^{+\infty} W(u, v) \, e^{+2\pi i(u\xi + v\eta)} \, du \, dv. \tag{1}$$

By use of the cascade relationship between measured and actual Wiener spectra; i.e.,

$$W'(u, v) = W(u, v)|T(u, v)|^2,$$

where $T(u, v)$ and $h(x, y)$ are Fourier transform pairs, the former denoting the transfer function of the scanning system, then:

$$C'(\xi, \eta) = \int\int_{-\infty}^{+\infty} W(u, v)|T(u, v)|^2 \, e^{+2\pi i(u\xi + v\eta)} \, du \, dv. \tag{2}$$

If for scanning in one direction we set $\eta = 0$,

$$C'(\xi) = \int_{-\infty}^{+\infty} \left(\int_{-\infty}^{+\infty} W(u, v)|T(u, v)|^2 \, dv \right) e^{+2\pi iu\xi} \, du. \tag{3}$$

By the Wiener-Khintchin theorem the measured one-dimensional autocorrelation function and the measured Wiener spectrum are Fourier transform pairs:

$$C'(\xi) = \int_{-\infty}^{+\infty} W'(u)\, e^{+2\pi i u \xi}\, du. \qquad (4)$$

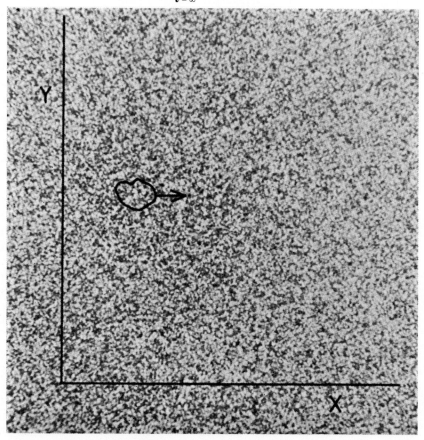

FIG. 1. Scanning a noise pattern with an aperture of arbitrary shape.

Comparison of equations (3) and (4) then yields a formula quoted by Jones[4]:

$$W'(u) = \int_{-\infty}^{+\infty} W(u, v)|T(u, v)|^2\, dv. \qquad (5)$$

This equation relates the measured one-dimensional spectrum, $W'(u)$, with the actual two-dimensional spectrum.

As an example it is supposed that the noise pattern is scanned with a slit which is infinitely long perpendicular to the scan and infinitesimally narrow in the scan direction. The scanning aperture can then be described by delta functions:

$$h(x, y) \equiv \delta(x); \qquad T(u, v) \equiv \delta(v).$$

In this case it follows that

$$W'(u) = W(u, 0),\qquad(6)$$

and the measured one-dimensional Wiener spectrum will be a section of the actual two-dimensional spectrum.

In practice a very long narrow slit is used, and we denote the length by L and the width by a. Assuming the transfer function of the measuring optics can be neglected:

$$T(u, v) = \text{sinc}\,(au)\,\text{sinc}\,(Lv),$$

and hence

$$W'(u) = \text{sinc}^2\,(au) \int_{-\infty}^{+\infty} W(u, v)\,\text{sinc}^2\,(Lv)\,dv.\qquad(7)$$

If the slit is sufficiently long that $W(u, v)$ is essentially constant over the spatial frequency range where the function $\text{sinc}^2\,(Lv)$ is appreciably non-zero, then

$$W'(u) = \text{sinc}^2\,(au)W(u, 0) \int_{-\infty}^{+\infty} \text{sinc}^2\,(Lv)\,dv.$$

The value of the integral is simply $1/L$, and hence

$$W'(u) = \frac{\text{sinc}^2\,(au)}{L} W(u, 0).\qquad(8)$$

Thus by scanning the noise pattern with a long narrow slit of known dimensions, a section of the actual two-dimensional Wiener spectrum can readily be obtained.

If the noise pattern is scanned in the x-direction by an infinitesimally small "point" aperture, which can be represented by

$$h(x, y) \equiv \delta(x)\delta(y); \qquad T(u, v) = 1;$$

it follows that

$$W'(u) = \int_{-\infty}^{+\infty} W(u, v)\,dv.\qquad(9)$$

In this case the measured one-dimensional Wiener spectrum is an integration over the v spatial frequency dimension of the actual two-dimensional spectrum.

Suppose that in practice the small point is a circular aperture of radius r. In this case it follows that equation (9) must be replaced by an equation which includes the transfer function of the aperture, and hence,

$$W'(u) = \int_{-\infty}^{+\infty} W(u, v) \left(\frac{2J_1(2\pi r\sqrt{u^2 + v^2})}{2\pi r\sqrt{u^2 + v^2}}\right)^2 dv.\qquad(10)$$

If the noise pattern is statistically isotropic equation (10) can in principle be solved for $W(u, v)$ if $W'(u)$ is known. However, in practice this solution is difficult, and a small aperture is rarely used for measuring Wiener spectra. A long narrow slit is preferable, and gives directly a section of the two-dimensional spectrum. Further discussions of the implications of one- and two-dimensional scans have been given by Klein and Langner[7] and Trabka[8].

The Wiener Spectrum and the Noise Parameter, G

The noise parameter G has been defined as

$$G = A\sigma_A^2. \tag{11}$$

The relationship between the measured value of G and the actual Wiener spectrum has been investigated by Jones[4] and Shaw[9], among others. By definition the total measured mean-square fluctuation is equal to the volume of the measured two-dimensional Wiener spectrum (see equation (46) of Chapter 6), and hence the value of G obtained with a scanning system having transfer function $T(u, v)$ will be

$$G = A \iint_{-\infty}^{+\infty} W(u, v)|T(u, v)|^2 \, du \, dv.$$

For a circular scanning aperture (i.e., with rotational symmetry) and for isotropic noise patterns, it follows by substitution of $\omega^2 = u^2 + v^2$ that

$$G = 2\pi A \int_0^\infty W(\omega)|T(\omega)|^2 \omega \, d\omega. \tag{12}$$

If the optical transfer function of the scanning system can be equated to the product of that of the circular scanning aperture (of area $A = \pi r^2$) and that of the optical system, $T_0(\omega)$, then

$$T(\omega) = \frac{2J_1(2\pi r\omega)}{2\pi r\omega} T_0(\omega). \tag{13}$$

Combining equations (12) and (13):

$$G = 2 \int_0^\infty W(\omega)|T_0(\omega)|^2 \frac{J_1^2(2\pi r\omega)}{\omega} \, d\omega. \tag{14}$$

If both the Wiener spectrum and the transfer function of the optical system are virtually constant over the spatial frequency range for which the function $(1/\omega)J_1^2(2\pi r\omega)$ is significantly non-zero, then equation (14) reduces to

$$G = 2W(0) \int_0^\infty \frac{J_1^2(2\pi r\omega)}{\omega} \, d\omega.$$

The value of the integral can be shown to be exactly one-half, and hence we arrive at the relationship

$$G = W(0). \tag{15}$$

Thus the Selwyn coefficient S will be defined by

$$S^2 = 2G = 2W(0). \tag{16}$$

The validity of equations (15) and (16) depends on the constancy of the functions $W(\omega)$ and $|T_0(\omega)|^2$ over a spatial frequency range defined by the dimensions of the scanning aperture. In terms of the new variable $x = 2\pi r\omega$,

$$\int_0^\infty \frac{J_1^2(2\pi r\omega)}{\omega} d\omega = \int_0^\infty \frac{J_1^2(x)}{x} dx = \tfrac{1}{2}.$$

In Fig. 2 the function $(J_1^2(x))/x$ is plotted as a function of x. The function is zero at $x = 0$ and again at $x = 3\cdot 83$, and more than 80% of the area of the

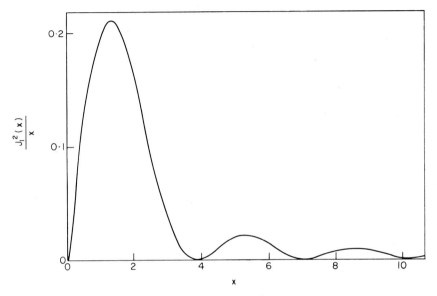

FIG. 2. The functions $\dfrac{J_1^2(x)}{x}$ plotted against x.

function lies between these two x-values. For this reason the spatial frequency range over which constancy of the product $W(\omega)|T_0(\omega)|^2$ is required is usually taken in practice as that corresponding to the x-range 0–3·83. This is interpreted in terms of practical scanning aperture sizes in Fig. 3. For an aperture 20 μm in diameter the first zero occurs at approximately 60 cycles/mm, while for a 40 μm diameter aperture it occurs at around 30 cycles/mm. Up to these spatial frequencies a practical microdensitometer system is almost certain to have a flat transfer function for the optics, although as we shall see, the Wiener spectrum may not be flat for very coarse-grained films, films exposed to electrons or X-rays, or for colour images.

Since the Wiener spectra of most film noise patterns and the transfer function of microdensitometer optical systems usually decrease with increase in spatial frequency, it follows from equation (14) that in general

$$G \leqslant W(0),$$

with the equality holding for very large scanning apertures. As a useful rule of thumb[9] the equality may be taken to hold for conventional silver halide processes when the scanning aperture has a diameter of around 50 μm.

In terms of the one-dimensional autocorrelation function the equivalent inequality will be

$$G \leqslant \int_{-\infty}^{+\infty} C(\xi)\, d\xi.$$

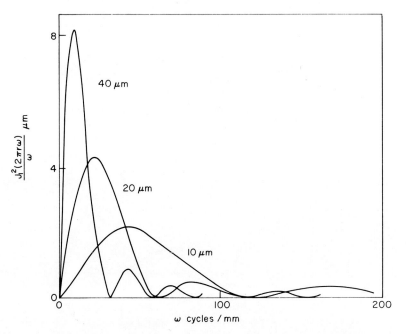

FIG. 3. The function $\dfrac{J_1^2(2\pi r\omega)}{\omega}$ plotted against ω, for apertures of diameter $(2r)$ equal to 10, 20 and 40 μm.

Detailed analyses of the relationship between the autocorrelation function and the Selwyn granularity coefficient have been carried out by Marriage and Pitts[10,11].

It is clear that the Wiener spectrum contains more information about the noise properties than either G or S. For conventional fine-grained films whose Wiener spectra are flat over a substantial spatial frequency range, the scale value, W(0), will then describe the spectrum, and will be fully defined by either G or S when they are suitably measured with large scanning apertures. In general the complete spectrum is required to define the noise, especially for colour images and for exposures to X-rays.

Other measures of the noise statistics have been proposed, an example of one of these being the so-called syzygetic granularity[12]. This may be related to both the Wiener spectrum and the autocorrelation function, although again it generally contains less information than either of these functions[5], and this and other similar measures are now mainly of historical interest.

Transmittance and Density Fluctuations

The definitions of the autocorrelation function and Wiener spectrum are usually in terms of density fluctuations. Sometimes however it is more relevant to define these functions in terms of transmittance fluctuations, and most analogue methods of measurement yield the transmittance functions directly. If the transmittance fluctuation is small compared with the mean transmittance, then

$$D = -\log_{10}T, \quad \sigma_D = -\log_{10}e \, \frac{\sigma_T}{T},$$

where σ_D and σ_T denote the respective rms density and transmittance fluctuations. Under these conditions the relationship between the Wiener spectrum of the density fluctuations, $W_D(\omega)$, and that of the transmittance fluctuations, $W_T(\omega)$, becomes

$$W_D(\omega) = \frac{(\log_{10}e)^2}{T^2} W_T(\omega). \tag{17}$$

If the transmittance fluctuations are not small, then the relationship between σ_D and σ_T will depend on the probability distribution function of the transmittance fluctuations, $p(T)$. From equation (36) of Chapter 6 the equation for σ_D^2 may be written as

$$\sigma_D^2 = \overline{D(x,y)^2} - D^2,$$

where $D(x, y)$ denotes the point-by-point density whose mean value is D. In the general case we may represent these values by the appropriate first and second moments of the probability distribution:

$$D = -\int_0^1 \log_{10}T \, p(T) \, dT$$

$$\overline{D(x,y)^2} = \int_0^1 (\log_{10}T)^2 \, p(T) \, dT. \tag{18}$$

To investigate the magnitude of the error obtained when using the linear relationship between σ_D and σ_T, Trabka[13] represented $p(T)$ by a beta-distribution, and his results are shown in Fig. 4. It can be seen that the error is generally quite small: for a mean transmittance of $T = 0.1$—corresponding to a mean density of $D = 1$—a relative transmittance fluctuation of 0.3—corresponding to a density fluctuation of approximately 0.13—produces an error of about 3%. In practice the density fluctuation would rarely be as high as in this example, except for scanning apertures which are smaller than those normally used.

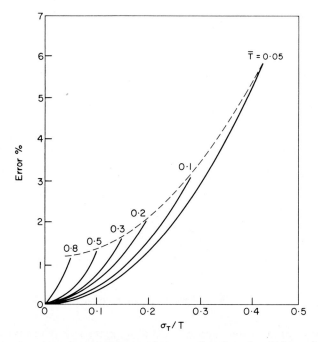

FIG. 4. Percentage error in measurement of σ_D as a function of σ_T/T for various average transmittances, assuming a linear relationship[13]. The abscissa corresponding to the intersection of the dashed curve and a particular solid curve gives the upper bound of σ_T/T for each value of T.

8.2 Measurement of the Wiener Spectrum

The Scale Value

The range of possibilities for measurement of the Wiener spectrum is illustrated in Fig. 5. These methods divide into those involving analogue and digital techniques, and these both subdivide into direct measurement of the Wiener spectrum or prior measurement of the autocorrelation function.

Of special interest to all methods of measurement is the scale value of the Wiener spectrum, W(0), which we have interpreted as the value of the spectrum for very low spatial frequencies. For Wiener spectra which are virtually flat this value may be sufficient to describe the complete spectrum. However, in the more general case it may still be appropriate to measure this value separate to the main experiment for the full spectrum. This is because the experimental design for the complete spectrum (for example the scanning aperture dimensions) may be such that the accuracy is insufficient at low spatial frequencies, and a side experiment may be appropriate to determine the low frequency (or scale) value of the spectrum. This problem did not arise in measurement of the MTF as considered in the previous chapter. This is because the MTF is not the signal as such, but rather a dimensionless signal transfer function by which the "scale value" of the signal is modulated during recording. However, the Wiener spectrum represents the absolute noise and hence must include the scale value.

Simple methods may be devised for the measurement of W(0), based on the simple relationship between W(0) and G of equation (15), whereby

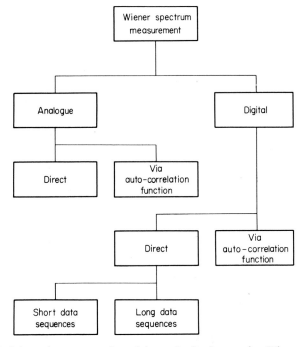

FIG. 5. Schematic representation of the methods of measuring Wiener spectra.

$W(0) = A\sigma_A^2$ when A is large enough to act as a low-pass filter of spatial frequencies. If a uniformly exposed and developed area of film is scanned using a micro-densitometer aperture typically 50 μm in diameter on the film, then σ_A may be calculated from the recorded density fluctuation.

If N statistically independent values of the density fluctuation, ΔD_i, are taken from the recorded trace, the mean-square density fluctuation may be calculated as

$$\sigma_A^2 = \frac{1}{N} \sum_{i=1}^{N} \Delta D_i^2. \tag{19}$$

In practice the calculation may involve the use of a false mean. The relative standard error in σ_A^2 is given by[14]

$$SE = \sqrt{\frac{2}{N}}. \tag{20}$$

and this is plotted in Fig. 6 as a function of the number of statistically independent readings. For N = 1000 the error in $\sigma_A{}^2$, and hence in W(0), is less

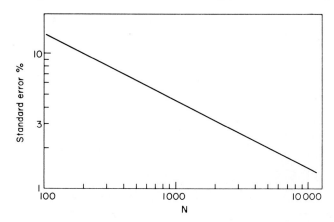

FIG. 6. The standard error in $\sigma_A{}^2$ for N statistically independent readings.

than 5%. We shall see that this is small compared with the errors involved in digital calculations of the complete spectrum.

A straightforward calculation of W(0) is simple when digital recording on paper or magnetic tape is available. However by making use of the fact that the density fluctuations often have a Gaussian probability distribution, the calculation of the standard deviation of a microdensitometer trace may be reduced to

FIG. 7. Illustration of the rapid evaluation of the *rms* density fluctuation of a microdensitometer trace.

an even simpler operation which can be carried out without computational aids. The technique is best described by example.

Figure 7 shows the microdensitometer output for a medium speed film at a mean density of approximately 1.0, scanned by a 50 × 50 μm aperture. Two lines are drawn at levels D_1 and D_2, the levels being in the region of the mean value ± σ_A, but their exact position is not critical. Sampling the fluctuations at suitable intervals (in this case 50 μm intervals are necessary to give statistically independent readings) the fractions of values falling below each level are determined. These fractions are the values of the cumulative probability distribution function for the levels D_1 and D_2. For a Gaussian process the cumulative distribution function is the error function, as plotted in Fig. 8. Locating the two fractional values, $P(D_1)$ and $P(D_2)$ on this curve gives the separation of the two levels as a multiple of the standard deviation. In this particular example $P(D_1) \simeq 0.19$ and $P(D_2) \simeq 0.78$. Reading from the curve,

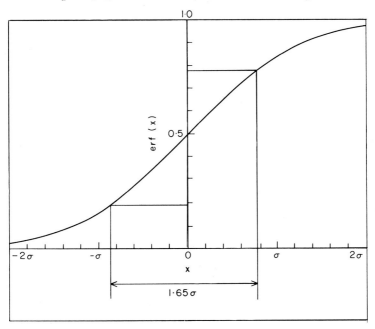

FIG. 8. Use of the error function curve for rapid evaluation of σ_A.

the separation $D_2 - D_1$ density units thus corresponds to approximately $1.65 \sigma_A$, and hence $\sigma_A \simeq 0.015$.

In practice it is found that for up to approximately 1000 data values the additional error introduced by this technique is of the same order of magnitude as the statistical error. This simple method, and variations of it, can reduce the computational load considerably in calculating W(0) from conventional microdensitometer traces.

If very large numbers of measurements of W(0) are to be made, it may be more conveninent to use an analogue method as routine. In one typical method[15] a film sample is rotated continuously using a specially constructed stage in a microdensitometer, and the photomultiplier output (which is proportional to the transmittance of the sample) is fed into a true rms voltmeter calibrated directly to read σ in transmittance units. The film sample must be free from dust and scratches, and must also be uniformly exposed and developed over a relatively large area (around 5 cm²). Difficulties in maintaining good focus can arise, but this problem is not as acute as in the corresponding method of measuring the complete spectrum. A novel method requiring only a conventional microdensitometer has been described by Riva[16].

Analogue Methods

An elegant method of measuring the autocorrelation function of the transmittance fluctuations was suggested by Fellgett[3]. Suppose that by some means it is possible to "superimpose" a noise pattern upon itself, but displaced by a distance (ξ, η), and that the total transmitted intensity, $I(\xi, \eta)$, is measured over a large area XY. In this case the intensity transmitted through the two patterns will be defined by:

$$I(\xi, \eta) = k \iint_{XY} T(x, y)T(x + \xi, y + \eta) \, dx \, dy,$$

where $T(x, y)$ denotes the transmittance distribution of the pattern (not to be confused with the optical transfer function, $T(u, v)$). If the transmittance distribution is written in terms of the mean value, T, then

$$T(x, y) = T + \Delta T(x, y),$$

and hence

$$I(\xi, \eta) = k \left(T^2 + T \iint_{XY} \Delta T(x, y) \, dx \, dy \right.$$
$$+ T \iint_{XY} \Delta T(x + \xi, y + \eta) \, dx \, dy$$
$$\left. + \iint_{XY} \Delta T(x, y) \Delta T(x + \xi, y + \eta) \, dx \, dy \right). \quad (21)$$

As the area XY over which the transmitted intensity is measured becomes large, the first two integrals in equation (21) tend to zero, leaving

$$I(\xi, \eta) = k \left(T^2 + \iint_{XY} \Delta T(x, y) \Delta T(x + \xi, y + \eta) \, dx \, dy) \right).$$

The remaining integral term is simply equal to the autocorrelation function, $C_T(\xi, \eta)$, of the transmittance fluctuations, and hence

$$C_T(\xi, \eta) = \frac{I(\xi, \eta)}{k} - T^2.$$

This method could be implemented by use of an arrangement such as shown in Fig. 9. The photographic sample is illuminated, and imaged back on itself by a lens and cube-corner reflector. The twice transmitted light corresponds to $I(0, 0)$. If one mirror of the cube corner is now tilted through an angle θ, and the lens has focal length f, the transmittance will now correspond to $I(\zeta)$, where

$$\zeta = \sqrt{\xi^2 + \eta^2} = 2f\theta \quad (22)$$

The main disadvantage of this method follows from examination of equation (22): for a lens with $f = 10$ cm and a 1 μm autocorrelation displacement, it is required that θ changes only by approximately 1 arc second, and both the

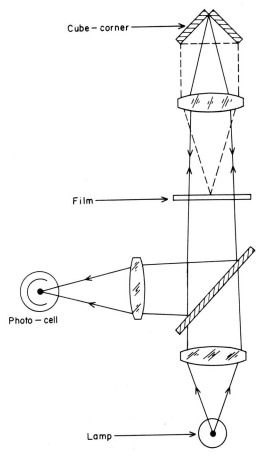

FIG. 9. Features of a possible analogue method of measuring the autocorrelation function.

optical quality and mechanical adjustment of the cube corners would have to be of a very high standard.

Another way of implementing this physical type of autocorrelation is to use a standard microdensitometer as the "camera" to obtain a photomicrograph of the noise pattern. This can usually be done by taking a photograph on a copying plate placed in the plane prior to the recording photomultiplier, which contains a focused and enlarged image of the noise pattern. The developed image is then replaced exactly in this plane, and the original sample is moved across the field of focus as in the normal scanning procedure, and the recorded trace shows the autocorrelation function as a "blip" centered on a maximum value corresponding to the case where the patterns are in exact registration.

If the photomicrograph is made on the straight-line region of the characteristic curve of the copying process, of slope γ, and the average transmittance levels of the original and the photomicrograph are T_1 and T_2 respectively, then it can readily be shown that for small transmittance fluctuations,

$$I(\xi) = T_1 T_2 - \gamma \frac{T_2}{T_1} C_T(\zeta).$$

Since these physical autocorrelation function methods yield a section $C(\zeta)$ through the two-dimensional autocorrelation function, it is necessary to determine the Hankel transform of $C(\zeta)$ to obtain a section $W(\omega)$ through the one-dimensional Wiener spectrum. An advantage of such methods is that the autocorrelation function does not contain the effect of any scanning aperture. The large area XY over which the autocorrelation function is measured simultaneously, determines the sampling accuracy of the measured function but not the shape of its profile.

The analogue method of measuring $W(0)$ described in the previous section may readily be extended to the measurement of the complete Wiener spectrum[4,7,16-18]. The noise pattern is moved at constant speed past the microdensitometer sampling aperture. The image density fluctuations become voltage fluctuations in the microdensitometer output, and it is assumed that these fluctuations are linearly related by the gain $G(f) = dV/dD$ of the measuring system. In general this gain will be a function of temporal frequency.

The power spectra of fluctuating voltages are easily measured using commercially-available spectrum analysers. Let $N(f)$ denote the output power spectrum measured in mean-square volts per unit temporal frequency bandwidth. If the film moves at a constant velocity s in the x-direction, then the measured voltage power spectrum will be related to the Wiener spectrum of the film noise by[4]:

$$N(f) = \frac{2}{s} \left| G(f) \right|^2 \int_{-\infty}^{+\infty} \left| T\left(\frac{f}{s}, v\right) \right|^2 W\left(\frac{f}{s}, v\right) dv,$$

where $f/s (= u)$ and v denote spatial frequencies, and $T(u, v)$ is the transfer function of the sanning system.

When the transfer function of the scanning system is dominated by that of the aperture, which is a long slit of length L and width a, then it follows that

$$W(\omega) = \frac{sLN(f)}{2|G(f)|^2 \text{sinc}^2(au)} \tag{23}$$

where $W(\omega)$ is a section of the two-dimensional Wiener spectrum.

It is also possible[16] to use a single band-pass filter plus an RMS voltmeter to analyse a single temporal frequency, f. Various spatial frequencies can then be investigated by changing the velocity of the film scanning as appropriate.

This type of method for measuring the complete spectrum has the same disadvantage as the corresponding method for W(0). If scanning is by means of a rotating stage, a fairly large noise sample is required and there may be problems in obtaining the necessary uniformity. Also the problem of maintaining focus may be a major one, to be overcome with special techniques such as the use of compressed air to maintain a constant distance between the microscope objective and the rotating sample. The focus position can then be adjusted by altering the air pressure. Apart from convenience the main advantage, as for all analogue methods, is that the statistical accuracy of the measurement can be high. It will be seen that when a circular scan of 6 mm radius is used ($X = 2\pi r \simeq 40$ mm) the error is approximately 5% for a measurement bandwidth of 10 cycles/mm. In a digital calculation as many as 20 000 sampling points would be required to achieve this same accuracy.

Just as coherent light may be used for measurement of the transfer function, it may also be used to measure the noise[19-22]. The amplitude in the Fraunhofer or Fourier plane is simply the Fourier transform of the amplitude transmittance of the diffracting sample. Thus if a film noise sample has an *amplitude* transmittance of $T_a(x, y)$, the intensity in the Fraunhofer plane, $I(u, v)$, will be

$$I(u, v) = k \left| \iint_{XY} T_a(x, y) \, e^{-2\pi i(ux+vy)} \, dx \, dy \right|^2, \qquad (24)$$

where k is a constant and XY denotes the area of the illuminated sample. If we write

$$T_a(x, y) = T_a + \Delta T_a(x, y),$$

it follows that

$$I(u, v) = k_1 \delta(u)\delta(v) + k \left| \iint_{XY} \Delta T_a(x, y) \, e^{-2\pi i(ux+vy)} \, dx \, dy \right|^2, \qquad (25)$$

where k_1 is another constant. Thus $I(u, v)$ will be proportional to the Wiener spectrum of the amplitude transmittance fluctuations of the noise pattern, except at very low spatial frequencies where there will be a contribution from the delta function representing the "straight through" beam. It must be stressed that this is not the same function as the Wiener spectrum of the *intensity* transmittance fluctuations, nor is it simply related to that function in the general case. If the noise pattern consisted of completely opaque and perfectly clear areas (i.e., with transmittance either 0 or 1) the two functions would then become identical. Photographic noise patterns may approximate to this at low and high density levels, where in effect there are a few opque areas on a clear background, or vice versa, and the effects of multiple scattering are small.

This type of method is used mainly to measure the noise of films designed for use in coherent light, as in holography, where the properties of the amplitude transmittance are then the relevant ones. Often a small amount of scattered light is detrimental to holographic reconstruction, and measurement of the Wiener spectrum of the complex amplitude functions can involve detecting low light levels. One measuring system due to Vilkomerson[21] is illustrated in Fig. 10.

It may also be difficult to measure the scattered light very close to the zero-order straight-through beam, due to the high intensity of this beam. One solution to this problem is to insert, for example, a π phase step across the

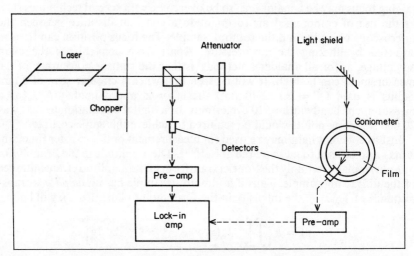

FIG. 10. System used for measuring light scattered by a photographic layer in coherent illumination[21].

diffracting aperture[20]. This removes the bright central order and re-distributes the energy over the other regions of the Fraunhofer plane.

Digital Methods

The Wiener spectrum may be computed directly from the recorded density fluctuations, or by Fourier transformation of the autocorrelation function. In practice there is usually little difference in computing time between the two, in spite of the advent of the "fast" Fourier transform which should in principle make the direct method the fastest. If for N data values approximately N values of the autocorrelation function are calculated (i.e., N^2 multiplications and additions), then clearly the calculation of the Wiener spectrum from the autocorrelation function would be slower than a direct calculation. However normally only the first few autocorrelation-interval values need to be calculated, and this tends to equalize the computing times required for each method.

If for example the final bandwidth of each estimate of the Wiener spectrum is to be 10 cycles/mm, then by the sampling theorem the autocorrelation function need only be calculated over the range -50 to $+50$ μm. Because the autocorrelation function is an even function the actual range of calculation is thus only 0 to 50 μm, say 26 points for a 2 μm sampling interval. Thus 26 × N multiplications and additions must be performed, and this is comparable with $2N \log_2 N$ operations required by the fast Fourier transform for typical values of N in the region of 2000. The literature on the digital measurement of photographic and other power spectra[16,23-30] is divided on which method to use, and of course computing time is not the only criterion since computing facilities are often available in excess of demands. A detailed description of the direct digital calculation of the Wiener spectrum will be given shortly, but first we shall briefly consider digital methods of calculating the autocorrelation function.

A uniformly exposed and developed film sample is scanned in a microdensitometer using a long narrow slit. A total of N density values, each spaced by δx, are read onto paper or magnetic tape. The value of δx is chosen in accordance with the sampling theorem (Nyquist formula). The density values, D_i, may be pre-processed to remove both very low and high spatial frequency components[29], yielding a set of density fluctuations, ΔD_i, about the mean density level. The first n points on the autocorrelation function are calculated using the formula

$$C'_j = \frac{1}{N-j} \sum_{i=1}^{N-j} \Delta D_i \Delta D_{i+j}; \qquad j = 0, 1, \ldots n. \tag{26}$$

The value of n is selected using the sampling theorem:

$$\frac{1}{n\delta x} = \delta\omega,$$

where $\delta\omega$ is the required bandwidth of the measurement.

The autocorrelation function may be multiplied by a so-called "lag window", typically of the form:

$$f(x) = \tfrac{1}{2}\left(1 + \cos\frac{x}{n\delta x}\right).$$

This operation results in a smoothing of the Wiener spectrum, and the Fourier transform of this function is known as the "spectral window". The choice of a suitable window is important if the autocorrelation function or the Wiener spectrum are likely to contain sharp peaks, but for most photographic noise cases the form of the window is not critical.

If the Wiener spectrum is required, it may be found by Fourier transformation of the autocorrelation function:

$$W'_k = L\delta x \sum_{j=0}^{n} C'_j e^{-\frac{2\pi i j k}{n}}; \qquad k = 0, 1, \ldots \frac{n}{2}, \tag{27}$$

where W'_k represents the estimate of the Wiener spectrum at the spatial frequency $\omega = k/n\delta x$, and L is the slit length. The statistical accuracy of these estimates will be considered shortly (see Fig. 13).

From equation (26) the basic arithmetic involved in the calculation of the autocorrelation function is seen to be very simple. If the number, n, of values required for the autocorrelation function is not too large, then it may be possible to carry out the calculation in real time on a special purpose computer. Commercial instruments are available for this type of calculation, and such instruments usually have additional facilities for measuring probability distributions of random signals.

The real time calculation can be simplified further by making use of the properties of Gaussian random processes, although only the shape of the autocorrelation function can be found when using this simplification. Suppose

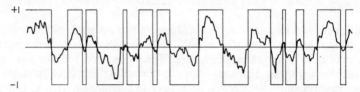

FIG. 11. Illustration of "clipping" of a random signal.

that each sampled value of the density fluctuation greater than the mean is called $+1$, and each of the value less than the mean is called -1: this procedure is known as "clipping", and is illustrated in Fig. 11. It can be shown[31] that for a Gaussian process the normalized autocorrelation function is related to that of the clipped process, $C_1(\xi)$, by the formula,

$$C(\xi) = \sin\left(\frac{\pi}{2}C_1(\xi)\right), \tag{28}$$

where both $C(\xi)$ and $C_1(\xi)$ are equal to 1 at $\xi = 0$. A modified form of this basic

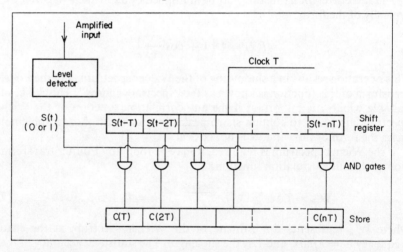

FIG. 12. Scheme for measuring the autocorrelation function of a clipped signal.

method has been used to measure autocorrelation functions in photon counting experiments[32].

A simple implementation of this method using standard digital circuitry is shown in Fig. 12, where the clipping levels are logical 1 and logical 0. The current value of the incoming signal is multiplied simultaneously by all the delayed values using logical AND gates, and the cumulative sums are stored in counting registers. This method yields only the shape of the autocorrelation function, since the magnitude of the fluctuations is lost in the clipping process. However the problem of recording data on tape does not exist since the calculations are carried out in real time, and it is therefore practical to use relatively large numbers of sample points and thereby to achieve good statistical accuracy. With a little modification the counting registers may also be used to measure the probability distribution of the density fluctuations, and hence the overall magnitude of the fluctuations and magnitude of the autocorrelation function.

For direct digital calculation of the Wiener spectrum the appropriate strict definition is:

$$W'(\omega) = \underset{X \to \infty}{\text{limit}} \left\langle \frac{L}{X} \left| \int_{-\frac{1}{2}X}^{+\frac{1}{2}X} \Delta D'(x) \, e^{-2\pi i \omega x} dx \right|^2 \right\rangle, \quad (29)$$

where L is the length of the measuring slit,
$\Delta D'(x)$ is the measured density fluctuation at the point x,
$W'(\omega)$ is the measured Wiener spectrum.

The measured Wiener spectrum is then related to the true spectrum by

$$W'(\omega) = W(\omega)|T(\omega)|^2, \quad (30)$$

where $T(\omega)$ is the transfer function of the measuring system. Although equation (29) indicates that both a space average and an ensemble average must be taken, in practice only a relatively small number of data values may be avilable. The main problem in digital Wiener spectrum analysis is to obtain maximum statistical accuracy per unit bandwidth from this available number of data values.

The spectrum may be calculated from either short or long data sequences, depending mainly on the method available for computing the Fourier transform. If the fast transform is available it is quicker to use one long sequence of data in the calculation (typically 2048 or 4096 values), whereas the cyclic method is useful for transforming many sets of short data sequences. However, in both cases practical decisions have to be made concerning:

(i) the sampling interval, δx;
(ii) the total number of data values;
(iii) the scanning-slit dimensions.

Before these quantities can be determined in a systematic manner, the statistical error which is involved must be considered and the implications of aliasing must be taken into account.

If the measured density fluctuations have a Gaussian probability distribution, then estimates of $W(\omega)$ will follow a chi-squared distribution, where under optimum measuring conditions there are approximately two degrees of freedom, regardless of the length of the trace. A longer set of data does not decrease the error of any individual value of the spectrum, but since the bandwidth is smaller it will in effect give a lower error per unit bandwidth. The arithmetic mean of M independent estimates of $W(\omega)$ obtained from M different noise records is equivalent to 2M degrees of freedom. For any fixed bandwidth, $\delta\omega$, there is no difference in stability between M pieces of data of unit length, and one piece of data of length M.

The chi-square distribution with two degrees of freedom is highly asymmetric, and it is meaningless to define a standard error. However for a greater

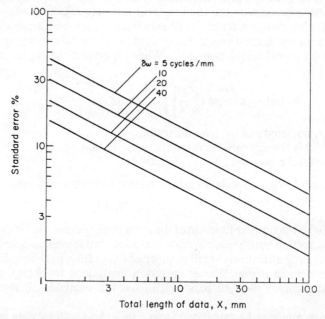

FIG. 13. The relative standard error of estimates of the Wiener spectrum as a function of the total sample length, for various values of the effective bandwidth.

number of degrees of freedom it will tend to a Gaussian distribution, which is symmetrical, and the standard error is given approximately by:

$$SE \simeq \frac{1}{\sqrt{X\delta\omega'}}, \qquad (31)$$

where X is the total length of the data, and $\delta\omega'$ is the *effective* bandwidth of the measurement. The standard error is plotted as a function of the total length of data for several typical spatial frequency bandwidths in Fig. 13.

As an example, suppose that the effective bandwidth is 10 cycles/mm, and that approximately 40 mm length of data is available. From Fig. 13 the standard error of each independent point on the Wiener spectrum will be $\simeq 5\%$. Anticipating a sampling interval of 2 μm, it is seen that 20 000 data values are required, and it is recalled that in the equivalent measurement of W(0) the same accuracy was obtained using only 1000 data values.

The phenomenon of aliasing also has important practical implications. If a waveform is sampled at intervals of δx, then by the sampling theorem (Nyquist formula), the maximum observable frequency is defined by $\omega_{max} = 1/2\delta x$.

However if spatial frequencies greater than ω_{max} are present in the sampled waveform (in the present case this will be determined mainly by the nature of the noise pattern and the width of the scanning slit), then they will contribute to the measured spectrum in the range 0 to ω_{max}. This is illustrated in Fig. 14, where as far as the values at the sample points are concerned, the two frequencies of 100 and 400 cycles/mm are indistinguishable.

The alias frequencies confused with some frequency ω are defined by

$$(2\omega_{max} - \omega), (2\omega_{max} + \omega), (4\omega_{max} - \omega), (4\omega_{max} + \omega), \text{etc},$$

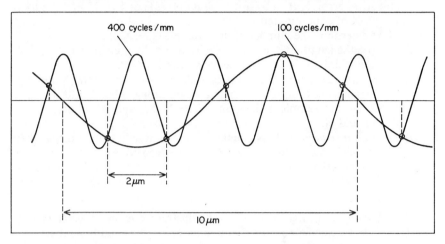

FIG. 14. Illustration of aliasing: the frequency ω is confused with the frequency $2\omega_{max} - \omega$, where in the case shown $\omega = 100$ cycles/mm and $\omega_{max} = 250$ cycles/mm.

and hence the measured Wiener spectrum is related to the actual spectrum by

$$W_1(\omega) = W(\omega) + W(2\omega_{max} - \omega) + W(2\omega_{max} + \omega) + \ldots, \quad (32)$$

where $W_1(\omega)$ denotes the spectrum including aliasing.

There are two main ways of dealing with this problem, the obvious one involving reduction of the sampling interval, and thereby increasing ω_{max}. However this is an inefficient procedure, because it is necessary to increase the number of sampled values in proportion to maintain statistical accuracy, since this is governed by the total length of the data. A better method involves preprocessing the data using either analogue or digital methods, to ensure that no high spatial ferquencies are present in the sampled trace. The simplest way of implementing this in the photographic case is by control of the width of the scanning slit, and here a compromise is necessary: a narrow slit is desirable so that low frequencies are not significantly attenuated, whereas a wide slit is required to attenuate frequencies greater than ω_{max}. The first zero of a slit of width a is at $1/a$ cycles/mm if a is in mm, and generally a is chosen such that

$$\frac{1}{a} \simeq \frac{3}{2}\omega_{max}. \quad (33)$$

We can now define an overall strategy for systematic selection of the measurement parameters of the sampling interval, the total number of samples, and the slit dimensions. The general aim is to select these quantities so as to give a maximum statistical accuracy for minimum computational effort. A simple procedure is as follows.

(1) Specify the maximum spatial frequency of interest, ω_{max}, and hence determine δx from the Nyquist formula.
(2) Specify the effective measurement bandwidth, $\delta\omega'$, and the required statistical error, s, then by use of equation (31) or Fig. 13 find the total length X of data required.
(3) Next determine the number of data values, $N \simeq X/\delta x$.
(4) Choose the length of the microdensitometer slit such that $L > 1/\delta\omega'$, and the width a is in accordance with equation (33).
(5a) *Short data sequences.* If M sets of n data values are used ($Mn = N$), each set must be long enough to give a measurement bandwidth $\delta\omega'$, and hence by the sampling theorem $\delta\omega' = 1/n\delta x$, and M is determined from n and N.
(5b) *Long data sequences.* If N points are sampled at intervals of δx, then the intrinsic measurement bandwidth is given by $\delta\omega = 1/N\delta x$. Since only an effective bandwidth of $\delta\omega'$ is required, this will still be obtained if blocks of M Wiener spectrum values are averaged, where $M = \delta\omega'/\delta\omega$.

The following numerical example will illustrate this strategy. A Wiener spectrum is to be calculated from digital data under the conditions that:

(i) $\omega_{max} = 250$ cycles/mm;
(ii) $\delta\omega' = 10$ cycles/mm;
(iii) $s \simeq 10\%$.

The measurement parameters are then as follows:

(1) Since $\omega_{max} = 250$ cycles/mm, therefore $\delta x = 2$ μm.
(2) Since $\delta\omega' = 10$ cycles/mm, and $s = 10\%$, it follows from Fig. 13 that $X \simeq 10$ mm.
(3) The number of data values will be $N = X/\delta x \simeq 5\,000$.
(4) Since $\delta\omega' = 10$ cycles/mm, L must be greater than 100 μm, say 300 μm. Since $\omega_{max} = 250$ cycles/mm, it follows that $a \simeq 2\cdot5$ μm.
(5a) For short data sequences, $\delta\omega' = (1/n\delta x)$, and thus $n = 50$, $M = 100$; so 100 sets of 50 values sampled 2 μm apart are required, and the scanning slit should be $2\cdot5 \times 300$ μm.
(5b) For long data sequences the intrinsic bandwidth is $1/N\delta x = 0\cdot1$ cycles/mm. To achieve an effective bandwidth of 10 cycles/mm blocks of 100 successive values of $W(\omega)$ should be averaged.

Practical Results

The Wiener spectrum of photographic image noise is not a unique function for each film type. However, fewer measurement conditions need to be specified when comparing measured Wiener spectra than when comparing MTFs, since problems due to non-linearities are rarely as important. Some of the most important conditions for a given film are:

(i) the mean exposure or image density level;
(ii) the nature and coherence of the exposing radiation;
(iii) the development conditions;
(iv) the microdensitometer illuminating and collecting optics.

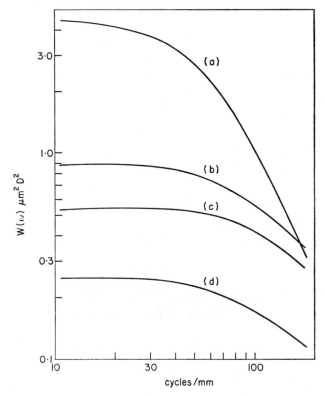

FIG. 15. Wiener spectra of some conventional silver halide films at a mean density of 0.8: (a) X-ray film; (b) fast film; (c) medium-speed film; (d) slow film.

Some typical Wiener spectra of conventional silver halide films are shown in Fig. 15. With the exception of the X-ray film, the spectra are relatively flat in the range 0–100 cycles/mm, and this is a characteristic of normal negative films. This degree of flatness of the spectrum implies that the noise parameter G may be measured to determine W(0) according to equation (15) with reasonable accuracy when using circular scanning spertures as small as 20 μm in diameter (see Fig. 3). The flatness of the spectrum is also of course why the parameter G is so important in many practical cases. A wide selection of other experimental measurements of photographic Wiener spectra may be found in references 6, 7, 16, 18 and 30.

It is found that colour images do not have such flat Wiener spectra, and some examples are shown in Fig. 16. The non-flatness is due to the relatively large size of the dye clouds, and the Wiener spectrum is a useful tool for analysing the effects of colour coupling on the structure of the resultant dye cloud[33].

The most important single factor influencing photographic Wiener spectra is the mean image density level. This was predicted for the model analysis of G

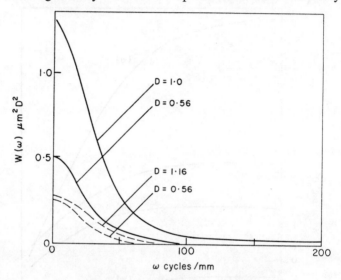

FIG. 16. Wiener spectra of the cyan dye image of two colour films[30]:
———— fast substantive film;
– – – medium-speed non-substantive film.

in Chapter 3, and is confirmed by practical results (see for example Fig. 25 of Chapter 2). The variation of the complete Wiener spectrum with mean image density is illustrated in Fig. 17 for a typical film-developer combination. The spectrum is seen to be approximately proportional to the density level at all spatial frequencies. A model which predicts these relationships will be considered shortly.

In Chapters 3 and 5 the way in which the statistics of the photon-grain interaction influences the noise was analysed in some detail. X-rays prove a special case, since one or more grains may be produced by each of these high-energy quanta. The so-called "quantum mottle" arising from this interaction is observed for both direct X-ray exposures, and for screened exposures in which the X-rays are absorbed by a pair of fluorescent screens placed in contact with the film. The Wiener spectrum of a coarse-grained X-ray film is shown in Fig. 18 for ordinary exposure to photons. The corresponding spectrum for an X-ray exposure has the effects of quantum mottle which show up as increased noise at low frequencies (less than 5 cycles/mm in this example). In

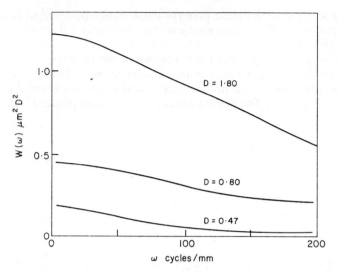

FIG. 17. Wiener spectra for several density levels of a medium-speed film[30].

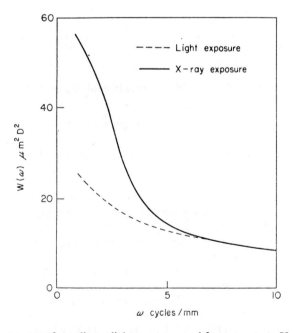

FIG. 18. Wiener spectra for ordinary light exposures and for exposure to X-rays (screened exposure).

this particular case the noise patterns could easily be distinguished by eye. The implications of quantum mottle on the transfer of noise will be discussed later in this chapter.

For visible-light exposures the coherence of the radiation at the exposure stage can influence the image noise. This effect is particularly marked for Lippmann-type films used in holography, and some experimental results are shown in Fig. 19[34]. The Wiener spectra as shown are those of the complex

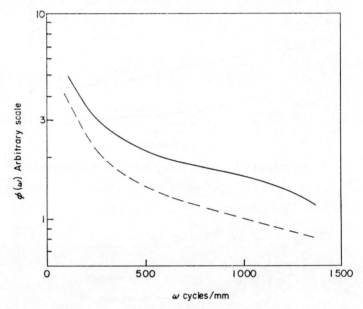

FIG. 19. Wiener spectra of the complex amplitude transmittance fluctuation, $\phi(\omega)$, of a holographic film[34]:
———— coherent light exposure;
– – – incoherent light exposure.

amplitude transmittance fluctuations of the image, and the increase in noise level is due to the formation of a coherent-light speckle pattern inside the photographic layer.

Since the composition of the developer is a main factor governing the physical structure of the image grain, it also plays an important role in defining the magnitude of the noise, and it is found experimentally that so-called "fine-grain" developers do in fact yield photographic images with lower Wiener spectrum values. Some recent results[35] suggest that the composition of the developer may not be as critical as the development time, and for a given film at a fixed image density level W(0) may vary by a factor of 2–4 for different development times. Of course other photographic parameters such as gamma are also influenced by such changes, and it then becomes relevant to assess the overall change by the DQE[35]. A full analysis of DQE which includes the MTF and the Wiener spectrum will be carried out later in this chapter.

When measuring the photographic MTF it is of little importance whether the image density is measured in terms of specular or diffuse density, since the MTF is normalized to unity at low spatial frequency. However, both the rms noise[36] and the Wiener spectrum depend critically on the illuminating and collecting optics of the microdensitometer, both of which determine the specularity of the measured density. This aspect of microdensitometry will be discussed in more detail in Chapter 9.

8.3 Analysis and Application of the Wiener Spectrum

Wiener Spectrum Models

Under the conditions for large scanning apertures that $G = W(0)$, the Siedentopf formula[37] as stated in equation (66) of Chapter 3 may be written as:

$$W(0) = \log_{10}e \; \bar{a}_D \; D \left(1 + \frac{\overline{\Delta a_D^2}}{(\bar{a}_D)^2}\right),$$

where D is the mean density level, $\bar{a}_D$ is the mean projection area of the developed grains at the density level D, $\overline{\Delta a_D^2}$ is the corresponding mean-square fluctuation.

In cases where the image grains are mono-sized, or where both $\bar{a}_D$ and $\overline{\Delta a_D^2}$ are independent of density, it follows that $W(0)$ will be proportional to the image density, and this is found experimentally for typical silver halide processes except in the higher density regions. The slope $dW(0)/dD$ is usually greater than that predicted by theory based on the crystal sizes, since it is determined by the structure of the developed image grains.

Despite the fact that the Sidentopf formula appears to account satisfactorily for experimental results with conventional photographic layers, there are a number of assumptions made in its derivation which appear to be physically unrealistic, and there have been many attempts to devise more realistic models[38-43]. The Siedentopf formula considers the image density fluctuations as arising solely from fluctuations in number and area of image grains, about the mean level which is proportional to the number and area of the grains as defined by the Nutting formula. In calculating these fluctuations it is assumed that the grains are Poisson-distributed in the photographic layer, whereas in real layers there will always be some degree of "crowding" which destroys the Poisson statistics. Also the two-dimensional overlapping of the three-dimensional layer configuration of grains as "viewed" by the scanning aperture would be expected to have some influence on the measured fluctuations, and this will depend on the exact scattering conditions within the developed layer during microdensitometry. The Nutting formula predicts that each image grain of a given size class makes an equal contribution to the image density, independent of position, due to scattering within the layer during densitometry. However, under the normal specular conditions in microdensitometry it is unlikely that each grain makes such a separate contribution to either the image density or fluctuations in density, and factors such as overlapping will have some influence.

In the Siedentopf formula the shape of the developed grains is unimportant, but if grain overlapping is to be taken into account the shape of the developed grains must be incorporated in the model. Bayer[38] has investigated the relationship between the *rms* density fluctuation and the mean density for a non-scattering layer with grains of circular cross-section and uniform size. The

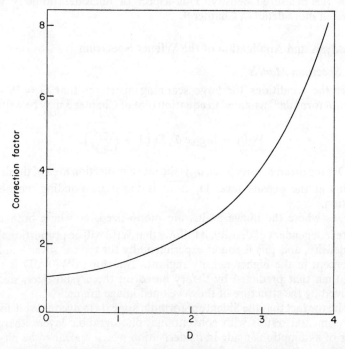

FIG. 20. The correction factor to be applied to the value of σ_A given by the Siedentopf formula to account for the overlapping of grains, shown as a function of density[38].

results of his analysis are summarized in Fig. 20, which shows as a function of density level the correction factor that should be applied to the rms density fluctuation; i.e. to the formula

$$\sigma_A = \sqrt{\log_{10} e \, D \, \frac{a}{A}}.$$

This correction factor is quite high at higher densities, but since the Siedentopf formula agrees quite well with experimental data the correction factor would in fact seem to make the agreement between theory and experiment worse. It is possible that the non-Poisson grain crowding reduces the predicted fluctuation at high density levels, and serves as a compensating factor. Crowded-layer models[40] do not however predict a significant decrease in the fluctuations for grains separated by typical average distances as in practical layers. The discrepancy between a realistic overlapping and crowded grain model and experimental results most probably lies in the nature of light scattering in the developed layer.

The analysis of the noise of the overlapping grain model has been extended to opaque grains of arbitrary shape, with some interesting results[41,42]. It can be shown that the low spatial frequency value of the Wiener spectrum of the transmittance fluctuations, $W_T(0)$, has upper and lower bounds defined by:

$$-T^2 \log_e T \leqslant \frac{W_T(0)}{K\bar{a}} \leqslant T(1 - T), \qquad (34)$$

where

$$K = 1 + \frac{\overline{\Delta a^2}}{(\bar{a})^2}$$

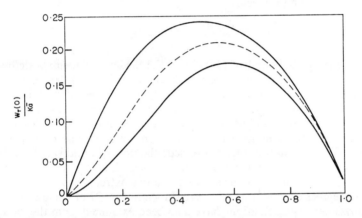

FIG. 21. Upper and lower bounds to the low spatial frequency value of the Wiener spectrum of the transmittance fluctuations for overlapping opaque grains of arbitrary shape[42]. The dashed line is the curve for circular grains of uniform size.

is the value corresponding to the mean transmittance, T. These bounds are shown as a function of T in Fig. 21, where the dotted curve corresponds to the overlapping circular grain model with constant grain area. The shape of this $W_T(0)$—T curve is quite different to the shape of the $W_D(0)$—D curve, and the value of $W_T(0)$ typically reaches a maximum in the range T = 0·4–0·6.

The upper and lower bounds have a special significance in terms of the models which we have already considered. The *lower* bound is simply a statement of the Siedentopf formula

$$W_D(0) = \log_{10} e \, \bar{a} \, D \, K, \qquad (35)$$

where by equation (17)

$$W_T(0) = \frac{T^2}{(\log_{10} e)^2} W_D(0). \qquad (36)$$

Combining equations (35) and (36):

$$\frac{W_T(0)}{K\bar{a}} = -T^2 \log_e T,$$

which is the lower bound of (34).

The *upper* bound is that corresponding to the close-packed or checkerboard model with a random spatial distribution of sensitivity classes[44]. In this case it can readily be shown that

$$W_D(0) = a(\log_{10} e)^2 \frac{1-T}{T}. \quad (37)$$

This equation holds for completely opaque image grains. Combination of equations (36) and (37) yields

$$\frac{W_T(0)}{a} = T(1-T),$$

which is the upper bound of (34). Thus the Siedentopf formula defines the lower bound for the noise, while the checkerboard model defines the upper bound. However, it must be emphasized that these upper and lower bounds are only valid for completely opaque grains. If the grain transmittance of the checkerboard model is adjusted to give equal maximum densities compared with the random grain model (see for example equations (25) to (29) of Chapter 1), the bounds are then reversed: the upper bound is defined by the Siedentopf formula and the lower bound is defined by the checkerboard "finite D_{max}" model. G-D curves for these two extreme cases and for intermediate degrees of randomness[43] were shown in Fig. 27 of Chapter 3.

The *rms* density fluctuations have also been examined from the viewpoint of continuous parameter Markov chains[45]. A layered model for colour granularity has also been described[59].

Models have been devised to explain the complete shape of the Wiener spectrum, in addition to its low frequency or scale value. For mono-sized grains which are circular in shape the Siedentopf formula can be extended to higher spatial frequencies[4,19] to give:

$$W(\omega) = \log_{10} e \, a \, D \left(\frac{J_1(2\pi\omega r)}{\pi\omega r}\right)^2, \quad (38)$$

where $J_1(\ )$ denotes the first-order Bessel function and $r = \sqrt{a/\pi}$ is the grain radius. At low density levels this expression has been shown to give a reasonable fit with experimental data[46].

The calculation of the Wiener spectrum of mono-sized but overlapping circular grains is more difficult[47,48]. It can be shown that the autocorrelation function of the transmittance fluctuations is then defined by:

$$\frac{C_T(\zeta)}{T^2} = \left(\frac{1}{T}\right) f\left(\frac{\zeta}{2r}\right) - 1, \quad (39)$$

where $C_T(\zeta)$ is a section of the rotationally-symmetrical autocorrelation function, and $f(\zeta/2r)$ is defined by:

$$f\left(\frac{\zeta}{2r}\right) = \frac{2}{\pi}\left(\cos^{-1}\frac{\zeta}{2r} - \frac{\zeta}{2r}\sqrt{1 - \left(\frac{\zeta}{2r}\right)^2}\right).$$

The one-dimensional Wiener spectrum (i.e., a section through the two-dimensional function) is then defined by the appropriate Hankel transform as

$$W_T(\omega) = 2\pi \int_0^\infty C_T(\zeta) J_0(2\pi\omega\zeta) \zeta \, d\xi. \tag{40}$$

In spite of ignoring the influence of light scattering in the image layer, equations (39) and (40) have been shown to give a fairly good fit with experimental data at low and moderate densities[46], provided that an additional scaling factor is used.

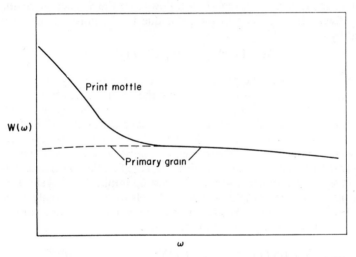

FIG. 22. Typical Wiener spectra of "primary grain" and "print mottle"[49].

Wiener Spectrum Transfer

An advantage of the MTF over the spread function was demonstrated to be the cascade property in systems analysis. Similarly an advantage of the Wiener spectrum over the autocorrelation function is in the analysis of the transfer of noise through a system. A common problem involving photographic noise transfer is that where a print is made from a negative, and the overall Wiener spectrum of the image noise is then considered as having two components, as illustrated in Fig. 22. There is a component known as "print mottle", although this in fact arises from the noise of the negative, combined with the normal noise due to the print material itself, which in this context is sometimes called "primary grain"[49].

Suppose that a uniformly exposed and developed negative with a transmittance Wiener spectrum denoted by $W_{T,n}(u, v)$ is transferred to a print material. If $M_p(u, v)$ represents the combined transfer functions of the printing system and the print material, and the print is made at unit magnification, then the Wiener spectrum of the effective exposure fluctuations in the print layer, $W_{E,p}(u, v)$, will be defined by

$$W_{E,p}(u, v) = W_{T,n}(u, v)M_p^2(u, v).$$

If the effective exposure is small, we may assume that

$$W_{\log E,p}(u, v) = W_{D,n}(u, v)M_p^2(u, v), \tag{41}$$

where the new suffixes, $\log E$ and D, denote the respective Wiener spectra of the input and output noise in terms of the fluctuations in $\log E$ and density.

If the exposure is made on the region of the characteristic curve of the print material such that $D = \gamma_p \log_{10} E$, it follows that the Wiener spectrum of the density fluctuations arising in the print due to the noise of the negative is defined by

$$W_{D,p}(u, v) = \gamma_p^2 W_{D,n}(u, v)M_p^2(u, v). \tag{42}$$

The noise statistics of the print and of the printed-on negative will not be strictly independent. Printing light forms the print image after transmission through the negative noise pattern, and hence there will be some correlation between the two sets of noise statistics. In other words the noise of the print depends on the mean print density level, and hence on the exposure level in the negative which fluctuates due to the negative noise. This is usually expressed by saying that the noise is signal dependent, and the fact that the image noise level thus conveys signal information has interesting implications in signal restoration from noisy photographic images[54,55]. However it is usually assumed in calculating the total noise, $W_{tot}(u, v)$, that the two noise patterns are statistically independent and additive in the sense that:

$$W_{tot}(u, v) = \gamma_p^2 W_{D,n}(u, v)M_p^2(u, v) + W_{D,p}(u, v), \tag{43}$$

where $W_{D,p}(u, v)$ denotes the Wiener spectrum of the density fluctuations due solely to the print material.

The justification for the assumptions which are implicit in equation (43) is that experimental data[50] is in good accord with predictions made by this equation. Figure 23 shows the Wiener spectra of a print of one film on another, and of the two films individually, with close agreement shown between theory and experiment.

A similar argument may be used to derive the total Wiener spectrum of the noise for screened X-rays[51], leading to the relationship at the exposure level, E, of

$$W_{tot}(u, v) = \frac{(\log e_{10})^2 \gamma_p^2}{E^2} W_E(u, v) + W_{D,p}(u, v). \tag{44}$$

The Wiener spectrum of the effective exposure fluctuation, $W_E(u, v)$, is related to the Wiener spectrum of the quantum exposure, $W_Q(u, v)$, by

$$W_E(u, v) = g^2 W_Q(u, v) M^2(u, v), \quad (45)$$

where $M(u, v)$ denotes the MTF of the film/screen combination, and g is the amplification factor associated with the screen. If the X-ray quanta obey Poisson statistics then the Wiener spectrum of the quantum exposure can be assumed to be flat within the range of spatial frequencies of practical interest,

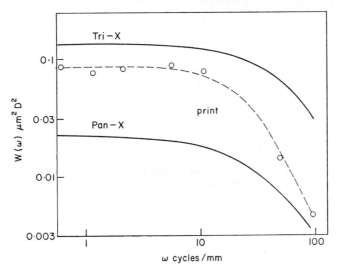

FIG. 23. Wiener spectrum of a contact print of Kodak Tri-X on Kodak Pan-X film, and the spectra of the separate films. The print curve is calculated from the other two, with circles indicating measured values[50].

and it can readily be shown (by considering the Fourier transform of the auto-correlation function, which will be a delta function) that it will be defined by

$$W_Q(u, v) = q,$$

where q denotes the average number of quanta per unit area. If the exposure after screen amplification is defined by $E \propto gq$, it follows that

$$W_{tot}(u, v) = \frac{k\gamma_p^2}{q} M^2(u, v) + W_{D,p}(u, v), \quad (46)$$

where k is a constant.

For an ideal film/screen system with $M(u, v) = 1$ within the spatial frequency range of interest,

$$W_{tot}(u, v) = \frac{k\gamma_p^2}{q} + W_D(u, v). \quad (47)$$

This situation is illustrated in Fig. 24. An ideal system would increase the Wiener spectrum of the total noise at all spatial frequencies compared with the

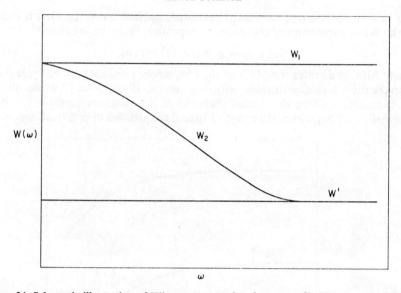

FIG. 24. Schematic illustration of Wiener spectra of grain patterns[51]: W_1—a perfect film/screen system (equation 47); W_2—a practical film/screen system (equation 46); W'—the "primary grain".

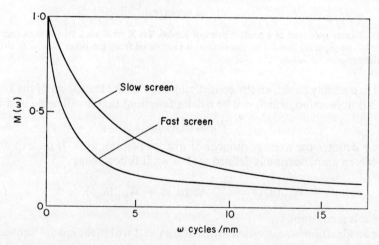

FIG. 25. MTFs of slow and fast screen/film systems, as measured from Wiener spectra[51].

practical system. It also follows that a "fast" screen which only needs to absorb a small average number, q, of X-ray quanta to produce sufficient image density will yield a higher noise level than a slower screen. This is essentially due to the fact that both fast and slow screen/film combinations may have the same DQE approaching 100% (see Section 3 of Chapter 5), and there is then mutual scaling up or down of both signal and noise. A full spatial frequency analysis of DQE will follow shortly.

In practice the transfer function of the screen/film system plays an important role in determining the Wiener spectrum of the total noise. If typical MTFs of slow and fast screens are taken into account, as illustrated in Fig. 25, then the visual appearance of the image noise may in fact be less objectionable for a fast screen, since only lower spatial frequencies will then be present, producing a more diffuse noise pattern (although there is of course also loss of signal). The MTFs of X-ray screens have been calculated by Swank[58].

The fact that the shape of the Wiener spectrum depends on the MTF of the screen/film combination can provide the basis for the measurement of these transfer functions, and those shown in Fig. 25 were determined in this way[52].

A comprehensive analysis of noise in screen/film systems has been given by Lubberts[53].

The Wiener Spectrum and DQE

Wiener spectrum analysis of image noise and noise transfer has permitted a more complete description of noise than that determined by the noise parameter, G, used in earlier chapters. By equation (15) we can now interpret G measured with a large scanning aperture (low-pass filter of spatial frequencies), as the low frequency value of the Wiener spectrum. Since the formulation of DQE—which compares the input and output fluctuations—has been based so far purely on G, it is similarly possible to interpret this value of DQE as relating to the comparison of low spatial frequency fluctuations in the input and output. In terms of the complete Wiener spectra of the input and output noise it now becomes possible to arrive at a more complete definition of DQE. The full expression for DQE in terms of spatial frequency is one of the most important functions for assessing the performance of an imaging detector, and in Chapter 10 we shall see that this function is intimately related to the total information transfer involved in imaging.

A full Fourier treatment of DQE has been given by Shaw[9], who considered S/N transfer as a function of spatial frequency, essentially by an extension of the analysis given in Chapter 5. However, a somewhat simpler but less rigorous argument can be based on the comparison of input and output fluctuations, as in Chapter 1. As in Chapter 1 (see equations (30) and (31)) the output fluctuations are referred back in terms of effective input fluctuations. If we consider image area A, with associated mean-square fluctuation σ_A^2 following a uniform exposure to an averate of q_A quanta, then the output noise referred back in terms of the input fluctuations will be:

$$\sigma_{out}^2 = \sigma_A^2 \left(\frac{dq}{dD}\right)^2 = \sigma_A^2 \left(\frac{q_A}{\gamma \log_{10} e}\right)^2 = GA \left(\frac{q}{\gamma \log_{10} e}\right)^2,$$

where q denotes the average quantum exposure per unit area. Interpreting this as the low spatial frequency $(0, 0)$ value, by equation (15):

$$\sigma_{\text{out}}^2(0, 0) = W(0, 0)A \left(\frac{q}{\gamma \log_{10} e}\right)^2. \tag{48}$$

It might appear from equation (48) that at higher spatial frequencies the appropriate (u, v) coordinates would be substituted for $(0, 0)$. However it can readily be seen that this would lead to an anomaly, since for conventional films we have seen that the trend is for $W(u, v)$ to decrease with increase in spatial frequency, thus predicting a decrease in the output fluctuations measured on the same scale as the input fluctuations. Of course the answer is that it is also necessary to modify the macro characteristic curve transfer when we are dealing with the micro case for high spatial frequencies. Assuming linear transfer it is thus also necessary to replace γ in equation (48) by $\gamma M(u, v)$, where in Chapter 7 this product was termed the contrast transfer function. Hence,

$$\sigma_{\text{out}}^2(u, v) = W(u, v)A \left(\frac{q}{\gamma M(u, v) \log_{10} e}\right)^2. \tag{49}$$

Since the square of the MTF is always a more rapidly decreasing function of spatial frequency than is the Wiener spectrum, equation (49) now correctly predicts an increase in noise in real terms at higher spatial frequencies.

Assuming the input quanta obey Poisson statistics and have a flat Wiener spectrum within the range of spatial frequencies of practical interest, then as assumed in our earlier treatment of X-rays,

$$\sigma_{\text{in}}^2(u, v) = q_A, \tag{50}$$

independent of spatial frequency. Combining equations (49) and (50),

$$\text{DQE}(u, v) = \frac{\sigma_{\text{in}}^2(u, v)}{\sigma_{\text{out}}^2(u, v)} = \frac{\gamma^2 (\log_{10} e)^2 M^2(u, v)}{q W(u, v)}. \tag{51}$$

Defining the Wiener spectrum by $W(u, v) = W(0, 0)n(u, v)$, i.e., as the product of the scale value and a modulation term, $n(u, v)$, where $n(0, 0) = 1$, it is also possible to express equation (51) as

$$\text{DQE}(u, v) = \frac{\gamma^2 (\log_{10} e)^2}{q W(0, 0)} \frac{M^2(u, v)}{n(u, v)} = \text{DQE}(0, 0) \frac{M^2(u, v)}{n(u, v)}. \tag{52}$$

For an isotropic imaging process this may be written as

$$\text{DQE}(\omega) = \text{DQE}(0) \frac{M^2(\omega)}{n(\omega)}. \tag{53}$$

In earlier chapters the value of DQE under discussion has essentially been that of DQE(0) in equation (53). The ratio $M^2(\omega)/n(\omega)$ provides a spatial frequency modulation term for DQE, which in practice will normally be dominated by the MTF component, and hence

$$\text{DQE}(\omega) \simeq \text{DQE}(0) M^2(\omega).$$

Figure 26 shows a typical DQE—spatial frequency curve for a fast panchromatic film. We have already seen that DQE is a strong function of exposure level, and thus for a complete description of the DQE characteristics a three-dimensional plot such as indicated in Fig. 27 is appropriate. Some practical three-dimensional plots of DQE have been given by Vendrovsky et al[56].

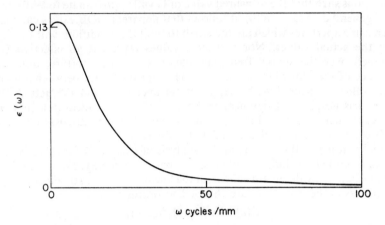

FIG. 26. The variation of DQE with spatial frequency[9] for a fast panchromatic film at a mean density of 0·58.

When measuring the low frequency DQE component, DQE(0), it follows that the noise value to be used must be strictly that of W(0). However, the practical

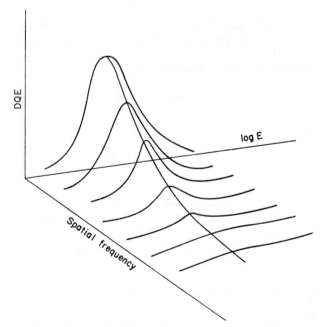

FIG. 27. The variation of DQE with log exposure and spatial frequency.

working relationships are usually expressed in terms of G. The exact relationship between G and W(0) was investigated earlier in this chapter in terms of scanning aperture dimensions, and rules of thumb were specified for these dimensions such that the measured value of G will approximate to W(0). Since in the general case G < W(0), it follows that apparent DQE values calculated when using apertures which are too small to fulfil the conditions will be higher than the actual values. Nor will such values represent an effective DQE averaged over the spatial frequency bandpass of the aperture, unless the influence of the film MTF is included. Thus the difference between apparent DQE values measured with inappropriate apertures, and the actual DQE values, has no physical significance. Some earlier DQE measurements were subject to such errors, and this general problem has been discussed generally by Zweig et al.[57], and analysed in detail by Shaw[9].

Finally, it is possible to extend the analysis of Chapter 4 for DQE transfer during printing to include the influence of spatial frequency. Equations (23) to (28) of Chapter 4 will remain valid if, at the linear printing magnification x, the parameters of the two stages are changed according to:

$$\gamma_1^2(q) \rightarrow \gamma_1^2(q) M_1^2(q, \omega);$$

$$\gamma_2^2(D) \rightarrow \gamma_2^2(D) M_2^2\left(D, \frac{\omega}{x}\right);$$

$$G_1(q) \rightarrow W_1(q, \omega);$$

$$G_2(D) \rightarrow W_2\left(D, \frac{\omega}{x}\right).$$

All spatial frequencies have been referred to those originally in the negative, and as before q represents the negative exposure level, and D the corresponding print density level.

Equation (28) of Chapter 4 now becomes:

$$\varepsilon_{1+2}(q, \omega) = \varepsilon_1(q, \omega) \frac{1}{1 + \frac{1}{f}}, \tag{54}$$

where

$$f = x^2 \frac{W_1(q, \omega)}{W_2\left(q, \frac{\omega}{x}\right)} \gamma_2^2(D) M_2^2\left(D, \frac{\omega}{x}\right).$$

In those cases where $f \gg 1$ it again follows that the overall DQE will be that of the first stage with negligible reduction due to the second stage, and this is now a general conclusion for all spatial frequencies and not only for DQE(0). For contact printing ($x = 1$) this can be achieved if the noise introduced by the second stage is much less than that of the first stage at all spatial frequencies. For large magnifications both the MTF and Wiener spectrum requirements of the printing stage will be relatively uncritical for maintenance of DQE.

References

1. Wiener, N. (1930). Generalized harmonic analysis. *Acta Math.*, **55**, 117.

2. Schade, O. H. Image gradation, graininess and sharpness in television and motion picture systems. I. The grain structure of motion picture images—an analysis of deviations and fluctuations of the sample number. *J. Soc. Motion Pict. Tel. Engrs.* **58,** 181 (1952); II. The grain structure of television images. *J. Soc. Motion Pict. Tel. Engrs.,* **61,** 97 (1953).
3. Fellgett, P. B. (1953). Concerning photographic grain, signal-to-noise ratio, and information. *J. Opt. Soc. Amer.,* **43,** 271.
4. Jones, R. C. (1955). New method of describing and measuring the granularity of photographic materials. *J. Opt. Soc. Amer.,* **45,** 799.
5. Zweig, H. J. Autocorrelation and granularity. Part I. Theory. *J. Opt. Soc. Amer.,* **46,** 805 (1956). Part II. Results on flashed black-and-white emulsions. *J. Opt. Soc. Amer.,* **46,** 812 (1956). Part III. Spatial frequency response of the scanning system and granularity correlation effects beyond the aperture. *J. Opt. Soc. Amer.,* **49,** 238 (1959).
6. Hacking, K. (1964). An analysis of film granularity in television reproduction. *J. Soc. Motion Pict. Tel. Engrs.,* **73,** 1015.
7. Klein, E. and Langner, G. (1963). Relations between granularity, graininess, and the Wiener spectrum of density deviations. *J. Photogr. Sci.,* **11,** 177.
8. Trabka, E. A. (1965). Wiener spectrum scans obtained from an isotropic two-dimensional random field. *J. Opt. Soc. Amer.,* **55,** 203.
9. Shaw, R. (1963). The equivalent quantum efficiency of the photographic process. *J. Photogr. Sci.,* **11,** 199.
10. Marriage, A. and Pitts, E. (1956). Relation between granularity and autocorrelation. *J. Opt. Soc. Amer.,* **46,** 1019.
11. Pitts, E. and Marriage, A. (1957). Relation between granularity and autocorrelation, II. *J. Opt. Soc. Amer.,* **47,** 321.
12. Jones, L. A., Higgins, G. C. and Stultz, K. F. (1955). Photographic granularity and graininess. VIII. A method of measuring granularity in terms of the scanning area giving a threshold luminance gradient. *J. Opt. Soc. Amer.,* **45,** 107.
13. Trabka, E. A. (1969). Relationship between rms density and transmittance fluctuations of photographic film. *J. Opt. Soc. Amer.,* **59,** 662.
14. Kendall, M. G. and Stuart, A. (1963). "The Advanced Theory of Statistics", Chapter 9, Volume I. Griffin, London.
15. Altman, J. H. (1964). The measurement of rms granularity. *Appl. Opt.,* **3,** 35.
16. Riva, C. (1969). Comparison of the various methods for measuring the granularity of photographic layers. *Phot. Korres.,* **105,** 111, 128, 143, 159.
17. Frieser, H. (1959). Noise spectrum of developed photographic layers exposed by light, X-rays and electrons. *Photogr. Sci. Eng.,* **3,** 164.
18. Frieser, H. and Findeis, G. (1965). Measurement of average noise and noise spectrum of photographic layers. *Phot. Korres.,* **101,** 85, 101.
19. Thiry, H. (1963). Power spectrum of granularity as determined by diffraction. *J. Photogr. Sci.,* **11,** 69.
20. Stark, H., Bennet, W. R. and Arm, M. (1969). Design considerations in power spectrum measurements by diffraction in coherent light. *Appl. Opt.,* **8,** 2165.
21. Vilkomerson, D. H. R. (1970). Measurement of the noise spectral power density of photosensitive materials at high spatial frequencies. *Appl. Opt.,* **9,** 2080.
22. Smith, H. M. (1972). Light scattering in photographic materials for holography. *Appl. Opt.,* **11,** 26.
23. Blackman, R. B. and Tukey, J. W. (1968). "The Measurement of Power Spectra". Dover, New York.

24. Jenkins, G. M. and Watts, D. G. (1968) "Spectral Analysis and its Applications", Holden-Day, San Francisco.
25. Otnes, R. K. and Enochson, L. (1972). "Digital Time Series Analysis". Wiley, New York.
26. Bendat, J. S. and Piersol, A. G. (1971). "Random Data: Analysis and Measurement Procedures". Wiley, New York.
27. Akcasu, A. Z. (1961). Measurement of noise power spectra by Fourier analysis. *J. Appl. Phys.*, **32**, 565.
28. Bingham, C., Godfrey, M. D. and Tukey, J. W. (1967). Modern techniques of power spectrum estimation. *I.E.E.E. Trans. Audio and Electroacoustics*, **AU-15**, 56.
29. De Belder, M. and De Kerf, J. (1967). The determination of the Wiener spectrum of photographic emulsion layers with digital methods. *Photogr. Sci. Eng.*, **11**, 371.
30. Wall, F. J. B. and Steel, B. G. (1964). Implications of the methods chosen for the measurement of the statistical properties of phtographic images. *J. Photogr. Sci.*, **12**, 34.
31. Van Vleck, J. H. and Middleton, D. (1966). The spectrum of clipped noise. *Proc. I.E.E.E.*, **54**, 1. (first written as a classified report, 1943).
32. Jakeman, E. (1970). Theory of optical spectroscopy by digital autocorrelation of photon counting fluctuations. *J. Phys. (A)*, **3**, 201.
33. Verbrugghe, R. and De Belder, M. (1967). Influence of color coupling process on granularity and sharpness in color films. *Photogr. Sci. Eng.*, **11**, 379.
34. Biedermann, K. (1970). The scattered flux spectrum of photographic materials for holography. *Optik*, **31**, 367.
35. Shaw, R. and Shipman, A. (1969). Practical factors influencing the signal-to-noise ratio of photographic images. *J. Photogr. Sci.*, **17**, 205.
36. Schmitt, H. C. and Altman, J. H. (1970). Method of measuring diffuse rms granularity. *Appl. Opt.*, **9**, 871.
37. Siedentopf, H. (1937). Concerning granularity, density fluctuations and the enlargement of photographic negatives. *Physik Zeit.*, **38**, 454.
38. Bayer, B. E. (1964). Relation between granularity and density for a random dot model. *J. Opt. Soc. Amer.*, **54**, 1485.
39. Picinbono, M. B. (1955). Statistical model for the distribution of silver grains in photographic films. *Comptes Rendus*, **240**, 2206.
40. Trabka, E. A. (1971). Crowded emulsions: granularity theory for monolayers. *J. Opt. Soc. Amer.*, **61**, 800. Lawton, W. H., Trabka, E. A. and Wilder, D. R. (1972). Crowded emulsions: granularity theory of multilayers. *J. Opt. Soc. Amer.*, **62**, 659.
41. Castro, P. E., Kemperman, J. H. B. and Trabka, E. A. (1973). Alternating renewal model of photographic granularity. *J. Opt. Soc. Amer.*, **63**, 820.
42. Benton, S. A. and Kronauer, R. E. (1971). Properties of granularity Wiener spectra. *J. Opt. Soc. Amer.*, **61**, 524.
43. Langner, G. (1963). Calculation of the granularity and contrast threshold of photographic layers. *Phot. Korres.*, **99**, 177.
44. Shaw, R. (1967). Image characteristics of model photodetectors. *J. Photogr. Sci.*, **15**, 78.
45. Yu, F. T. S. (1969). Markov photographic noise. *J. Opt. Soc. Amer.*, **59**, 342.
46. Berwart, L. (1969). Wiener spectrum analysis of experimental emulsions with cubic homogeneous grains, comparison of the spectra with the Wiener spectrum of commercial emulsions. *J. Photogr. Sci.*, **17**, 41.

47. Savelli, M. (1958). Practical results of a study of a three-parameter model for the representation of the granularity of photographic films. *Comptes Rendus*, **246**, 3605.
48. O'Neill, E. L. (1963). "Introduction to Statistical Optics". Chapter 7. Addison-Wesley, Reading, Mass.
49. Stultz, K. F. and Zweig, H. J. (1959). Relation between graininess and granularity for black-and-white samples with nonuniform granularity spectra. *J. Opt. Soc. Amer.*, **49**, 693.
50. Doerner, E. C. (1962). Wiener spectrum analysis of photographic granularity. *J. Opt. Soc. Amer.*, **52**, 669.
51. Rossmann, K. (1963). Spatial fluctuations of X-ray quanta and the recording of radiographic mottle. *Amer. J. Roent.*, **90**, 863.
52. Rossmann, K. (1962). Modulation transfer function of radiographic systems using fluorescent screens. *J. Opt. Soc. Amer.*, **52**, 774.
53. Lubberts, G. (1968). Random noise produced by X-ray fluorescent screens. *J. Opt. Soc. Amer.*, **58**, 1475.
54. Roetling, P. G. (1965). Effects of signal-dependent granularity. *J. Opt. Soc. Amer.*, **55**, 67.
55. Lohmann, A. W. (1965). Image formation and multiplicative noise. *J. Opt. Soc. Amer.*, **55**, 1030.
56. Vendrovsky, K. V., Veitzman, A. I. and Ptashenchuk, V. M. (1972). The relationship of signal-to-noise ratio and quantum efficiency to resolution in photographic layers. *Zh. Nauch. Prikl. Fotogr. Kinematogr.*, **17**, 426.
57. Zweig, H. J., Stultz, K. F. and MacAdam, D. L. (1961). Effect of complex granularity patterns on the determination of quantum efficiency. *J. Photogr. Sci.*, **9**, 273.
58. Swank, R. K. (1973). Calculation of modulation transfer functions of X-ray screens. *Appl. Opt.*, **12**, 1865.
59. Trabka, E. A. and Lawton, W. H. (1974). Colour granularity: a layered model. *J. Photgr. Sci.*, **22**, 131.

Exercises

(1) At an average density level of $D = 1$ a certain film is found to have the following proportion of image grains in the various size classes:

Area μm^2	Proportion
0·25	0·04
0·70	0·10
0·75	0·20
1·00	0·18
1·25	0·14
1·50	0·08
1·75	0·06
2·00	0·04
2·25	0·04
2·75	0·02
	1·00

Calculate the mean and variance of the grain area. Estimate the mean number of image grains per unit image area, and the low frequency value, $W(0)$, of the Wiener spectrum of the image noise. Also calculate $W(0)$ for image grains of the same mean area, but without a spread of sizes, and comment on the comparison with the previous value.

(2) Explain why the error in each independent point in an estimate of the Wiener spectrum is very much greater than the error in the variance, σ^2, calculated using the same number of data values.

How many data values are required in each case to yield a standard error of 2%, where for the Wiener spectrum the bandwidth is 10 cycles/mm?

(3) The two-dimensional autocorrelation function of a sample of photographic image noise is to be measured, and for practical reasons based on the recording light level required in microdensitometry, the scanning aperture must have a minimum area of 100 μm^2 on the image.

Describe how you would choose the optimum shape of the scanning aperture. Indicate also the outline of a computer programme to calculate the two-dimensional autocorrelation function from the measured data.

(4) Indicate the relationship between the Wiener spectrum, the autocorrelation function, and the Selwyn granularity coefficient. Discuss the advantages and disadvantages of each of these as measures of photographic noise.

Wiener spectra of photographic noise patterns are often measured by scanning the pattern in a microdensitometer using a long slit of width a. Assuming that:
    (i) the optical system of the microdensitometer has a flat MTF for spatial frequencies of practical interest;
    (ii) the Wiener spectrum of the noise pattern is also flat;
explain the nature of the measured Wiener spectrum and also of the measured autocorrelation function, and discuss their practical significance.

(5) In the table below the mean-square density fluctuation, σ^2, as measured with a 50 μm diameter scanning aperture, is shown as a function of average density level for films A, B and C, each value having been calculated from 2000 density readings

	$\sigma^2 \times 10^3$ (in density units)		
Density	a	b	c
0·20	0·75	0·120	0·175
0·32	1·20	0·170	0·220
0·46	1·55	0·285	0·280
0·60	1·90	0·370	0·335
0·80	2·25	0·465	0·380
1·00	2·60	0·605	0·455

Plot this data and draw a smooth curve through each set of points, taking into account any statistical error. Discuss critically the form of these curves, and from them estimate the mean undeveloped grain area of each film.

(6) Give a general discussion of the practical and theoretical problems which are involved in the determination of the Wiener spectrum of photographic noise. State the Wiener-Khintchin theorem and explain its relevance in this context.

8. IMAGE NOISE ANALYSIS

When using methods involving numerical calculation, explain the comparative merits of computing the Wiener spectrum: (a) directly; or (b), via the autocorrelation function.

(7) Make use of the approximate method of estimating σ (based on Fig. 8) to find the variance of the random processes given in Appendix I.

Compare your results with values calculated in the usual manner, and thereby comment on the accuracy of this method for rapid evaluation of $W(0)$.

(8) The working equation $\text{DQE} = \dfrac{(\log_{10} e)^2 \gamma^2}{G \, q}$ is used to determine the DQE of an imaging process at exposure level q quanta per unit area. A circular scanning aperture ($A = \pi r^2$) of radius $r = 5 \, \mu$m is used to measure G, according to $G = A \sigma_A^2$.

If the actual Wiener spectrum of the image noise approximates to a triangle function which reaches zero at 400 cycles/mm, and the optics of the scanning system has MTF which approximates to a triangle function which reaches zero at 600 cycles/mm, use a graphical method to determine the error that would be made in assuming that the value obtained by use of the working equation is the value of DQE(0).

If the MTF of the imaging process itself is 0·4 at 100 cycles/mm, what is the DQE at this spatial frequency, expressed as a fraction of DQE(0)?

(9) A Wiener spectrum is to be calculated numerically under the conditions that:

$$\omega_{\max} = 750 \text{ cycles/mm};$$
$$\delta\omega' = 25 \text{ cycles/mm};$$
$$s = 10\%.$$

Devise appropriate scanning and data-processing parameters for calculation involving short data sequences.

(10) Describe the measurements which are necessary for the measurement of the low spatial frequency value of DQE, and discuss the main sources of error.

The working equation of question (8) is used to determine DQE(0), with a scanning aperture of dimensions which will give a systematic error of $+15\%$. The other experimental errors are:

$$\gamma: \pm\ 2\%;$$
$$A: \pm\ 2\%;$$
$$\sigma_A: \pm\ 5\%;$$
$$q: \pm 15\%$$

Discuss the overall experimental error spread. If the systematic error is ignored, estimate the probability that the true value lies within the apparent error spread.

9. Microdensitometry

9.1 Design of Conventional Microdensitometers

Single and Double Beam Microdensitometers

The measurement of micro-image properties such as the MTF and the Wiener spectrum of noise requires some form of microdensitometer for analysis of the developed image. The optical, mechanical and electronic components of the microdensitometer will all influence its performance, but the design of the optical system is usually most crucial in determining systems linearity and the existence of a unique microdensitometer transfer function.

The optical system of a typical single beam microdensitometer[1] is shown schematically in Fig. 1. A light source A_1 is focused by a field lens on the aperture stop A_2, which coincides with the ocular of the "illuminating" microscope, which in turn forms an image of the field stop B_1 on the sample B_2 to be scanned. A "pick-up" microscope projects an enlarged image of the sample onto the scanning aperture B_3: the planes B_1 to B_3 are conjugate. A condenser located immediately after the scanning aperture images the exit pupil of the pick-up microscope on the cathode of the photomultiplier tube, A_4, and all the A-planes are also conjugate. This system eliminates the effects of spatial variations in cathode sensitivity, since the same area of the cathode is illuminated regardless of the size of the scanning aperture. In some instruments a piece of diffusing glass is used in place of the pick-up condenser[2].

Standard optical components can be used in either of the microscopes. The pick-up microscope usually has a magnification in the range 20–200, depending on application. The illumination system is essentially the same as that for Kohler illumination which is commonly used in photomicrography, except that a compound microscope is substituted for the usual substage condenser. This ensures that a sharp, aberration-free image of the field stop is obtained in the sample plane.

To simplify focusing and to locate the test area on the sample, a hinged "focusing" mirror is mounted in the light path after the pick-up microscope to reflect the image onto a ground-glass screen. This screen is parfocal with the scanning aperture, and the position of the aperture is shown on it with accuracy. After the sample is located, focused and aligned, the mirror is swung out of the light path. This system has the disadvantage that it is not possible to view the sample while it is being scanned. However no stray light is introduced, as in some methods of aligning and focusing involving, for example, a whitened area surrounding the scanning slit. The optical systems of other single beam

9. MICRODENSITOMETRY

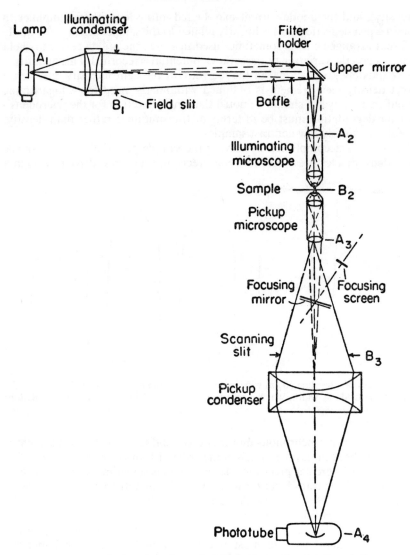

FIG. 1. Optical system of a typical single beam microdensitometer[1].

microdensitometers are essentially the same as the one described, the main differences being in the viewing system[2-5].

If the anode current is small compared to that in the dynode chain the output of the photomultiplier will be linearly related to the incident light intensity, and in practice linearity is usually obtained over an intensity range of around 1000:1. The photomultiplier anode current will be proportional to the transmittance of the image sample (if it is assumed that the optical system is linear in intensity), and the anode current may be fed into a logarithmic amplifier to obtain a pen deflection that is linearly related to the sample density. A slightly different arrangement[1] is sometimes used in which the anode current is held constant and the variation of the cathode potential is recorded. This quantity is approximately proportional to the logrithm of the light intensity, and hence direct density measurement is obtained without the need for a logarithmic amplifier. However it should be noted that any correction for the microdensitometer degradation must be in terms of transmittance rather than density, except for suitably low-contrast samples.

Scanning is accomplished by driving the sample past the aperture. In the microdensitometer described above, a precision lead screw driven through a

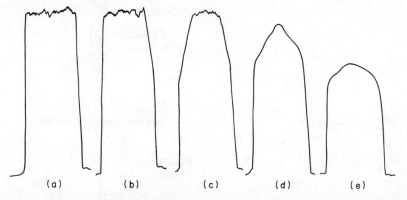

FIG. 2. Traces of a test line 70 μm wide, made with a scanning slit 1 μm wide, showing the effect of the field slit width, which was[1]: (a) 6·5 μm; (b) 20 μm; (c) 40 μm; (d) 82 μm; (e) slit removed.

gear reducer by a synchronous motor is used, and this type of arrangement is adequate if the accuracy required is of the order of 1 μm over a distance of a few millimeters. For greater precision a lever system is sometimes used[3,4], with or without a position transducer for monitoring the smoothness of the motion of the sample stage. This stage usually runs on roller bearings on ways formed by two ground and polished stainless steel rods.

Shortly we shall discuss the problems that arise from the partial coherence of the optical systems used in most microdensitometers. Another important problem concerns the reduction of flare light in the optical system, and this is usually achieved by painting metallic mountings matt black and using light baffles. Microdensitometer traces are shown in Fig. 2 of a test lines 70 μm wide, which were made for a range of field slits with a scanning slit of 1 μm. It is clear that (subject to coherence restrictions) the smallest possible field slit should be used when maximum accuracy is required.

Flare light also gives rise to errors in the measurement of the density of very small points or narrow lines less than approximately 2 μm wide and with a central density greater than approximately 1·0, and this is called the Schwarzschild-Villiger effect[6]. This effect may also be important in noise measurements[7]: in Fig. 3 the *rms* density fluctuation measured with a small scanning aperture (15 μm^2 in area) is plotted as a function of the illuminated area of the sample. Due to light scattered within the image layer the effective scanning area increases as the illuminated area increases, and hence the rms fluctuation is decreased.

Single beam microdensitometers are widely used in photographic image analysis, partly because of their relatively simple construction. However, care

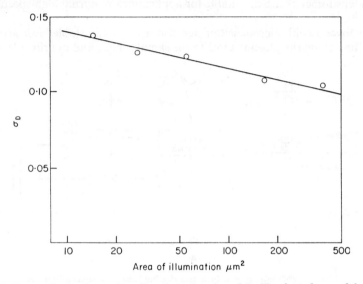

FIG. 3. The *rms* density fluctuation plotted as a function of the illuminated area of the film[7], for a scanning aperture of area 15 μm^2.

has to be taken to stabilize the power supplies and amplifiers. It is customary to chop the light beam so that AC amplification is used, since this is generally preferable to a DC system. Instability of the light source and power supply can only be overcome in a statisfactory manner by using a double beam instrument.

A schematic diagram of the optical system of the Joyce-Loebl double beam microdensitometer is shown in Fig. 4. The optical system shown on the left hand side of the diagram is identical with that shown previously for the single beam instrument, except that the illuminating and pick-up microscopes are replaced by less elaborate lens systems. The second beam in the microdensitometer passes through a reference wedge whose position is governed by the control electronics. The intensities of the reference and sample beams are sequentially compared, and the electronic system causes the reference wedge to be moved in or out of the reference beam until the two intensities are equal. The position of the reference wedge is then a measure of the density of the sample, and this is recorded on a plotting table by a pen having mechanical linkage to the wedge. Instabilities in the light source, power supplies and electronic components do not affect this type of measuring system. The reference wedge is usually linear in density along its length, and consequently readings are obtained directly in density units without the need for special circuitry. Wedges may be interchanged so that the scale of the recorded trace can be selected in the range 0·02–0·25 density units per cm. The main disadvantage of the double-beam arrangement is that the response time is fairly slow since the reference wedge has to be moved mechanically into the beam. This means that double beam microdensitometers are unsuitable for applications requiring high-speed data acquisition.

The Joyce-Loebl microdensitometer also uses an adjustable lever arrangement for linking the plotting table to the sample stage, and in principle this is

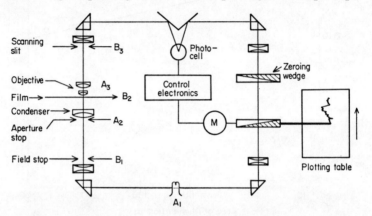

FIG. 4. Schematic illustration of a double beam microdensitometer.

more stable than driving the two independently. Meaningful linear magnifications of up to 1000 are possible with this system, with 1 mm on the plotting table then being equivalent to 1 μm on the sample.

Digital Systems

It is often desirable to record density data on paper or magnetic tape for numerical analysis by computer. Two distinct types of instrument have been developed for this purpose: the first, used for precision analysis when the rate of data recording is of secondary importance, consists of a standard microdensitometer with an analogue-to-digital converter and a recording unit. The second type, of which the drum scanner is typical, is a completely redesigned and more complex instrument for very high-speed measuring rates.

The Joyce-Loebl instrument described earlier can readily be adapted for digital recording. The sample stage is decoupled from the plotting table, and is driven by programmable stepper-motors in the x and y directions. The basic interval is between 1 and 5 μm in the x direction and 5 μm in the y direction. The reference wedge is linked mechanically to a precision linear potentiometer which is used by the analogue-to-digital converter. The rate of digitizing is limited by the time taken by the reference wedge to reach the null position, and this depends on the density differences between neighbouring sample points, but is typically around 20 readings per second. The output is produced either on paper or magnetic tape, the advantage of magnetic tape in this case is its high storage density (around 40 readings per cm of standard tape), rather than its fast recording rate. This type of system retains the accuracy features of the standard microdensitometer which are essential for measuring photographic micro-image properties.

In drum scanners[8,9] the image sample, which must of course be on a flexible base, is placed over an opening in a cylindrical drum and clamped to it such that the film forms part of the circumference of the drum. The drum rotates continuously, and after each revolution the optical carriage containing the illumination and pick-up optics is stepped along the drum axis. An opening in the drum opposite to the film position allows the system to measure zero density once per revolution, and compensation for any short term drift. The density data is converted from analogue to digital form and recorded directly on magnetic tape at up to 20 000 readings per second. This type of instrument usually samples at a minimum interval of 25 μm, and is therefore unsuitable for analysis of micro-image properties unless it is modified. It is mainly used for picture processing applications where high speed data recording is of primary importance.

9.2 Theory of Imaging in Microdensitometers

Effects of Partial Coherence

The illumination system used in most microdensitometers is shown reduced to its essential components in Fig. 5, from which it can be seen that light from a

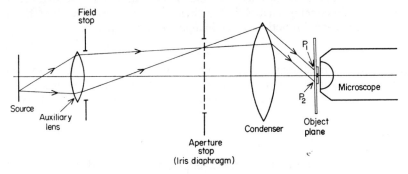

FIG. 5. Kohler illumination in a microscope[10].

single point on the source is incident on the object plane at a certain angle. Suppose that the light source is monochromatic. If the source consisted only of one very small point the illumination in the object plane would be coherent, since the complex amplitude at a point such as P_1 would always have a fixed relationship with that at P_2. Practical sources are relatively large, and the diameter of the aperture stop governs the actual area of the source used to illuminate the object. When the aperture stop is very small, fairly coherent illumination may be obtained, but if the aperture stop is large the points P_1 and P_2 are in effect illuminated by many point sources and the total amplitudes and phases will not have any fixed relation in time. The latter represents incoherent illumination.

When the illumination is incoherent the object and image intensities are related by a simple convolution formula for regions of the image field over which the aberrations and focusing vary only by a small amount (optical systems are not isoplanatic in this sense as has been assumed for the imaging properties of the photographic process). Within this limination an optical transfer function can be used to describe the microdensitometer system, and measured photographic properties such as the MTF and Wiener spectrum can be corrected for microdensitometer degradation as appropriate. Unfortunately the illumination in any practical microdensitometer system is neither totally coherent or incoherent, and is termed partially coherent. Optical systems that are partially coherent are not linear in either complex amplitude or intensity, and simple linear filter theory, such as outlined in Chapter 6, cannot be applied. However, for an imaging system of given numerical aperture it is possible to produce illumination that is effectively incoherent by a suitable choice of the size of the field stop and the numerical aperture of the substage condenser.

It will be recalled from Chapter 6 that we may regard an image formed in incoherent light as being the sum of a large number of weighted intensity point spread functions, the weighting factors being the intensity at each corresponding point in the object. This addition of weighted intensities can be carried out only if the illumination of neighbouring resolution cells is indeed incoherent. A resolution cell has an area of the same order as the point spread function of the imaging lens, and thus for effectively incoherent illumination we require that the radius of a coherence patch in the object, r_{coh}, must be much less than the radius of the spread function of the imaging lens, $r_{\text{objective}}$:

$$r_{\text{coh}} \ll r_{\text{objective}}. \tag{1}$$

To proceed further than this general requirement it is necessary to examine the factors that govern the coherence of illumination systems in more detail. Consider two points, 1 and 2, of intensity I_1 and I_2 on a screen illuminated by a monochromatic light source. The mutual intensity between the two points, I_{12}, is defined as

$$I_{12} = \langle V_1(t) V_2^*(t) \rangle,$$

where $V_1(t)$ and $V_2(t)$ are the complex amplitudes of the field at points 1 and 2, and $\langle \rangle$ denotes the time average. As the average intensity, I, is defined by

$$I = \langle |V(t)|^2 \rangle,$$

it follows that $I_{11} = I_1$ and $I_{22} = I_2$. The degree of coherence, μ_{12}, is defined as

$$\mu_{12} = \frac{I_{12}}{\sqrt{I_1 I_2}}.$$

For completely coherent illumination (i.e., when V_1 and V_2 are perfectly correlated), $\mu_{12} = 1$, and for completely incoherent illumination (i.e., when V_1 and V_2 are uncorrelated), $\mu_{12} = 0$.

For an incoherent source of known dimensions and at a known distance from the object plane, as in Fig. 5, the degree of coherence at the object plane can be found from the van Cittert-Zernike theorem[10], which is stated as follows.

> The complex degree of coherence, which describes the correlation of vibrations at a fixed point P_2 and a variable point P_1 in a plane illuminated by an extended quasimonochromatic primary source, is equal to the normalized complex amplitude at the corresponding point P_1 in a certain diffraction pattern centred on P_2. This pattern would be obtained on replacing the source by a diffracting aperture of the same size and shape as the source, and on filling it with a spherical wave centred on P_2, the amplitude distribution over the wavefront in the aperture being proportional to the intensity distribution over the source.

As an example, suppose the exit pupil of the condenser in Fig. 5 may be treated as a uniform incoherent source. The complex degree of coherence between two points in the object plane separated by a distance x_{12} is simply equal to the Fraunhofer diffraction pattern of a circular aperture (the Airy disc function):

$$\mu_{12} = \frac{2J_1(z)}{z}, \tag{2}$$

where:

$$z = \frac{2\pi}{\lambda} NA_c x_{12};$$

$J_1(\ )$ denotes the 1st order Bessel function;
NA_c denotes the numerical aperture of the condenser.

It should be noted that this expression defines the degree of coherence regardless of the aberrations of the condenser lens.

The degree of coherence has been measured experimentally[11] for the Joyce-Loebl microdensitometer described earlier, and some results are shown in Fig. 6, where the agreement between theory and experiment is seen to be good.

The inequality given in equation (1) for effectively incoherent illumination may now be restated. If the imaging objective in the pick-up microscope is aberration free, its intensity point spread function is given by

$$h(r) = \left(\frac{2J_1(z')}{z'}\right)^2,$$

where

$$z' = \frac{2\pi}{\lambda} NA_0 r;$$

NA_0 denotes the numerical aperture of the objective. It can be seen that both the degree of coherence in the object plane and the spread function of the imaging system have a similar form. The coherence patch must be much smaller than the spread function area, which is equivalent to the condition:

$$NA_c \gg NA_0, \text{ or } s = \frac{NA_c}{NA_0} \gg 1. \tag{3}$$

A further condition must be imposed. The condenser exit pupil cannot be an incoherent source since it is illuminated by a field stop whose area governs the degree of coherence at the condenser. Thus for effectively incoherent illumination it is also required that the field stop is "large".

Detailed investigations of partially coherent imaging in microdensitometers have been made by Swing[12,35] and Kinzly[13]. Although partially coherent systems are not linear in either the complex amplitude or the intensity, they are

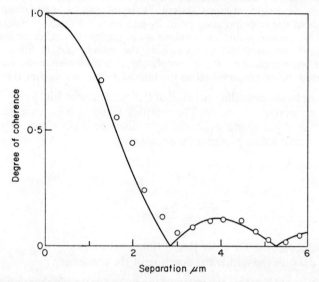

FIG. 6. The degree of coherence, μ_{12}, between two points in the object plane of a Joyce-Loebl microdensitometer, plotted as a function of the separation between them[11]. The condenser lens has a nominal numerical aperture of 0·08.
———— theoretical curve;
○○○○○ experimental values.

linear in mutual intensity, and the analysis of such a system can be reduced to characterization of the propagation of mutual intensity from source to image. Having calculated the image intensity in this way, some criterion of effectively incoherent illumination is required. For a system linear in intensity, the image intensity spectrum will be equal to the product of the object intensity spectrum and the optical transfer function. Therefore a partially coherent system will be regarded as effectively incoherent if for the spatial frequency range of practical interest:

$$G(\omega) = F(\omega)T(\omega), \qquad (\omega < \omega_{max});$$

where:

$G(\omega)$ is the Fourier transform of the image intensity;
$F(\omega)$ is the Fourier transform of the object intensity;
$T(\omega)$ is the incoherent optical transfer function;
ω_{max} is the maximum spatial frequency of interest.

In this way Swing[12,35] showed that two conditions must be satisfied for effectively incoherent illumination:

(1) $$s \geq 1 + \frac{\omega_{\max}}{\omega_0}, \qquad (4)$$

where: $s = NA_c/NA_0$, and ω_0 is the maximum spatial frequency transmitted by the objective lens;

(2) $$W \geq \frac{4\lambda}{NA_c}, \qquad (5)$$

where W is the width of the field slit referred to the object plane.

The value of the parameter s must lie between 1 and 2, and is typically around 1·5. Although equation (5) gives no upper bound for the field slit-width, it must not be so great that the effects of flare light become serious.

As an example, suppose that a microdensitometer is used with an imaging objective of $NA_0 = 0.25$ for MTF measurements up to 300 cycles/mm, and that we wish to determine the minimum values that should be used for the condenser numerical aperture and the reduced field slit width for linear operation of the instrument, assuming $\lambda = 500$ nm. For an aberration-free objective the cut-off spatial frequency is given by

$$\omega_0 = \frac{2NA_0}{\lambda},$$

and it follows that $\omega_0 = 1000$ cycles/mm. From inequality (4), $s \geq 1.3$, and therefore $NA_c \geq 0.33$. From inequality (5), $W \geq 6$ μm, and thus the numerical aperture of the condenser must be greater than 0·33 and the field slit must be wider than 6 μm in the object plane.

A practical solution to the problems caused by partial coherence is provided by so-called linear microdensitometers[11,14,15,36–38], which have a basic optical system as shown in Fig. 7. The sample is still illuminated by the image of the source aperture, but all the light transmitted by the sample is collected. It can be shown that the output of the microdensitometer is given by the convolution of the intensity transmittance of the sample with the projected source aperture. If this aperture is unresolved by the condenser lens, then the microdensitometer transfer function is simply equal to that of the condenser lens.

One disadvantage of this system for MTF measurement is that the scanning aperture is so small that the influence of image noise may be excessive. This can be reduced by using a line scanning aperture produced either by a cylindrical condenser lens or by an unresolved slit at the source plane. The output is independent of the coherence properties of the source, and a laser may be used[15]. There are also fewer focusing and alignment problems.

A useful feature of this type of microdensitometer is that singly diffuse density is measured, as with most large-area densitometers. Conventional microdensitometers measure more specular density, although this varies with

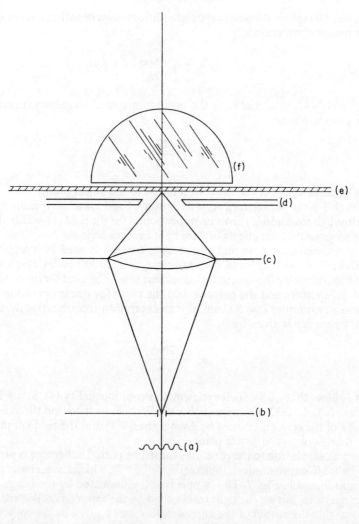

FIG. 7. The linear microdensitometer[14]: (a) source; (b) slit; (c) high-quality condenser; (d) baffles; (e) sample; (f) light collector.

the numerical apertures of the condenser and objective, as shown in Fig. 8[16]. It is extremely important to standardize density measurement when combining quantities, as in the expression for DQE, which involve both macro- (γ) and micro- (σ) parameters. Recent developments in linear microdensitometers are given in references 36 to 40.

The Microdensitometer Transfer Function

We now assume that effectively incoherent illumination is used in the microdensitometer, and it is therefore meaningful to refer to a unique OTF (for a given field angle, focus position, wavelenth, etc.). The total OTF of the microdensitometer, $T_M(u, v)$ is the product of the transfer function of the objective lens (pick-up optics), $T_0(u, v)$, and that of the scanning aperture, $T_A(u, v)$:

$$T_M(u, v) = T_0(u, v)T_A(u, v). \qquad (6)$$

The electronic system may also contribute a further multiplicative transfer

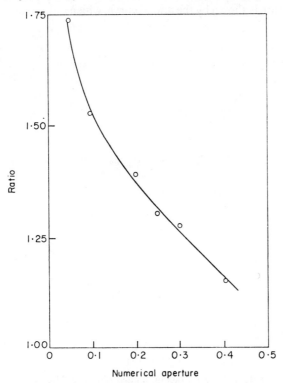

FIG. 8. Variation of the ratio of the measured density to singly diffuse density with the numerical apertures of the illuminating and pick-up optics[16] (which were equal in this case).

function, but in practice this can usually be neglected. A good microdensitometer should approximate to a diffraction-limited optical system, for which the transfer function is given by:

$$T_0(\omega) = \frac{2}{\pi} \cos^{-1} \frac{\omega}{\omega_0} - \frac{\omega}{2\omega_0} \sqrt{1 - (\omega/\omega_0)^2}, \qquad (7)$$

where

$$\omega = \sqrt{u^2 + v^2}, \text{ and } \omega_0 = \frac{2NA_0}{\lambda}.$$

This function is shown in Fig. 9 for $NA_0 = 0.25$ and $\lambda = 500$ nm.

The transfer function of a slit of width W is given by its normalized Fourier transform:

$$T_A(\omega) = \frac{\sin \pi W \omega}{\pi W \omega}, \tag{8}$$

and this function is also shown in Fig. 9, for $W = 2$ μm. The total system transfer function is shown by the dotted curve, which is the product of the other two curves.

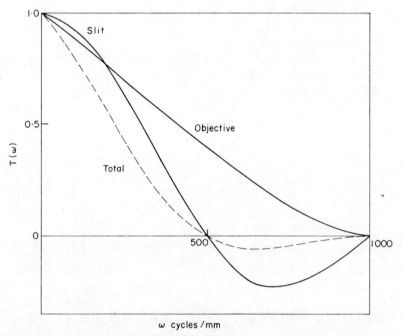

FIG. 9. Transfer functions for a diffraction-limited objective, $NA_0 = 0.25$, $\lambda = 500$ nm; a slit of width 2 μm; and the total system.

The transfer function of an optical system free from aberrations, as defined by equation (7), is only applicable to the imaging of infinitely thin objects at the paraxial focus of the lens. However, a typical photographic layer may have a thickness of 5–20 μm, and for lenses of high numerical aperture the transfer function changes significantly over this distance. The optical transfer function of a system free of geometrical aberrations is shown in Fig. 10 for various degrees of defocus. The number on each curve refers to the parameter m, defined by

$$m = \frac{\pi}{2\lambda}(NA_0)^2 x,$$

where x is the defocus distance. For example, if $\lambda = 500$ nm and $NA_0 = 0.25$

(in air), then $m \simeq 0.2\, x$, so that the curve for $m = 1$ corresponds to a defocus of only 5 μm in air, or about 7·5 μm in a layer of refractive index of 1·5. This result means in effect that the microdensitometer transfer function depends on the thickness of the image sample, and this is an important factor in studies of

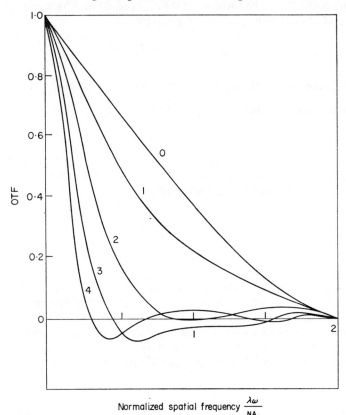

FIG. 10. Transfer functions for a defocused lens system that is free from geometrical aberrations[17].

images through their depth[18]. It also makes it necessary to set precise tolerances for the mechanical construction of the sample stage and scanning mechanism.

One of the simplest methods of measuring the transfer function of a microdensitometer is to scan a knife-edge. A suitable edge with a finite density difference can be prepared from an aluminium or silver coated glass slide[19,20]. A razor blade edge may also be suitable if selected carefully. The reflecting nature of these edges tends to introduce errors at the high-density side of the edge trace, and photographically-produced edges on special Lippmann-type emulsions may be preferable in this respect. The density difference across any test edge should not be so great as to introduce electronic or mechanical factors (such as mechanical overshoot) which are not present when recording normal image fluctuations. The edge trace is converted into intensity units, differentiated and Fourier transformed, to yield the microdensitometer transfer function.

A possible source of error in this method of measurement, as in all methods involving the scanning of one-dimensional patterns with a slit, is due to slit misalignment[21]. It can be shown that the effect of slit misalignment is to introduce an additional transfer function given by

$$T_{skew}(\omega) = \frac{\sin \pi L \phi \omega}{\pi L \phi \omega}, \qquad (9)$$

where L is the slit length and ϕ is the angle of misalignment.

The fractional error due to misalignment is given by

$$F = 1 - T_{skew}(\omega). \qquad (10)$$

For small fractional errors (corresponding to small $\pi L \phi \omega$), the term $\sin \pi L \phi \omega$ may be approximated by the first two terms of its series expansion, and equations (9) and (10) solved for ϕ, leading to

$$\phi \simeq \frac{\sqrt{3!F}}{\pi L \omega}. \qquad (11)$$

Suppose for example that a fractional error of less than 0·01 is required for a microdensitometer transfer function at 150 cycles/mm, when using a slit length of 150 μm. Substitution of these values in equation (11) yields a maximum permissible misalignment of approximately 0·0035 radians, or 12 mins of arc. Such accuracy is only attainable if great care is taken in alignment.

Another method of measuring the microdensitometer transfer function[22] makes use of the fact that the measured Wiener spectrum of photographic image noise, $W'(\omega)$, is related to the true Wiener spectrum, $W(\omega)$, by

$$W'(\omega) = W(\omega)|T_M(\omega)|^2.$$

If a fine-grained film is used it may usually be assumed that $W(\omega)$ is constant up to several hundred cycles/mm, and consequently the measured Wiener spectrum is equal to the square of the MTF of the microdensitometer. This method eliminates the problems involved in scanning a knife-edge, but has the usual statistical error associated with Wiener spectra measurements.

Correcting for Microdensitometer Degradation

Correcting the measured transmittance distribution for microdensitometer degradation is straightforward in the absence of noise. If the transmittance distribution of photographic image, excluding noise, is denoted by $f(x, y)$, and the measured transmittance is $g(x, y)$, then assuming that the microdensitometer has effectively incoherent illumination,

$$g(x, y) = f(x, y) \circledast h(x, y),$$

where $h(x, y)$ denotes the microdensitometer spread function, and $\circledast$ again denotes convolution. The Fourier transforms are then related by

$$G(u, v) = F(u, v)T_M(u, v).$$

It follows that

$$f(x, y) = \mathscr{F}^{-1}\left(\frac{G(u, v)}{T_M(u, v)}\right), \quad (12)$$

where $\mathscr{F}^{-1}$ denotes inverse Fourier transformation.

To correct for the degrading effect of the microdensitometer the measured transmittance distribution, $g(x, y)$, is in effect passed through a linear filter whose transfer function, $T_{fil}(u, v)$, is defined by

$$T_{fil}(u, v) = \frac{1}{T_M(u, v)}. \quad (13)$$

Similar filters are used to restore images degraded by camera lenses or image motion. The restoration is exact in the sense that the output of the filter is identical to the original image (unless $T_M(u, v)$ contains zero values), at least up to the cut-off frequency of the microdensitometer system. It should be noted that in general this correction must be made in transmittance units, and can only be made in density units for low contrast images. When correcting for the microdensitometer in a Wiener spectrum measurement the appropriate equation will be

$$W(\omega) = \frac{W'(\omega)}{T_M^2(\omega)},$$

where $W(\omega)$ is the actual Wiener spectrum and $W'(\omega)$ is the measured spectrum.

Often however we do not want restoration in this exact sense. For example, the problem may be to restore a signal that is partly obscured by noise, as is normal when measuring film MTFs. Several authors[23-25] have shown that the filter defined by equation (13) is not optimum in this respect. Furthermore, the exact form of the optimum filter depends to some extent on the precise definition of "optimum". If $S(x, y)$ is the original (noiseless) signal and $S'(x, y)$ is the restored (noisy) signal, then the optimum filter in the least-squares sense is the one that minimises

$$\int\int_{-\infty}^{+\infty} |S(x, y) - S'(x, y)|^2 \, dx \, dy.$$

If both signal and noise are treated as stationary random processes with Wiener spectra $W_S(u, v)$ and $W_N(u, v)$ respectively, then the optimum least-squares restoration filter turns out[25] to be:

$$T_{fil}(u, v) = \frac{T_M^*(u, v)}{|T_M(u, v)|^2 + \dfrac{W_N(u, v)}{W_S(u, v)}}. \quad (14)$$

This can be seen to reduce to the simpler filter as in equation (13) when the noise is zero. Other forms of optimum filter are discussed in the literature[25]. The use of such filters is important when noise cannot be reduced by other means, although it is always preferable to reduce noise experimentally, for example by lengthening the scanning slit or using multiple scanning slits, and then to apply the simple correction of equation (13).

It should be noted that even when noise is negligible, the simple correction applied only to the transmittance distribution of the image. The MTF of a film is often measured in terms of the effective exposure, and this is related to the transmittance in a nonlinear manner. A correction of the type defined by equation (13) in terms of spatial frequency (u, v), ignores the nonlinear transformation and generally gives overcorrection at high spatial frequencies. Strictly

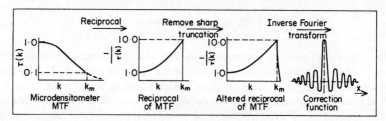

FIG. 11. The steps involved in the generation of a microdensitometer correction function[27].

speaking the correction should be made in terms of distance (x, y), by convoluting the measured transmittance with a suitable correction function[26,27], as illustrated in Fig. 13 of Chapter 7.

One method of producing the correction function is shown in Fig. 11. The function is essentially the inverse Fourier transform of the reciprocal of the microdensitometer MTF, and in the absence of image noise the correction is exact. However, there are problems that make it unattractive for general use, arising from the highly oscillating nature of the correction function and the necessity of numerical convolution. In practice correction is usually carried out simply by dividing the measured MTF by that of the microdensitometer (or the measured Wiener spectrum by the square of this function), and for a wide range of values of effective modulation and gamma the remaining error is less than 5%[26,34].

9.3 Special Features and Instruments

Automatic Focusing and Alignment

Routine measurements of quantities such as the MTF and Wiener spectrum generally require more automatic instruments than ones described so far for special research purposes. The focusing and alignment of the sample can be time-consuming, and require patience and skill on the part of the operator. Here we shall consider a method that has been used by Eastman Kodak for automatic focusing[28], and also some features of the Perkin-Elmer microdensitometer[29] relevant to general alignment problems.

Automatic focusing systems are usually based on the fact that the measured noise of a sample is greatest when the sample is in focus, and a feedback circuit adjusts the focus to maintain maximum noise level. One such system is shown schematically in Fig. 12. A reflex viewer with a beamsplitting prism is added to

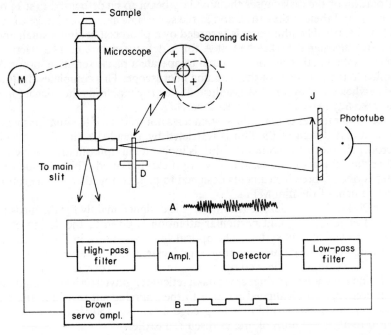

FIG. 12. A method of automatic focusing[28]. The disk D is composed of four segments, alternately positive and negative, cut from lenses as indicated by lens L. When the microscope is out of focus the phototube current has the form shown at A; after rectification it has the form shown at B, and will operate the servomotor M to restore focus.

a conventional scanning system, and the auxiliary image at J is used for the automatic focusing of the instrument.

The focusing system is based on two simple principles. First, if a weak positive lens were placed in the beam the image would move nearer the microscope, while a weak negative lens of the same power would move the image an approximately equal distance further from the microscope. If the image without these extra lenses were in focus on J, the degradation produced by the two lenses would be equal. If the image did not lie on J, one lens would produce more degradation than the other, and it follows that the direction of defocus can be found by determining which lens produces the better image.

Secondly, by moving either the positive or negative lens laterally, the auxiliary image can be made to move, since the lens introduces a certain amount of prismatic power when it is decentred, and this power varies with the degree of decentering. Therefore by moving the lens through the beam an image can be obtained that sweeps across the auxiliary focal plane. The scanning disk that is used is shown in the upper part of Fig. 12. It consists of four quadrants; two cut from +25 cm focal length lenses and two from −25 cm focal length lenses. The optical axes of these lenses lie in positions indicated by the + and − signs near the edge of the disk.

Rotation of the disk causes the auxiliary beam to be defocused first in one direction and then in the other, and at the same time causes the image to scan back and forth. The film noise is recorded by a photocell behind a small aperture, and this signal is sketched at A for a defocused image. The electronic control system makes use of the signal to operate a phase-sensitive motor, M, coupled to the fine focus of the scanning microscope. This completes the loop, so that when the scanning disk is adjusted to a proper phasing the microscope is automatically driven to the correct focus.

Since scanning takes place the system does not lend itself to situations where the illuminated area of the film is restricted by a field slit. In this particular microdensitometer the scanning "slit" is in fact a multiple-slit arrangement to reduce the influence of noise, and so a large field slit is used. This microdensitometer is one of three instruments designed to be used together for the routine measurement of the film MTFs [28].

The Perkin-Elmer microdensitometer[29] is designed mainly for the measurement of edge images, and particular attention is given to the design of the viewing and alignment system. The optical system is shown in Fig. 13 and the optical layout of the main measuring beam is essentially the same as that shown in Fig. 1. Auxiliary optical components are provided for the viewing mode. The $5\frac{1}{2}$ inch diameter viewing screen has a reticule engraved on it enabling a high resolution edge to be aligned parallel to the scanning slit to within 0·2 mm. Based on this skew, a 0·15 mm long edge magnified 150 times can then be aligned to about 12 mins of arc, as discussed earlier.

Another feature of this microdensitometer is the oscilloscope display of the transmittance distribution of the edge which is accomplished without moving the film sample. Directly above the measuring microscope there is a rotating prism, which when in the optical path produces a circular image motion on the measuring slit without introducing an image rotation. The oscilloscope beam sweeps in synchronism with the image motion, and the film transmittance or its first derivative (which in this case is the line spread function) is displayed. The profile display permits the operator to see the interrelated effects on the output transmittance due to illumination coherence, image noise, focus, alignment errors, etc. The operator can also use this profile to select the most desirable area of the sample, prior to scanning to produce digital output on magnetic tape. The gradient display peaks sharply when an image is properly focused and aligned.

The design of the microdensitometer also takes into account possible difficulties in sample positioning. Sample alignment is accomplished after approximate positioning by rotating an 11 inch diameter ring, the x and y sample adjustments being mounted on top of the ring. Rotational motion does not cause the image to swing across the measuring slit when the scanning stage is centered. Sample alignment can be very difficult in the absence of independent rotational adjustment.

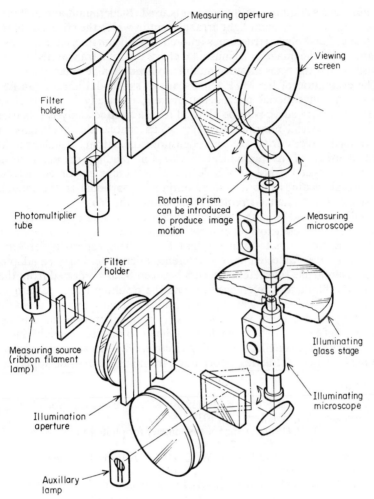

FIG. 13. Optical system of the Perkin-Elmer microdensitometer[29].

Other Special Features

Image noise is often a problem when measuring the MTF of a film. In conventional microdensitometers the effect of noise can be reduced using a multiple scanning slit[28]. The arrays of slits must be designed for a particular sinusoidal test chart, although a single array of slits can be used for several spatial frequencies. For example, an array of slits whose spacing is such that the slits fall on each cycle of a 50 cycles/mm pattern can also be used for a 100 cycles/mm pattern, since the slits will then coincide with every second cycle. The *rms* noise however decreases only as the square root of the number of slits, and the improvement is not very great. Coherent light processing methods reduce the effects of noise, but in general are sensitive to phase structures on the film, unless liquid-gate immersion is used. Biedermann and Johansson[30] have suggested a coherent light arrangement in which the effects of film noise are reduced and phase structure is ignored. Two-beam interference fringes are scanned over the image of a sinusoidal chart, the total transmitted intensity is recorded, and the transmittance modulation can then be determined.

The evaluation of the MTF from scans of sinusoidal images can be quite tedious in conventional microdensitometers, especially if the macroscopic density-exposure curve is to be allowed for. In routine analysis it is useful to construct a special plotting table[28,31] which enables the MTF curve to be determined directly from the microdensitometer output. Usually the instrument consists of a mechanical device for reading the data from the microdensitometer trace, a simple analogue computer for converting the mechanical readings into a voltage output proportional to the modulation transfer factor, and an x-y plotter for recording the MTF.

Finally it should be noted that the quantity measured by a "microdensitometer" need not relate to light absorption of the image layer, and for example it may be advantageous to examine the relief image[32], especially for very high resolution films. Reflected light interference microscopes may be adapted for this measurement. Interference occurs between beams reflected from the surface of the film and a reference flat, and the resulting fringe pattern can be examined at high magnification. Such an instrument may also be useful for estimating the density of very fine lines (using a calibration curve of density against relief height), which would be difficult to measure directly due to the Schwarschild-Villiger effect. Scanning electron microdensitometry[33] may also be a useful way of extracting information from images in some applications.

References

1. Altman, J. H. and Stultz, K. F. (1956). Microdensitometer for photographic research. *Rev. Sci. Instr.*, **27**, 1033.
2. Herrnfeld, F. P. (1961). A microdensitometer for photographic research. *J. Soc. Motion Pict. Tel. Engrs.*, **70**, 904.
3. Southwold, N. K. and Waters, W. G. (1959). A microdensitometer for photographic research. *J. Photogr. Sci.*, **7**, 174.
4. Wall, F. J. B. and Steel, B. G. (1964). Implications of the methods chosen for the measurement of the statistical properties of photographic images. *J. Photogr. Sci.*, **12**, 34.
5. Gorokhovskii, Yu. N., Grigor'yev, A. G., Ivanov, A. M. and Stepochkin, A. A. (1972). New sensitometric instruments. *Soviet J. Opt. Tech.*, **39**, 161. (Page number refers to *Opt. Soc. Amer.* translation).
6. Schwarzschild, K. and Villiger, W. (1906). On the distribution of brightness of the ultra-violet light on the sun's disk. *Astrophys. J.*, **23**, 284.
7. Kraus, W. (1968). The illumination of the emulsion for photographic granularity measurements. *Photogr. Sci. Eng.*, **12**, 217.
8. Abrahamsson, S. (1966). A computer controlled film scanner. *J. Sci. Instr.*, **43**, 931.
9. Xuong, N. (1969). An automatic scanning densitometer and its application to X-ray crystallography. *J. Sci. Instr.*, Series **2**, 485.
10. Born, M. and Wolf, E. (1970). "Principles of Optics". (4th Ed.) Chapter 10. Pergamon Press, Oxford.

11. Reynolds, G. O. and Smith, A. E. (1973). Experimental demonstration of coherence effects and linearity in microdensitometry. *Appl. Opt.*, **12**, 1259.
12. Swing, R. E. (1972). Conditions for microdensitometer linearity. *J. Opt. Soc. Amer.*, **62**, 199.
13. Kinzly, R. E. (1972). Partially coherent imaging in a microdensitometer. *J. Opt. Soc. Amer.*, **62**, 386.
14. Grimes, D. N. (1971). Linear microdensitometry. *J. Opt. Soc. Amer.*, **61**, 1263.
15. Weingartner, I. (1971) Laser-microdensitometer. *Optik*, **32**, 508.
16. Schmitt, H. C. and Altman, J. H. (1970). Method of measuring diffuse RMS granularity. *Appl. Opt.*, **9**, 871.
17. Hopkins, H. H. (1955). The frequency response of a defocused optical system. *Proc. Roy. Soc.*, A231, 91.
18. Berg, W. F. (1969). The photographic emulsion layer as a three-dimensional recording medium. *Appl. Opt.*, **8**, 2407.
19. Kleinsinger, I. J., Derr, A. J. and Giuffre, G. F. (1964). Optical analysis of a microdensitometer system. *Appl. Opt.*, **3**, 1167.
20. Charman, W. N. (1965). Practical tests of a microdensitometer system. *Appl. Opt.*, **4**, 289.
21. Jones, R. A. (1965). The effect of slit misalignment on the microdensitometer modulation transfer function. *Photogr. Sci. Eng.*, **9**, 355.
22. Doerner, E. C. (1962). Wiener-spectrum analysis of photographic granularity. *J. Opt. Soc. Amer.*, **52**, 699.
23. Blackman, E. S. (1969). Letter to the Editor. *Photogr. Sci. Eng.*, **13**, 382.
24. Becherer, R. J. and Geller, J. D. (1968). Optimum shaded apertures for reducing photographic grain noise. Proceedings of the S.P.I.E. Seminar on "Image Information Recovery", **16**, 89.
25. Kinzly, R. E., Haas, R. C. and Roetling, P. G. (1968). Designing filters for image processing to recover detail. Proceedings of the S.P.I.E. Seminar on "Image Information Recovery", **16**, 97.
26. Langner, G. and Müller, R. (1967). The evaluation of the modulation transfer function of photographic materials. *J. Photogr. Sci.*, **15**, 1.
27. Jones, R. A. and Coughlin, J. F. (1966). Elimination of microdensitometer degradation from scans of photographic images. *Appl. Opt.*, **5**, 1411.
28. Lamberts, R. L., Straub, C. M. and Garbe, W. F. (1969). Equipment for the routine evaluation of the modulation transfer function of photographic emulsions. *Photogr. Sci. Eng.*, **13**, 205.
29. Galburt, D., Jones, R. A. and Bossung, J. W. (1969). Critical design factors affecting the performance of a microdensitometer. *Photogr. Sci. Eng.*, **13**, 205.
30. Biedermann, K. and Johansson, S. (1972). Evaluation of the modulation transfer function of photographic emulsions by means of a multiple-sine-slit microdensitometer. *Optik*, **35**, 391.
31. Leistner, K. (1964). Plotting table with grid plate for reading modulation transfer factors from microdensitometer recordings. *Photogr. Sci. Eng.*, **8**, 87.
32. Altman, J. H. (1966). Microdensitometry of high resolution plates by measurement of the relief image. *Photogr. Sci. Eng.*, **10**, 156.
33. Kofsy, I. L., Geller, J. D. and Miller, C. S. (1972). Scanning electron microdensitometry. *Appl. Opt.*, **11**, 2340.
34. Johansson, S. and Biedermann (1974). On the compensation for the transfer function of the microdensitometer in measurement of the MTF of photographic emulsions. *Photogr. Sci. Eng.*, **18**, 151.

35. Swing, R. E. (1973). The optics of microdensitometry. *Opt. Eng.*, **12**, 185.
36. Cronin, D. J. and Reynolds, G. O. (1973). Optical design considerations and test results for a linear microdensitometer, *Opt. Eng.*, **12**, 201.
37. Fallon, J. P. (1973). Design considerations for a linear microdensitometer, *Opt. Eng.*, **12**, 206.
38. Kinzly, R. E. (1973). Experimental evaluation of efflux optics for linear microdensitometry, *Opt. Eng.*, **12**, 218.
39. Special issue of *Optical Engineering*, 12(6), Nov-Dec, 1973.
40. Proceedings of SPSE Seminar on Microdensitometry, Boston, April-May, 1974.

Exercises

(1) A microdensitometer is to be used for measuring the MTF of a film. Discuss in detail the factors that govern the choice of:

(a) the field slit width;
(b) the numerical aperture of the objective;
(c) the overall magnification;
(d) the scanning slit width.

(2) The MTF of a microdensitometer is found by scanning a white-noise pattern, and thus is free from misalignment error. In the measurement of a film MTF by an edge trace method, a slit length of 200 μm is used. If the average error of alignment is 15 mins of arc, calculate the fractional error in the measured film MTF at 100 cycles/mm.

(3) A microdensitometer has a condenser of numerical aperture 0·1, and the mean wavelength of the illumination/detection system is 550 nm. Calculate the approximate radius of the coherence patch in the sample plane.
To what extent will the use of this condenser influence:

(a) measurement of the Selwyn granularity coefficient using a 50 × 50 μm^2 scanning aperture;
(b) MTF measurements using a 2 × 200 μm^2 scanning aperture?

(4) Using equation (14) for the least-squares restoration filter, and assuming that the microdensitometer MTF is dominated by that of the 3 μm scanning slit, calculate and plot the form of the optimum filter for values of $W_N(u, v)/W_S(u, v)$, assumed independent of (u, v), equal to: 0; 0·1; 0·5; 1·0.

(5) A microdensitometer has an aberration-free objective of numerical aperture 0·25 and a scanning slit of 2 μm width. Plot the MTF curves for the objective and slit, and for the combination, assuming that the illumination is effectively incoherent and that $\lambda = 500$ nm. Determine the spatial frequency at which the total MTF falls to 0·80.

(6) Show that for the linear microdensitometer represented in Fig. 7, the measured output is given by the convolution of the intensity transmittance of the film and the intensity point spread function of the condenser lens.

9. MICRODENSITOMETRY

(7) A microdensitometer is to be used to examine photographic images up to 250 cycles/mm. What values would you select for the numerical apertures of the condenser and objective, and for the field and scanning slit widths?

(8) In a microdensitometer the x, y adjustment of the sample may be positioned on top of, or below, the rotational adjustment. Which of these two systems is preferable in practice?

Suppose that the rotational alignment is set visually on the image of a slit 1×100 μm at a magnification $\times 100$. What must the minimum angular sensitivity of the rotation adjustment be to make full use of the resolution of the eye (approximately 0·1 mm at 25 cm)?

(9) It is found in practice that the imaging quality of a microdensitometer depends on the numerical aperture of the condenser (for a given objective). Discuss possible reasons for this.

10. Image Assessment by Information Theory

10.1 The Information Theory Approach

Introduction

The theorems around which the general name information theory has been adopted date back to 1948 and are due to Shannon[1], although the need for such a theory was discussed prior to this by Hartley[2], who also indicated how information might be measured on a logarithmic scale. These theorems were developed largely within the context of electrical communication channels, but can readily be adapted to any other type of communication system, as in the optical transmission or photographic recording of information.

In some ways the name information theory is unfortunate, since the term information is generally used in an empirical and subjective way, while the major emphasis of Shannon's theorems is that information can be given a precise and unique scientific measure. For this reason the names communication theory and entropy theory are sometimes used: the former as used by Shannon and for reasons which are obvious; the latter due to an interesting similarity between the measure of information (or the "state of disorder of knowledge") and the measure of entropy (or state of molecular disorder) in statistical thermodynamics. Whereas the name used is of little importance, such a general term can give the impression that the theory concerns subjective quality judgements, but in most applications it leaves off where human interpretation takes over. For example, it might be used in the design of a telephone system optimized for the transmission of typical speech forms, but not to measure the utility of messages spoken over the system, unless a purely symbolic and mathematical language were to be used.

Although the object of Shannon's work was to meet the immediate practical needs of communication engineers, subsequent examination of the theorems by Gabor[3], Brillouin[4] and others has demonstrated that it was more or less inevitable that Shannon arrived at his particular measure of information. This is because at a conceptual level the Shannon analysis of information and the definition of thermodynamic entropy have common origins, and differ only by choice of units to suit the application. The scale factor adopted for entropy in thermodynamics involves Boltzmann's constant, 1.38×10^{-23} J/°K, and entropy is then expressed in units of Joules per degree. For convenience the scale factor adopted for information is the "bit", which is the amount of information conveyed by the answer to a yes-or-no question when both answers had equal *a priori* probability.

This close relationship between thermodynamic entropy and information has been considered in general terms by Tribus and McIrvine[5], with some interesting and important implications for photographic systems. Considering a single molecule as the simplest thermodynamic system to which information theory can be applied, and by assuming that it has two states of equal probability, leads to the conclusion that the smallest entropy change associated with the acquisition of one bit of information is approximately 10^{-23} J/°K. Calculations for the energy/information ratio for practical systems based on character or digitally-encoded messages show that this varies from a few joules per bit to a few hundreths of a joule per bit, while much lower ratios are relevant for audio or graphic systems. Estimates of Tribus and McIrvine for pictorial systems are shown in Table I.

TABLE I. Information and energy estimates for various pictorial activities.[5]

Pictorial information activity	Information content (Bits)	Energy (Joules)	Energy (Joules per bit)
Telecopy one page (telephone facsimile)	576 000	20 000	·03
Projection of 35 mm slide (one minute)	2 000 000	30 000	·02
Copy one page (xerographic copy)	1 000 000	1 500	·002
print one high quality opaque photographic print (5″ × 7″)	50 000 000	10 000	·000 2
Project one television frame (1/30 s)	30 000	6	·000 02

Relevance to Photography

This brings us in a natural way to the question of why it is relevant to apply information theory to photographic systems and processes. It is estimated that the total amount of printed information alone is in excess of 10^{16} bits per year, and that this figure doubles about each decade. With numbers such as these involved it is important that the problems of storing and retrieving information on such a scale should be analysed from first principles as a basis for the technological solutions. Maimon[6] has presented interesting data in this respect, as prepared according to technology forecasting principles used by the U.S.A. Post Office. This shows the large number of possibilities for graphic data systems and devices, microforms and related equipment, which may result from the invention of new technologies or the development of existing ones during the 1970s. It is clear from studies such as these that the efficient storage of information by imaging processes, and objective evaluation of their capacities is becoming increasingly vital.

A similar exercise has been carried out by Nakamura and Sakata[7], who surveyed the potential information storage capacities of various technologies in view of predicted needs for the year 1990, and analysed the relative merits of existing types of conventional and unconventional photographic processes. Pierce[8] has also given similar general considerations to the information-handling capabilities of existing and predicted communication systems, during the course of a clear introductory description of the concepts of information theory.

In view of these present and future large-scale information-handling problems, the need for a fundamental analytical approach as provided by information theory is self-evident. There are also other classes of problem in applied photography which can benefit from the information theory approach. These come under the general description of experimental design in precision information-collecting experiments where the photographic process is used as the recording medium, and where it is important to achieve the highest information recording rate. Here the problem is often one of coding: the statistical structure of the incoming signal is fairly well known, and the question becomes one of how best to present the signal to the photographic recording element. As we shall see, it may then be desirable to match the Wiener spectrum of the signal to the spatial frequencies in the photographic image which yield the highest information capacity and rate: these frequencies will be determined by the MTF and the Wiener spectrum of the noise. This approach demonstrates the very close relationship between information capacity and recording rate, and DQE, and extends the principle discussed in Chapter 5 whereby input and output statistics should be matched in the DQE sense.

The original papers of Shannon were published in text-book form in 1949, with additional contributions by Weaver[9]. The decade of the 1950s saw the origins of the application of information theory to optical rather than electrical communication systems[10-14]. In particular, Fellgett and Linfoot[11] laid the foundations for the analysis of optical systems with the influence of noise included. Such work led the way for detailed considerations of the photographic process as a communication channel, and in the 1960s a series of papers followed concerning photographic assessment by the information theory approach. Some of these papers will be discussed later in this chapter, but first there follows an introductory account of the main theorems of Shannon. This is necessarily simplified and condensed, the intention being to stress the general principles and results which are useful in the context of imaging processes. A simple introduction to information theory has been given by Woodward[15].

10.2 Shannon's Theorems

Measure of Information

Suppose that a communication channel of unspecified nature is capable of transmitting only 27 distinct symbols, one at a time, consisting of the 26 letters of the alphabet plus a word-space. The addition of a second channel transmitting independently and in parallel with the first one would permit 27 × 27 = 729 combined symbols to be transmitted, since each symbol over the first channel could be combined with each over the second. Although the number of symbols is squared, intuition tells us that the information rate would only be doubled. Reasoning along these lines leads to Hartley's original proposal[2] for the information capacity of a communication channel as the logarithm of the number of distinct symbols, n, capable of transmitting:

$$\text{information capacity} = \log n. \tag{1}$$

In practice the probability of occurrence of a message is usually of more significance than the possible number of distinct messages. If the probability of one of the 27 symbols is denoted by p, then Shannon demonstrated that the information gained when the symbol occurs is defined by:

$$\text{information gain} = \log \frac{1}{p}. \tag{2}$$

In the special case where all 27 symbols are equally probable,

$$p = \frac{1}{27} = \frac{1}{n},$$

and the information gained on receiving one of them will be

$$\log \frac{1}{p} = \log 27 = \log n.$$

However, this equality will not hold in general, and this is why the log n measure of information is inadequate.

If logs are taken to base 2, a convenient unit of information follows, for if the probability of a symbol occurring is $p = \frac{1}{2}$ the associated information is then $\log_2 2 = 1$. Thus the amount of information yielded by a yes or no response when both had equal *a priori* probability is 1 unit. This again fits is well with intuitive ideas, and such a binary scale needs no elaboration to scientists. The unit of information is then called the "bit", which is an abbreviation of binary digit, although it will be shown that a binary digit only conveys one bit of information in the special case where each binary state had equal *a priori* probability. The units of information on a decimal scale will be equivalent to $\log_2 10 = 3.32$ bits on a binary scale, and the two scales can be easily converted.

Returning to our example, the information gain on receiving one of 27 equiprobable symbols would be $\log_2 27 = 4.76$ bits. In general however the letters of the alphabet do not occur with equal probability, as is evident in the written form of the English language. Also the information associated with sequences of symbols (words, sentences, complete messages) is often more relevant than for a single symbol. Further analysis is necessary to allow both for the different symbol probabilities and for correlations in sequences of symbols.

The Zero-Memory Information Source

Suppose a source S emits a sequence of symbols of a given alphabet, where the symbols occur according to a fixed probability law, and for which there is no statistical correlation between successive symbols. Such a so-called zero-memory source is completely described by the source alphabet,

$$S_1, S_2, S_3, \ldots, S_i, \ldots, S_n;$$

and the probabilities of occurrence, denoted by

$$p(S_1), p(S_2), p(S_3), \ldots, p(S_i), \ldots, p(S_n).$$

If the symbol S_i is received, the associated information gain will be $\log_2 p(S_i)$, and since the probability of this symbol occurring is $p(S_i)$, the *average* information gain per symbol, H(S), will be:

$$\mathrm{H}(S) = \sum_i p(S_i) \log_2 \frac{1}{p(S_i)},$$

where the summation is taken over the entire source alphabet. This last equation is usually written as

$$\mathrm{H}(S) = -\sum_i p(S_i) \log_2 p(S_i). \tag{3}$$

The quantity H(S) can be interpreted either as the average amount of information per symbol provided by the source, or the average uncertainty (or missing information) of the observer before receiving the source output, and is referred to as the *entropy* of the source.

Shannon gave a basic proof for the inequality

$$\mathrm{H}(S) \leqslant \log_2 n,$$

with the equality holding only in the special case:

$$p(S_1) = p(S_2) = p(S_3) = \ldots = p(S_n) = \frac{1}{n}.$$

So for a zero-memory source with an n-symbol alphabet the maximum value of H(S) will be when all the symbols are equiprobable.

The simplest example of a zero-memory source is a binary source with respective probabilities p and $1 - p$ for the binary states. This source will have entropy:

$$\mathrm{H} = -p \log_2 p - (1 - p) \log_2 (1 - p).$$

If this entropy is plotted as a function of p, then a curve such as shown in Fig. 1 results. H reaches a maximum of 1 when both symbols are equiprobable, and is zero when either of them is absolutely certain. This demonstrates the difference between a binary digit and 1 bit of information. If $p = \frac{1}{2}$ prior to observation and $p = 1$ after observation (i.e., one of two states of equal *a priori* probability has been observed), then H has changed from 1 to 0 and has decreased, while 1 bit of information has been gained. Decrease in entropy on observation is thus equivalent to gain in information.

The symbol frequencies of written English can be estimated from the analysis of large samples of typical text. This gives approximate probabilities of 0·2 for the word space; 0·1 for E; 0·05 for N, I, R and S; and so on. The least common symbols are J, Q and Z, with probabilities round 10^{-4}. By carrying out the summation process of equation (3) for the whole alphabet, the entropy is estimated as around 4 bits per symbol, compared with the value of 4·76 bits for 27 equiprobable symbols.

A typical message generated by such a zero-memory source with appropriate typical probabilities for the symbols would be: CGRAI EFO RSIUPBRN

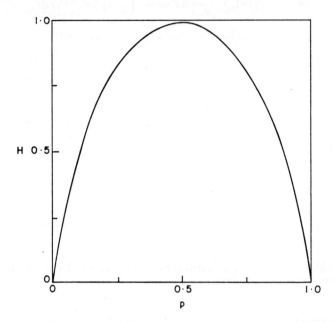

FIG. 1. Entropy of a zero-memory binary source as a function of symbol probability.

SDAEENURI TLAO ELTRTM RE HBE GXUNTWEFY. This is still not recognizable because the practical correlations between successive symbols have not been taken into account. However, before introducing the effects of such correlations, a more general definition of entropy is given for continuous functions rather than discrete symbols.

When the information is in the form of continuous functions the approach is very similar to that above. Consider for example a physical quantity, x, for which the initial knowledge is represented by a continuous probability distribution, $p_1(x)$. After gaining additional information about x, the new state of knowledge is represented by a new probability distribution, $p_2(x)$. These before-and-after situations are represented in Fig. 2.

By dividing the x-range into small intervals and using a limiting process, the summations in the discrete case may be replaced by integrations in the continuous case. The gain on receiving the information will be represented by the change in entropy in going from the initial to the final state, i.e.,

$$\text{information gain} = \int p_1(x) \log_2 \frac{1}{p_1(x)} \, dx - \int p_2(x) \log_2 \frac{1}{p_2(x)} \, dx.$$

So the information gain is measured by the decrease in H, where

$$H = \int p(x) \log_2 \frac{1}{p(x)} \, dx = - \int p(x) \log_2 p(x) \, dx. \tag{4}$$

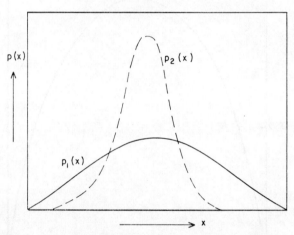

FIG. 2. Probability distributions representing the states of information of a physical quantity, x, before and after observation.

H is now called the continuous entropy of the observed quantity relative to the coordinate system x. This more general interpretation is of significance here, since in applied photography the signals to be recorded are often in the form of continuous intensity distributions.

Markov Information Sources

The zero-memory source of discrete symbols is inadequate to deal with the situation where correlations exist between successive symbols and sequences of symbols. Markov sources are then appropriate, and an mth-order Markov source is one for which the occurrence of a symbol depends on a finite number, m, of preceding symbols, but which is completely independent of those occurring $m + 1$ symbols (or more) previously.

The probability of the ith symbol, S_i, occurring after some particular sequence of m previous symbols, can now be denoted as $p(S_i/m)$. This conditional probability will have a defined value for each possible previous sequence of m symbols, and similar conditional probabilities will be defined for all the other S symbols. Thus the entropy of the source will be obtained by summing for all symbols over all previous m sequences:

$$H(S) = \sum_m \sum_i p(m, S_i) \log_2 \frac{1}{p(S_i/m)}$$

$$= \sum_{m+1} p(m, S_i) \log_2 \frac{1}{p(S_i/m)}, \quad (5)$$

since summing over all m and i for sequences of m followed by S_i, denoted by (m, S_i), is equivalent to summing over all possible $m + 1$ sequences.

Shannon demonstrated that

$$\sum_{m+1} p(m, S_i) \log_2 \frac{1}{p(S_i/m)} \leqslant \sum_i p(S_i) \log_2 \frac{1}{p(S_i)},$$

with the equality occurring only when the probability of a symbol is completely independent of all previous ones; i.e., when the source reduces to a zero-memory one. The result of equation (5) is in fact restricted to "well-behaved" alphabets and symbol transition probabilities, such that the ergodic property applies to generated messages. This is of little practical concern here, but for example the analysis would not apply if it were possible for a sequence to become "stuck" in any one state, or when the transitional probabilities do not have well-defined limiting values.

The analysis of English language as a Markov source is simple in the first-order case, which involves only the frequency of occurrence of letter pairs. Using the approximate probabilities for each possible pair and applying equation (5) gives a value of about 3·3 bits per symbol. The statistics for higher order cases are not so easy to calculate, and there is little available data. However whole words can equally be considered as the basic source symbols as can letters. Assuming a dictionary of say 10^4 words and an average word length of 5·5 letters, leads to a value of $\log_2 10^4 = 12·7$ bits per word, which is equivalent to 12·7/5·5 = 2·3 bits per letter. This is for a zero-order word source where all words have equal probabilities of occurrence, and allowing for the probabilities observed in typical texts reduces this value to about 2 bits per letter.

The correlation between adjacent words can also be considered in the mth-order sense. A typical second-order word sequence is: THE NEXT ONE OF THOSE WILL PERHAPS GO TO SEE WHAT BETTER THAN MANY ALTERNATIVE THEORIES FOR EVER. A limiting value of around 1·6 bits per letter has been estimated for the information associated with typical text, as compared with the value of 4·76 bits for the zero-order case with equal probabilities. It is thus concluded that only about one-third of the possible information is achieved in the practical case, and in effect the information in every message is conveyed three times over. However such so-called redundancy can be an advantage (and even sometimes a necessity) when the message is perturbed by noise during transmission, as for example during a telephone conversation. However, when there is no appreciable noise present it must be appreciated that the information capacity of say an information storage system which has been optimized for printed text will be under-utilized by a factor of three. This factor could in principle be removed by re-coding to eliminate the redundancy, and in many applications of information theory the difficult problems are those involving coding rather than calculations of information rates and capacities.

Devising minimum redundancy codes for picture and alphanumeric storage has been discussed by Rosenfeld[16] and Andrews[17] among others. The general principle of using redundancy to combat noise can be an important one when noise is present.

Communication Channels

An overall representation of the process of communicating, from original source to final destination, is shown in Fig. 3. Attention is now turned to the

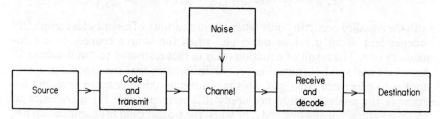

FIG. 3. General representation of a communication process.

channel over which the suitably-coded message is to be transmitted prior to reception and de-coding. All practical communication channels impose their own restrictions on the messages which can be transmitted over them.

Suppose that a string of symbols is coded and transduced into a time series for transmission over a communication channel. If M(T) represents the number of distinct messages which can be transmitted over the channel in time T seconds, Shannon demonstrated that a measure of channel capacity is given by

$$C = \lim_{T \to \infty} \frac{\log_2 M(T)}{T} \text{ bits/second.} \qquad (6)$$

If the entropy of the source is denoted by H bits per symbol, and if m is the average number of symbols per second produced by the source, then H' = m H, where H' measures the average amount of information from the source per second. Shannon's theorem for a noise-free channel states that for a source with entropy H bits per symbol and a channel with a capacity C bits per second, it is possible to encode the source output in such a way as to transmit over the channel at the average rate of C/H − r symbols per second, where r can be arbitrarily small.

It is not possible to transmit at an average rate greater than C/H, and this converse part of the theorem follows readily from intuitive reasoning. The entropy of the source cannot exceed the channel capacity, and hence C ⩾ H', so the number of symbols per second, m = H'/H, will be governed by the inequality C/H ⩾ H'/H.

The source as seen through the transducer which performs the coding should have the same statistical structure as the source which maximiset the entropy of the channel. The theorem then states that whereas an exact match is not generally possible, it can be approached as closely as desired. This leads to the definition of the coding efficiency α as the ratio of the actual rate of transmission to the capacity, C, i.e.,

$$\alpha = \frac{\text{actual rate}}{C},$$

and the redundancy β is then defined as $\beta = 1 - \alpha$. For example, for non-coded text as estimated above, $\alpha = \frac{1}{3}, \beta = \frac{2}{3}$.

Although the general problems of coding are complicated, rules of thumb are fairly simple to define. For example, so far as time series are concerned a good rule-of-thumb is that the logarithm of the reciprocal probability of a symbol or message must be proportional to the duration of the corresponding transmitted signal (i.e., a frequently-occurring symbol should occupy a short time interval, as exemplified by Morse code).

The Influence of Noise

Communication channels usually introduce noise to the message during transmission, and the received message will not then necessarily be identical with that originating from the transmitter. In this context it is important to distinguish between distortion and noise. The former is the name given to the effect whereby a particular transmitted signal always produces the same received signal. If this distortion has an inverse (i.e., no two different transmitted signals produce the same received message), then it may be corrected by performing the inverse process on the received message. The term noise is retained for those effects whereby the signal does not normally undergo the same predictable change during transmission. In photographic terms the MTF represents distortion while the Wiener spectrum represents noise.

If a received signal E is a function of the transmitted signal S and a further noise variable, N, then $E = f(S, N)$. The noise is usually a chance variable governed by the statistical processes at work during transmission, so there are two statistical processes involved in transmission; those of the signal and those of the noise. When noise is present it is not generally possible to reconstruct the original message from the received signal with certainty, so there will be missing information which can be measured by the entropies associated with signal and noise statistics.

If the entropy of the source, or input, is $H(x)$, and the entropy of the output is $H(y)$, then in the noiseless case $H(x) = H(y)$. When noise is present there are two conditional entropies, $H_x(y)$ and $H_y(x)$, namely the entropy of the output when the input is known, and *vice versa*. The joint entropy is defined as

$$H(x, y) = H(x) + H_x(y) = H(y) + H_y(x), \quad (7)$$

and the rate of transmission of information will be

$$R = H(x) - H_y(x); \quad (8a)$$

i.e., the amount of information sent minus the missing information. The conditional information sent is called the equivocation, and it measures the average ambiguity in the received message. This situation is illustrated in Fig. 4, from which it is seen that R is also defined by:

$$R = H(y) - H_x(y) = H(x) + H(y) - H(x, y). \qquad (8b)$$

A direct intuitive interpretation of equivocation is as follows. Suppose that an external observer can see both the transmitted and received signals. This observer notes the errors in the received message and transmits sufficient data to the receiver over a correction channel to enable the received signal to be corrected. From the previous theorem it will be possible to code the correction data to send it over the correction channel and correct all but an arbitrarily small fraction of errors, if the correction channel has a capacity equal to

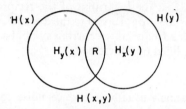

FIG. 4. Illustration of the relationship between input and output entropies, and the information rate, R.

$H_y(x)$. It will not be possible if the correction channel capacity is less than $H_y(x)$.

The channel capacity, C, will be given by the maximum rate of transmission which will occur when the source is properly matched to the channel, and we may write this as

$$C = \text{Max}\,(H(x) - H_y(x)). \qquad (9)$$

The maximization is for all possible sources which might be used as the input to the channel.

It may seem surprising that a definite capacity can be ascribed to a noisy channel, but the significance of the channel capacity is that by proper encoding it is possible to send information at the rate C with as small a frequency of errors as required. The problem of encoding thus becomes all-important. An approximation to the ideal would have the property that when the signal is perturbed by noise in a typical manner the original signal can still be recovered. In other words the perturbation by noise will not bring it closer to any other normal signal than the original one. This can be accomplished at the cost of a certain degree of redundancy in the coding, and this redundancy must be introduced in the best way to combat the noise statistics that are involved. As we have already noted, redundancy that is already present in the source can be helpful in this respect. For example, the redundancy in printed text allows a considerable amount of noise in a channel over which it is transmitted, without any significant loss of information.

Shannon's theorem for the case with noise expresses the maximization of equation (9) in the form

$$C = \lim_{T \to \infty} \frac{\log_2 M(T, r)}{T}, \qquad (10)$$

which demonstrates the close connection with the result for the noiseless case of equation (6). $M(T, r)$ represents the maximum number of signals such that the probability of an incorrect interpretation is less than or equal to r. In other words, no matter how the limits of reliability are chosen, about CT bits can be transmitted in T seconds when T is sufficiently large.

Suppose that we are concerned with a continuous function such as a time series, and that transmitted and received signals are limited to a certain frequency bandwidth, Δf. By the sampling theorem as discussed in Chapter 6, such a function is completely determined by its ordinates at discrete points $1/2\Delta f$ seconds apart, so in time T the function can be specified by $2T\Delta f$ coordinates, and thus is said to have $2T\Delta f$ degrees of freedom in time T. By analogy with equation (4) the entropy of an n-dimensional continuous source can be written as

$$H = -\lim_{n \to \infty} \frac{1}{n} \int\int \cdots \int p(x_1, x_2, \ldots, x_n) \log_2 p(x_1, x_2, \ldots, x_n)\, dx_1, dx_2, \ldots, dx_n, \qquad (11)$$

and this defines the continuous entropy relative to the coordinate system $x_1, x_2, \ldots, x_n$. Thus H is the entropy per degree of freedom, and we can also define the entropy per second, H', by dividing by the time T necessary to transmit n samples, and since $n = 2T\Delta f$,

$$H = 2\Delta f H'. \qquad (12)$$

Denoting the probability associated with the transmitted signal by

$$P(x) = p(x_1, x_2, \ldots, x_n),$$

and the noise by the conditional probability

$$P_x(y) = p_{x_1, x_2, \ldots, x_n}(y_1, y_2, \ldots, y_n),$$

then by analogy with the discrete channel

$$R = H(x) - H_y(x).$$

The channel capacity is then defined as the maximum rate when $P(x)$ is varied to maximise R, and by consideration of the conditional entropies it may be shown that this leads to the general expression:

$$C = \lim_{T \to \infty} \mathop{\text{Max}}_{P(x)} \frac{1}{T} \int\int P(x, y) \log_2 \frac{P(x, y)}{P(x)P(y)}\, dx\, dy. \qquad (13)$$

Power-Limited Signals, Gaussian Noise

Suppose the noise is additive and independent of the signal. In this case the equivocation will be a function only of the difference $z = y - x$, and we can express the noise as

$$P_x(y) = Q(y - x),$$

and an entropy $H(z)$ of the distribution $Q(z)$ can be assigned to the noise. In this case the information rate will be:

$$R = H(y) - H(z),$$

i.e., the entropy of the received signal less the entropy of the noise. Since $H(z)$ is independent of $P(x)$, maximizing R implies maximizing the entropy of the received signal, $H(y)$.

If there are physical constraints on the transmitted signals, then the entropy of the received signals must be maximized subject to these constraints. Suppose for example the noise is white gaussian, and that the transmitted signals are limited to an average power $P + N$, where N is the average noise power. To achieve the maximum information rate it is necessary to maximize the inegral

$$-\int p(x) \log_2 p(x) \, dx,$$

subject to two conditions. The first is by definition

$$\int p(x) \, dx = 1;$$

and the second is based on the fixed mean-square value:

$$\int x^2 p(x) \, dx = \sigma^2.$$

A solution may be obtained by the use of "undetermined multipliers" (see for example page 25 of Woodward[15]), and leads to the expression:

$$p(x) = \frac{1}{\sigma\sqrt{2\pi}} e^{-\frac{x^2}{2\sigma^2}}. \tag{14}$$

Substitution in the integral equation for entropy leads to:

$$H(x) = \log_2(\sigma\sqrt{2\pi e}). \tag{15}$$

Thus maximum entropy for the received signal occurs when it also has a gaussian form, and for a given mean-square value of x the gaussian distribution can be said to be the most random of all in this sense.

The entropy per second will be

$$H' = 2\Delta f H = \Delta f \log_2(2\pi e \sigma^2). \tag{16}$$

Since

$$\sigma^2 = P + N,$$

the entropy of the received message will be

$$H(y) = \Delta f \log_2 (2\pi e(P + N)) \text{ bits/second},$$

and similarly the noise entropy will be

$$H(z) = \Delta f \log_2 (2\pi eN) \text{ bits/second}.$$

The channel capacity will thus be

$$C = H(y) - H(z) = \Delta f \log_2 \frac{P + N}{N} \text{ bits/second}. \tag{17}$$

This is one of the most important of Shannon's theorems, and it signifies that with appropriate coding binary digits can be transmitted over the channel at this rate with an arbitrarily small frequency of errors.

To approximate to this limiting rate of transmission the transmitted signals would need to have the statistical properties of white noise. Suppose for example that at the transmitter the message is coded into one of M white noise samples. At the receiver the received message is tested against each of the M samples, and the sample with the least rms discrepancy is chosen as the transmitted signal, and de-coding then gives the transmitted message. The number M will depend on the tolerable frequency of errors, r, but in general the rate will be

$$\lim_{r \to \infty} \lim_{T \to \infty} \log_2 \frac{M(T, r)}{T} = \Delta f \log_2 \frac{P + N}{N},$$

so that no matter how small r is chosen, by making T sufficiently large it is possible to transmit, as nearly as required, $T\Delta f \log_2 ((P + N)/N)$ bits in time T.

This solution for a mean-square, or "power", constraint and white gaussian noise represents one of the simplest of the continuous-channel cases. Shannon considered other types of noise statistics and other constraints, (e.g., a peak-signal limitation which may correspond more closely to the minimum and maximum density limitations of the photographic process), but the solutions are not so exact and straightforward and these may be referred to in Shannon's original papers.

10.3 Photographic Applications: Discrete Signals

Information Storage Cells

Although the preceding analysis has demonstrated that the problems of optimizing information systems may involve both practical and theoretical difficulties, the problems may reduce to a much simpler form in some applications. When a specified type of signal (for example, in the form of binary digits) is to be stored photographically with highest information storage per unit area, then simplified models of the photographic process as a storage medium may prove adequate. Altman and Zweig[18] gave a method of analysis based on a unit storage cell in the image, using a simple model for the influence of noise due to Levi[19].

The model assumes that the photographic image consists of a mesh of individual cells each of area A, and in each of which information is stored in the form of one of a number of distinguishable density levels. Since each density level will be subject to uncertainty due to image noise, these levels must be separated by a sufficient density interval according to some practical criterion, and the question arises as to how the number of recording levels, M, may be determined. Suppose that $2k$ represents the number of standard deviations by

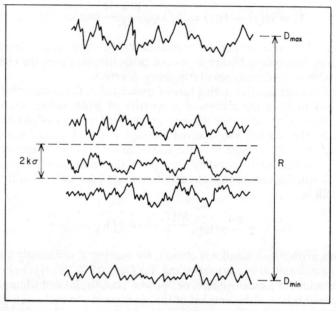

FIG. 5. Illustration in one-dimension of the available density range and separation of recording levels.

which adjacent levels are separated, and that $R = D_{max} - D_{min}$ represents the total available density range for recording, as illustrated in Fig. 5. Assuming that R is a multiple of $2k\sigma_A$, it follows that

$$M = \frac{R}{2k\sigma_A} + 1. \tag{18}$$

In this equation σ_A denotes the mean-square image density fluctuation as measured with a scanning aperture A. To make it possible to express equation (18) in such a simple form, it has been assumed that σ_A is the value measured at, say, mid-way in the range R, since in fact σ_A will generally increase with density level. This assumption means in effect that the level-separation criterion will be met with greater ease than suggested by the adopted value of k when the density is low, and with less ease when high.

If the ratio $R/2k\sigma_A$ is not an integer, equation (18) will produce a value of M which must be rounded down to the integer below (e.g., if $R/2k\sigma_A = 7.75$,

10. IMAGE ASSESSMENT BY INFORMATION THEORY

then since $7{\cdot}75 + 1 = 8{\cdot}75$, it follows that 8 recording levels would be appropriate).

Assuming the cell-size A is big enough for the relation $A\sigma_A{}^2 = G$ to hold, with G identical to the low-frequency value of the Wiener spectrum of the noise as investigated in Chapter 8, it follows that

$$M = \frac{RA^{\frac{1}{2}}}{2kG^{\frac{1}{2}}} + 1. \tag{19}$$

If the information is stored in such a way that each of the recording levels is

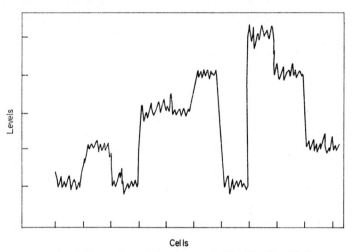

FIG. 6. One-dimensional illustration of signals recorded in 10 cells with 5 storage density levels.

equiprobable, a typical example being shown in Fig. 6, then the information capacity will be $\log_2 M$ bits per cell. If there are N cells per unit area, the information capacity per unit area will be

$$C = N \log_2 M. \tag{20}$$

Since $N = A^{-1}$, combining equations (19) and (20):

$$C = \frac{1}{A} \log_2 \left(\frac{RA^{\frac{1}{2}}}{2kG^{\frac{1}{2}}} + 1 \right). \tag{21}$$

The parameters R and G may be assumed to be constants for a given photographic process, and so for a specified separation criterion, k, only the variable A remains at our disposal, and equation (21) can be written in the form:

$$C = \frac{1}{A} \log_2(cA^{\frac{1}{2}} + 1), \tag{22}$$

where c is a constant. Investigation of equation (22) reveals that C increases as A decreases, so A should be as small as possible for maximum information capacity. However, at least two recording levels are necessary, so we conclude that in principle binary recording will give optimum information capacity. From equation (21) the cell area for binary recording will be defined by

$$A = \left(\frac{2k}{R}\right)^2 G. \qquad (23)$$

Whether or not binary recording is possible will depend on the practical conditions associated with the use of a cell size as defined by equation (23). This will depend on the size of the spread function, for if A is less than the area of the spread function there will be "blurring" of information into adjacent cells. So it is reasonable to specify that A must be of the same order or greater than the spread function.

Binary and Multilevel Recording

In order to make practical estimations of the information capacity according to this simple model, it is necessary to specify the criterion, k, for the separation of recording levels. This cannot be done outside the practical context of a given type of information storage-and-retrieval problem, and will depend on the practical accuracies associated with read-in and read-out. If the limiting factor is the criterion for distinguishing between the adjacent levels during read-out, then as pointed out by Altman and Zweig[18], a separation of $\pm 5\sigma_A$ would give an error rate of about 1 in 10^6 assuming the image noise fluctuations are normal-distributed. An error rate of 1 in a million might be tolerable for most applications, but to allow further practical tolerance a separation of $\pm 10\sigma_A$ should be reasonable. It is noted that doubling the separation will approximately half the number of recording levels, and hence reduce the information capacity of each cell by 1 bit.

As an example suppose that $k = 10$ and the available density range is $R = 2$ density units. From equation (23)

$$A_2 = 100\ G,$$

where the suffix 2 denotes binary recording. From equation (21) we deduce that the minimum possible cell sizes for multilevel recording would be:

$$A_3 = 400\ G;\ A_4 = 900\ G, \text{ etc.}$$

The procedure to determine the optimum storage of information must now take into account the size of the spread function as measured experimentally. The cell-size might be defined as the area for which the point spread function is reduced to, say, 10% of its peak value. This would give some information "cross-talk" between adjacent cells, but this need not be significant when the criterion for level separation is based on a comfortable tolerance. This area is then compared with the appropriate values of A_2, A_3, A_4 etc, such as calculated in the above example. The next value (A_M) above the spread function area is then chosen for the cell-size, and multilevel recording based on M is appropriate. It might happen that $M - 1$ level recording with smaller cell-size, equal to the spread function area, would have approximately the same storage capacity, as shown by a continuation of the above example.

10. IMAGE ASSESSMENT BY INFORMATION THEORY

Suppose that $G = 1\ \mu m^2 \times D^2$, as for a typical medium speed film at medium image density. In this case:

$$A_2 = 100\ \mu m^2,\ A_3 = 400\ \mu m^2,\ A_4 = 900\ \mu m^2,\ \text{etc.}$$

If the practical spread function defines a lower limit of $15 \times 15 = 225\mu m^2$ for the cell-size, then by equation (20) 3-level recording would give a storage capacity of

$$C = \frac{10^8}{400} \log_2 3 = 0.42 \times 10^6\ \text{bits cm}^{-2}$$

If on the other hand binary recording were used at spread function cell-size, the capacity would then be

$$C = \frac{10^8}{225} \log_2 2 = 0{\cdot}44 \times 10^6\ \text{bits cm}^{-2},$$

which is marginally higher than for 3-level recording.

As a further example we consider a fine-grained film with the same available density range but with $G = 0{\cdot}05\ \mu m^2 \times D^2$. In this case

$$A_2 = 5\ \mu m^2,\ A_3 = 20\ \mu m^2,\ A_4 = 45\ \mu m^2,\ A_5 = 90\ \mu m^2,\ A_6 = 125\ \mu m^2,\ \text{etc.}$$

Assuming a spread function of $10 \times 10 = 100\ \mu m^2$, then 6-level recording would be associated with

$$C = \frac{10^8}{125} \log_2 6 = 2{\cdot}1\ \text{bits cm}^{-2},$$

while binary recording would permit a capacity of

$$C = \frac{10^8}{100} \log_2 2 = 1{\cdot}0\ \text{bits cm}^{-2}.$$

The general conclusion is that the gain in information capacity by using multilevel rather than binary recording is not a substantial one. It follows from equation (20) that an increase in the number of cells, N, gives a linear increase in capacity, while an increase in M gives only a logarithmic increase, and therefore a small spread function is the most important factor. A spread function allowing a storage area of $1 \times 1\ \mu m^2$ as opposed to $100 \times 100\ \mu m^2$ would give an increase in capacity of 10^4. Whereas binary messages and codes are commonplace, recording levels higher than two may involve coding complications and difficulties. For all these reasons multilevel recording may offer little practical advantage over the binary case. This conclusion is confirmed by the results of Altman and Zweig[18] for a series of Kodak films which are summarized in Table II.

It is seen from Table II that the highest M-value for this wide range of film-types is only 8, representing at the most a loss of only $\log_2 8 = 3$ if binary recording is always used. This data emphasizes the benefit of a small spread function, since the film with a spread function as low as $1\ \mu m$ in diameter has much the highest storage capacity. There may of course be practical difficulties

TABLE II. Information storage parameters for a series of film types[18]. The M-levels were determined according to a 10σ criterion.

	Spread function diameter μm	Available levels M	$\log_2 M$	Bit capacity for an image area of 10 × 10 μm².		
				M-level	Binary	Experimental
Kodak Fine Grain Cine Positive	—	8	3	0·33	0·11	—
Recordak Fine Grain Type 5454	12·5	6	2·6	1·60	0·64	1·1
Recordak Fine Grain Type 7456	15	4	2	0·88	0·44	1·1
Kodak Plus-X	—	3	1·6	0·33	0·21	—
Kodak Pan-X	15	3	1·6	0·70	0·44	0·50
Kodak Royal-X Pan	27	2	1	0·14	0·14	0·05
Kodak High Resolution Type 649	1	2	1	160	160	160

associated with the quality of the optics for image read-in and read-out if such a small cell-size is used, and even maintaining the film free from dust and abrasions (which could lead to a much greater error-rate than predicted by the nominal level-separation criterion) may then pose a severe practical problem.

The final column of Table II shows the capacity as determined experimentally by Altman and Zweig, using a binary grid pattern. The theoretical and experimental results are quite close, the biggest discrepancy being for the most coarse-grained film of the series, Royal-X, where a cell-size more than twice the area determined by the spread function was found necessary even for binary recording. It is probable that this discrepancy is at least partly due to the simplification of assuming a constant value for the noise level, G, when making theoretical predictions. For binary recording it may be more appropriate to insert the value of G at maximum density in equation (19), and for coarse-grained films this value may be very high.

Altman and Zweig gave pictorial illustrations of binary-coded patterns having marginally-acceptable quality, and these demonstrated the advantages of a small spread function and low noise in a convincing manner. It must be stressed however that the analysis and conclusions for this simple cell model relate to the maximum information capacity as expressed in terms of bits per unit image area. So far no account has been taken of any exposure criterion for information read-in. For example it might be that in some practical applications the speed of Royal-X would outweigh its disadvantages of capacity. To calculate this it would be necessary to specify the exposure criterion, but it is simple to convert, using the input/output characteristics, the information in bits cm^{-2} to that in bits s^{-1}, or bits erg^{-1}, or even bits $cm^{-2} s^{-1}$, as practice demands. However in straightforward applications of information storage the packing-density is usually the most important criterion, along with that of access of read-out. This will not usually be the case in the application of photography to classes of experiment where the information collection-rate must be optimized, as will be discussed later in this chapter.

Problems of Coding

In considering Shannon's theorems, and as demonstrated by the above simple analysis and examples, the problem of coding the information is an important one, and may be the most difficult operation involved in the overall communication chain. The full information capacity will only be achieved if the information is coded in the optimum manner according to entropy considerations. Even with binary coding the distinction must be made between a binary digit and a bit, as discussed earlier. For non-optimum coding the information will be reduced even further than is the case when cell-size and recording levels are not chosen according to the full practical possibilites. For these reasons it may be useful to distinguish between the *information capacity*, as the limiting value with optimum coding and utilization of imaging parameters, and the *information content*; the latter signifying what is actually achieved under all the practical conditions of use.

The problem of coding was taken a practical stage further by Altman and Zweig, in a comparison between binary and alphanumeric data storage. For this they compared the images of letters 1·5 μm high, and binary coding with a density of about 40×10^6 bits cm^{-2}. Since the average bit-value for written text with normal redundancy has been estimated as about 1·6 bits per letter, this corresponds to a bit capacity of about 40×10^6 bits cm^{-2} for letters of 1·5 μm size. The corresponding images were of equivalent marginal quality, but it was found that a slight increase in letter size to about 2 μm gave a greatly increased alphanumeric image quality, with almost every letter then being distinguishable. Since the normal redundancy of about 3 in written text would then be unnecessary for accuracy in decoding, a value of around 60×10^6 bits cm^{-2} would then be possible in principle for alphanumeric text in this particular example, but coding would be necessary to achieve this. However, the conclusion is that alphanumeric text and binary data can be packed in the image with the same order of bit-density. Riesenfeld[20] confirmed this conclusion in a study of the relationship between information capacity and resolving power. This conclusion is a reassuring one for those applications in which it is inconvenient to convert letters and decimal numbers into binary form.

For colour films as opposed to black-and-white, there is the additional possibility of colour-coding of information prior to storage. Taking into account the number of distinguishable colours per cell as well as the number of distinguishable grey-scale density levels should give an overall increase in the number of distinguishable recording levels. This would seem to offer considerable scope for information storage, but it has been argued that only a factor of 2 or 3 might be gained in practice, whether based on optical read-out or visual impression[21]. Lehmbeck[22] made detailed measurements on a film which uses three colour layers to obtain an extremely high exposure latitude, but found information capacities of the same order as for normal latitude black-and-white films.

The coding of pictorial information has been discussed by Bartletson and Witzel[23] from the viewpoint of the luminance scale in the original and the visual impression in the photographic print. They found that about 6 bits per cell corresponds to a "continuous" tone image.

Novel approaches to the problem of coding and information storage have included the study of a system whereby each bit is recorded as the image of a small diffraction grating[24]. Multiplexing by using superposed exposures was found to give 8 simultaneous bits for one such system, with a limiting packing of about 3×10^6 bits cm^{-2}. Fast retrieval becomes possible for such systems due to the low error rates when the information is spread out over large areas for which the importance of dust and scratches and other imperfections is proportionally reduced. This use of so-called carrier frequencies for alphanumeric storage has also been investigated by Bestreiner et al.[25], who reported results of experiments with superpositions of up to 60-fold for printed matter. Further analysis of alphanumeric storage capacity has beeh carried out when using Fourier holograms[26].

In these latter examples of practical approaches to coding, we are diverging from the simple cell model. The use of this simple model allows many calculations to be made, and such calculations may be the ones of most practical relevance in some common types of application. However another approach is possible by continuing this divergence from the case of discrete signals and storage cells, and analysing the photographic process as a continuous communication channel perturbed by noise. We shall see that it then becomes convenient to work in the spatial frequency domain and to make use of Fourier analysis, as in Chapters 6 to 8.

10.4 Photographic Applications: Continuous Signals

Spatial Frequency Analysis

It is important to realize that the answers obtained for the information capacity when using this continuous channel approach will not be the same as those found by the previous discrete approach, although in practice they may turn out to be quite close. The answers are different because they concern different questions, or types of information input. In principle we would expect from Shannon's theorems that this new approach will yield the upper limit to the information capacity, although it may be very impractical to achieve this limit, and we just remember the distinction between information capacity and information content. To make full use of the information capacity might involve coding the input as a series of random noise samples.

The great benefit of this approach is that it allows the information capacity to be expressed as a close, if not exact, function of spatial frequency, and in turn the close relationship between information transfer rate and DQE then becomes apparent. For applications in scientific photography where overall systems—including the photographic recording element—must be designed to achieve the highest information rate about an incoming signal, this spatial frequency analysis of information may be essential. Without too much loss of mathematical rigour we shall see that it is possible to arrive at information relationships of great significance to the experimentalist, although they may not strictly define the exact bit-rate in any particular experiment.

During the early 1960s this spatial-frequency approach was the topic of a series of papers[27-33] which were based largely on the earlier analysis of Fellgett and Linfoot[12] for optical and photographic systems, and these included the first estimations of numerical values for the information capacity of practical photographic processes. The main features of some of these approaches will be discussed here.

The basis for the spatial frequency approach is usually taken as the result for the continuous channel with average mean-square limitation and gaussian noise. In fact the photographic process is more nearly a peak-limited channel, with its operating region between fog density and D_{max}. However there is no exact analytical solution for the information capacity of such a peak-limited channel, only upper and lower bounds as analysed by Shannon[1]. A further difficulty is due to the non-linearities of the photographic process and the wide variation of its imaging properties over the range of its operating limits. The spatial frequency implications of these have been discussed in detail in Chapters 7 and 8.

The result for the information capacity of a continuous channel with an average "power" limitation is, according to equation (17):

$$C = \Delta f \log_2 \frac{P+N}{N}.$$

As well as the usual assumption about optimum coding, the other main assumption implicit in this equation is that both the signal and the noise are gaussian, stationary and ergodic, and the signal and noise are statistically-independent and additive in the mean-square sense. The bandwidth may have any origin of reference, and width, so long as these conditions apply, but the average signal and noise "powers" must be invariant over the region Δf. If the signal and noise powers vary over Δf but are smooth continuous functions, then it is possible to carry out a summation over the frequency range. For example if the signal and noise powers are approximately constant over two adjacent regions A and B, each of width $\tfrac{1}{2}\Delta f$, then the capacity for the total bandwidth approximates to

$$C = \tfrac{1}{2}\Delta f(\log_2 (1 + P_A/N_A) + \log_2 (1 + P_B/N_B)). \qquad (24)$$

If we write P and N as $W_S(f)$ and $W_N(f)$ respectively, indicating that we are concerned with the Wiener spectra of signal and noise, then in the limit:

$$C = \int_0^\infty \log_2\left(1 + \frac{W_S(f)}{W_N(f)}\right) df. \tag{25}$$

Equation (25) is still that applicable to the case where signal and noise are one-dimensional functions of time, as for electrical time series. Fellgett and Linfoot[12] showed that this equation could be modified when signal and noise are two-dimensional functions of space, as for photographic images, to give

$$C = \tfrac{1}{2}\int_{-\infty}^{+\infty}\int_{-\infty}^{+\infty} \log_2\left(1 + \frac{W_S(u,v)}{W_N(u,v)}\right) du\, dv. \tag{26}$$

In this equation the Wiener spectra of signal and noise are expressed in terms of image density units, and if as usual u and v are expressed in terms of cycles/mm the information capacity will be in bits mm^{-2}. An image of A μm^2 will have a capacity of AC $\times$ 10^{-6} bits. Whereas it is inelegant to use a mixture of distance units such as mm and μm, this merely reflects common usage in specifying spatial frequencies on the one hand, and micro-image dimensions comparable with grain size on the other, and these come together in single equations relating to information. If an exposure time of t seconds was necessary to record the image, then the bit-rate per area A μm^2 will be AC/t $\times$ 10^{-6} bits s^{-1}. Similarly we can express the bit-rate as AC/J $\times$ 10^{-6} bits per Joule, or AC/q $\times$ 10^{-6} bits per photon, as appropriate, where both J and q relate to the area A.

Since the statistical properties of the photographic process, including image noise, may be assumed to be isotropic, and since for optimum coding the signal will also have the nature of an isotropic noise pattern, it is convenient to work in terms of the one-dimensional spatial frequency, ω, where $\omega^2 = u^2 + v^2$, leading to

$$C = \pi \int_0^\infty \log_2\left(1 + \frac{W_S(\omega)}{W_N(\omega)}\right) \omega\, d\omega. \tag{27}$$

Equation (27) illustrates the dilemma of evaluating the information capacity of the photographic process. Due to non-linearities the S/N ratio in terms of power spectra will only be constant over a limited input/output, exposure/density range. To keep equation (27) "exact" it is necessary to restrict it to small signals, but this "exactness" no longer relates to the information capacity as such. Shortly we shall investigate some of the ways which have been devised to deal with this dilemma, but first it is profitable to express equation (27) in terms of other photographic parameters.

Information and DQE

The ratio $W_S(\omega)/W_N(\omega)$ in equation (27) represents the output S/N ratio in terms of Wiener spectra, both relating to fluctuations in the photographic image and hence in image density units. If we denote this ratio by $(S/N)_{OUT}(\omega)$, then

$$(S/N)_{OUT}{}^2(\omega) = \frac{W_S(\omega)}{W_N(\omega)}, \tag{28}$$

where as in Chapter 5 the S/N ratio is defined in terms of density units, and the squared term arises since the Wiener spectra are in terms of the square of density.

In Chapter 8 we saw that the "black-box" DQE operator relating input and output S/N ratios may be replaced by a fuller spatial-frequency definition[34] of DQE, denoted here by $\varepsilon(\omega)$, where in terms of previously-defined photographic parameters:

$$\varepsilon(\omega) = \varepsilon(0)\frac{M^2(\omega)}{n(\omega)}; \quad \varepsilon(0) = \frac{(\gamma \log_{10} e)^2}{G\,q}. \tag{29}$$

It follows that the equivalent expression to equation (8) of Chapter 5 is now

$$(S/N)_{OUT}{}^2(\omega) = \varepsilon(\omega)(S/N)_{IN}{}^2(\omega). \tag{30}$$

This is illustrated in Fig. 7, and it is recalled that this equation will hold only when the input noise is due purely to quantum fluctuations.

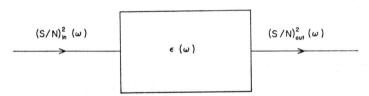

FIG. 7. Photographic process as a "black-box" DQE operator on the input S/N ratio as a function of spatial frequency, when the input noise is due to quantum fluctuations.

It remains to define the input S/N ratio in terms of the signal and noise Wiener spectra of the exposure fluctuations, $\Delta E/E$, or $\Delta q/q$:

$$(S/N)_{IN}{}^2(\omega) = \frac{W_S'(\omega)}{W_N'(\omega)}. \tag{31}$$

A convenient way of doing this[32] is to define

$$W_S'(\omega) = \alpha W_N'(\omega); \tag{32}$$

i.e., we express the mean-square signal fluctuations in the exposure as a multiple, α, of the inherent photon fluctuations in the signal. It is assumed that both Wiener spectra are flat within the range of spatial frequencies of practical interest, and hence that α is independent of spatial frequency.

It follows by combination of equations (28), (30), (31) and (32), that

$$\frac{W_S(\omega)}{W_N(\omega)} = \alpha\varepsilon(\omega), \tag{33}$$

and hence equation (27) now becomes

$$C = \pi \int_0^\infty \log_2 (1 + \alpha \varepsilon(\omega)) \, \omega \, d\omega. \tag{34}$$

Equation (34) contains only two practical variables: α measures the strength of the signal, and $\varepsilon(\omega)$ measures the efficiency of the imaging device. If equation (34) is to represent the full information capacity as such, α will in fact be defined by the strength of the signal which will just "fit" within the assumed "power" limits of the recording process. However apart from the unrealistic nature of assuming that the photographic process is power- rather than peak-limited, the question of linearity restricts the size of α. Photographic non-linearity now reduces to the question of the dependence of DQE on exposure level, and the output S/N ratio will only be defined by the product $\alpha \varepsilon(\omega)$ for low α values where $\varepsilon(\omega)$ may be assumed invariant over the corresponding exposure range.

Since the only measure of the performance of the imaging process included in equation (34) is the imaging DQE, then it might be argued that from the viewpoint of knowledge of the process the information capacity adds nothing that is not already provided by DQE[33]. But it has been argued that the converse is also true[35], and the question is one of practical utility and intrepretation. The information capacity defines the amount of information conveyed about a particular type of signal, in the presence of a S/N ratio determined by the DQE of the recording system.

Small Signals

In addition to avoiding the complications of non-linearities, a further simplification of equation (34) follows if the analysis is restricted to small signals. This approach was applied by Linfoot[27] and Shaw[29-32], though as already discussed does not yield the information *capacity* as such. The ratio

$$\frac{C}{\alpha} = \frac{\pi}{\alpha} \int_0^\infty \log_2 (1 + \alpha \varepsilon(\omega)) \, \omega \, d\omega$$

takes its maximum value when α tends to zero, and this maximum value is defined by

$$\frac{C}{\alpha} = \pi \log_2 e \int_0^\infty \varepsilon(\omega) \, \omega \, d\omega \tag{35}$$

Following Linfoot[27], we define this as the upper bound for the information recording rate per unit signal, α, for random small signals. When the Wiener spectrum of the image noise is flat ($n(\omega) = 1$), or at least invariant over the spatial frequency range for which the MTF is non-zero, as is often the case in practice, then by equation (29),

$$\frac{C}{\alpha} = \pi \log_2 e \, \varepsilon(0) \int_0^\infty M^2(\omega) \, \omega \, d\omega. \tag{36}$$

To compare two photographic processes, A and B, it would thus be appropriate to measure the ratio

$$\frac{\varepsilon_A(0)}{\varepsilon_B(0)} \frac{\int_0^\infty M_A^2(\omega)\,\omega\,d\omega}{\int_0^\infty M_B^2(\omega)\,\omega\,d\omega}.$$

If the interest is only in signals within a certain spatial frequency range, ω_1 to ω_2, then it would be reasonable to measure the above ratio with the integration between these limits rather than from zero to infinity. but again this is only if we are prepared to abandon conceptual and mathematical exactness. The ratio is no longer that of two bit-rates, but a useful comparison of the most relevant imaging parameters for information-gathering problems.

Similarly we might compare the performance of an imaging process with an ideal process specified by a low-frequency DQE of 100% and MTF of 1 for all spatial frequency values of interest. This ratio might usefully be termed the informational efficiency, and expressed as

$$I_{\text{eff}} = \varepsilon(0) \frac{\int_\omega M^2(\omega)\,\omega\,d\omega}{\int_\omega \omega\,d\omega}. \qquad (37)$$

To complete the analysis for small signals, we consider the importance of the integral over the spatial frequency range, and the scale factor, from three informational aspects.

(a) When dealing with problems where noise is of negligible concern, as for example in the transmission of information by a lens, then the integral

$$I_1 = \int_\omega M^2(\omega)\,\omega\,d\omega, \qquad (38)$$

will be of significance.

(b) When the noise is important, as in photographic images, and the information *rate* of recording is the criterion, I_1 will be scaled by the DQE, and we define the relevant quantity as

$$I_2 = \varepsilon(0)\,I_1. \qquad (39)$$

(c) There may be cases where image noise is important, but the interest is in the maximum information-packing in the image; i.e., an extension of the discrete case already discussed, to the continuous case. The relationship between the integral I_3 which is now of relevance, and the integral I_2, will be analogous to that of DQE and the noise-equivalent number of quanta (NEQ), detailed in Chapter 5 for essentially low spatial frequencies. Denoting the NEQ by $q'(0)$:

$$q'(0) = \frac{(\gamma \log_{10})e^2}{G} = q\,\varepsilon(0),$$

at the exposure level q quanta per unit area, and it follows that

$$I_3 = q'(0)\,I_1 = q\,I_2. \qquad (40)$$

The wider context in information problems of the concepts of DQE and NEQ, as analysed for detection problems at low spatial frequencies in Chapter 5, is now apparent. The integral I_1, supplemented by the DQE or NEQ values at low spatial frequency, provide the useful parameters of the image for information recording and storage. Since all are functions of exposure level, a three-dimensional plot of these functions as q and ω vary provides a useful overall picture of the informational possibilities of an imaging process.

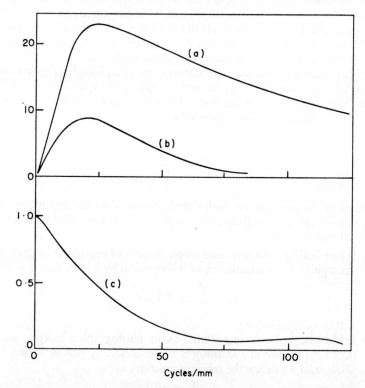

FIG. 8. (a) The function $\omega M_f^2(\omega)$ for a film; and (b) as cascaded with an optical system defined by $M_o^2(\omega)$, as shown below (c). From results calculated by Linfoot (reference 27, Figs 1, 2 and 3), with film identified as Agfa IFF, and optics as for typical fast camera lens.

Examples of such three-dimensional plots, and emphasis of this close relationship between DQE, NEQ, MTF, Wiener spectrum and information capacity, were given in a review by Shaw[36].

Linfoot[27] calculated the function $M^2(\omega)\,\omega$ for four film-types, and his results for one of these are shown in Fig. 8. He also considered the influence of the MTF of a typical fast camera lens, and cascaded lens with film for the overall systems performance. The result shown demonstrates that the major contribution to the integral I_1 for film alone may be from higher spatial frequencies than for the system as a whole.

This example does not include the influence of DQE. A comparison between films with typical DQE and MTF values shows the importance of DQE when the information rate is of relevance. Films A and B have MTFs as shown in

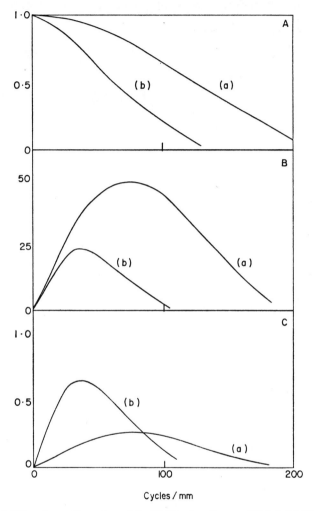

FIG. 9. (a) MTFs of two films, A and B. (b) The functions $\omega M^2(\omega)$. (c) The functions $\varepsilon(0) \omega M^2(\omega)$, compared for DQEs of $\varepsilon_A(0) = 0.5\%$; $\varepsilon_B(0) = 3.0\%$.

Fig. 9(a), with A having a higher MTF at all spatial frequencies, which reflects strongly in the respective functions $M^2(\omega) \omega$ as shown in Fig. 9(b). However, if $\varepsilon_A(0) = 0.5\%$ while $\varepsilon_B(0) = 3.0\%$, then Fig. 9(c) shows that film B now dominates in the spatial frequency range 0 to 90 cycles/mm, although A still has advantages above 90 cycles/mm due to its superior MTF. The application of either of these films would thus depend on the practical spatial frequency range of interest. Further plots of the type of Fig. 9(c), based on values as measured for various film types, have emphasized the role of DQE, and also of the amplitude of the signal and the importance of correct utilization of the available spatial frequency range in information recording[30,32,60].

Maximum Information Capacity

In an attempt to calculate the maximum information capacity, Shaw[30] considered increases in the amplitude of the signal until it was estimated that it would "fill" the available exposure latitude, but without allowance for the variation of DQE over this high exposure range. Figure 10 shows a summary of

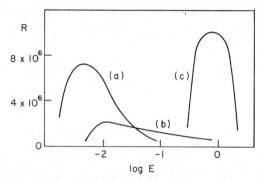

FIG. 10. Estimation of bit rates, R, calculated[30] for Ilford films identified as: (a) HPS; (b) HP3; (c) MICRO-NEG PAN. R is in bits erg^{-1}, and the log E axis is in terms of ergs cm^{-2}.

the results obtained in this way for three film types. By integrating over the spatial frequency range and obtaining I_2, the information was calculated in terms of bits erg^{-1}, and plotted as a function of the average signal level.

A similar attempt to calculate the information capacity of the photographic process was made by Jones[28]. He considered the operating characteristics of the photographic process as constrained between fog density and D_{max}, and although in this sense the photographic process is peak-limited he used Shannon's theorem for a power-limited channel, and then made various *ad hoc* corrections to take account of this. By using experimental data on four different films, and making some allowance for non-linearities, he arrived at estimates for the respective information capacities. His results, along with other comparative values of interest, are summarized in Table III.

The first column of numbers shows the information capacities of four films as identified, in terms of the number of bits that can be packed per cm^2. The values for Royal-X, Plus-X and Pan-X are about 4–6 times higher than those due to Altman and Zweig[18]—which were shown in Table II—for the simple cell model with a 10σ level-separation criterion.

The next column shows the equivalent film area to store 1 bit. Next, by taking into account the exposure energy required for recording, the bit-rate is expressed in terms of bits per erg. In this case Royal-X, the film with highest

TABLE III. Information capacities of four films, and various comparative values, as estimated by Jones[28].

Film	Information capacity bits cm^{-2} ($\times 10^{-6}$)	Area for 1 bit μm^2	Information rate bits erg^{-1} ($\times 10^{-6}$)	Exposure for one bit photons ($\times 10^{-3}$)	Comparative time for Hi-Fi system sec cm^{-2}	Film area equiv to one TV frame cm^2/frame
Royal-X	0·449	200	26·5	8·18	3·01	2·98
Tri-X	0·845	118·4	7·35	29·4	5·1	1·76
Plus-X	1·86	53·8	6·45	33·6	11·2	0·80
Pan-X	2·85	35·0	7·45	29·2	17·2	0·52

speed and noise of the four, has easily the highest rating, whereas it has much the lowest storage capacity per unit image area. The next column shows the energy criterion converted to the number of photons required to yield 1 bit, the number involved ranging from around 8000 to 33 000 photons.

The final two columns show two comparative numbers estimated by Jones which are of general interest. First there is the time in seconds need by a high-fidelity channel to convey the same information as 1 cm^2 of film. For this the channel was assumed to have a bandwidth of 10^4 cycles per second and a S/N power ratio of 10^5 (i.e., 50 decibels). The numbers estimated are seen to be of the order of a few seconds. Finally there is the area of film equivalent to a single TV frame in the information capacity sense. For the latter a bandwidth of 4.5×10^6 cycles per second was assumed, with a S/N ratio equivalent to 30 decibels. It is seen that 1 cm^2 film is equivalent to about a single TV frame.

Other descriptions of methods and approximations for measuring the information capacity of the photographic process are due to Biedermann and Frieser[37], and Langner[38], and the problems have been reviewed by Lebedev[39].

The practical and theoretical difficulties associated with the definition, measurement and utilization of photographic information capacity have already been expounded. It is useful however to consider the origins of some of these difficulties, which are due largely to the form of the input/output measurements, for example in terms of ergs and image density. DQE, MTF and the Wiener spectrum all depend on the image recording level, and all are statistical representations. A more fundamental approach to information capacity would be to use a quantum-mechanical approach in terms of the discrete input/output photon/grain states. Such an approach would have its origins in the information capacity of the exposure light in terms of the photon statistics.

Jones[40] calculated the information capacity of a beam of light based on the Poisson statistics of nondegenerate ensembles of Bose entities, and also made general comparisons with the capacity of various types of radiation detector[41]. Hisdal calculated the information content of photon beams for given degeneracies[42]. The beam was assumed to fall on an object, and the reflected beam which conveys the information about the object was measured by a photo-emissive type of detector or detector array. Takahasi[43] has given a detailed analysis of the information conveyed by "noisy" quantum-mechanical channels, including interpretation by Markov chains for the conditional probabilities, or correlations, which are involved.

There is no reason in principle why similar methods should not be applied to the photographic process, with calculations being made of the information conveyed per photon by an image grain. As was demonstrated in Chapter 5, applying basic fluctuation theory leads to a more fundamental understanding of DQE in terms of the photographic inefficiencies which operate on the input photons and the photon noise than when working in terms of image density. Of course the conditional probabilities involved in the photographic recording of quanta (which can be considered as the introduction of degeneracies) are many, and an analytical solution for information capacity would be difficult. In previous chapters we have seen that Monte Carlo methods have been necessary to specify the fates of quanta prior to absorption by a grain (to yield the MTF) and after absorption by a grain (to yield the quantum sensitivity). However perhaps more fundamental studies of this nature at the photon/grain level will eventually give a more satisfactory solution to the problem of the information conveyed by the photographic process.

Practical Applications

In spite of difficulties, the application of information assessments in applied photography has proved useful in many diverse fields. Some of these not already mentioned will be reviewed here.

Linfoot[44] used information theory to study the design of astronomical spectrographs. He analysed the influence of noise and resolution on the information in photographic spectra, and related these to the instrumental slit-width. Reverting back to examples for discrete signals, Levi[45] has considered the photographic process for computer output storage, and concluded cluded that photography is a highly efficient storage medium in this respect. McCamy[46] considered the packing of information in legible microphotographs. He found the information capacity of Kodak High Resolution Plates to be around 3.25×10^6 bits mm^{-2}, and showed that legible alphanumeric text could be stored at about 2×10^6 bits mm^{-2}. He described how the King James version of the Old and New Testaments could thus be reduced to a square of 4×4 mm^2. This brings us back to the question of redundancy in written text. Arps[47,48] has considered the entropy of printed matter at limiting legibility, and some general principles involved in page-coding to reduce redundancy.

Clark and Zarem[49] analysed the information rate of high-speed camera systems, including spatial-frequency bandwidth, shutter speed and S/N ratio, and compared several ultra-high-speed electronic cameras. Trinder[50] applied statistical entropy methods to the classification of typical terrain features in aerial photography. Bershad[51] considered the relationship between resolution and information capacity and rates of optical communication channels for a number of conditions, including that for a given number of uniformly-distributed point sources in a given image region, with additive white gaussian noise as background. Frieden[52] considered the restoration of object information from a noisy image using a minimum mean-square restoration error-criterion, showing the relationship with Shannon's measure of information.

Of the many comparisons between conventional and unconventional photographic process, an interesting one from the information viewpoint concerns the application of photochromic materials[53]. Beiser[54] considered high-speed scanning of high-resolution media over a wide signal bandwidth, and compared electron-beam and laser-beam scanning with conventional silver-halide and

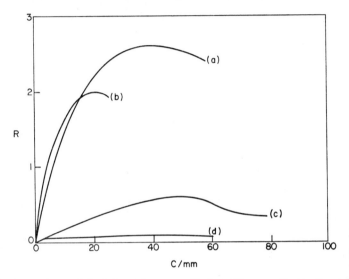

FIG. 11. Comparison of the information rates, R, in arbitrary units, for Spectracon intesifier plus: (a) medium speed emulsion; (b) fast emulsion; (c) slow emulsion. (d) is for unaided Kodak IIa-0 emulsion. Results due to Kahan and Cohen[58].

thermoplastic recording. Znamensky et al.[55], investigated the use of holographic techniques to increase microfilm storage. Tubbs[56] has discussed the possibilities of thermochromic and photochromic processes as reversible storage and display materials in holographic recording, leading to estimates of read-write-erase optical information stores of 10^9 bits, and with very low access times.

Beckman[57] measured DQE as a function of spatial frequency in the comparison of two image intensifier tubes, showing higher values of DQE at low frequency compared with the photographic process, but with more rapid loss of DQE with increase in frequency. He also compared information capacities by plotting against spatial frequency various functions closely related to those of equations (38) to (40). Kahan and Cohen[58] used similar techniques to compare the information with the Spectracon image tube plus film with that of unaided photography. Some of their results are summarized in Fig. 11, showing the substantial information gains to be made with this type of image tube. Measurement of gains by multistage image converters in terms of real informational sensitivity have also been calculated by Fanchenko and Frolov[59].

References

1. Shannon, C. E. (1948). A mathematical theory of communication. *Bell Syst. Tech. J.*, **27**, 379, 623.
2. Hartley, R. V. L. (1928). Transmission of information. *Bell Syst. Tech. J.*, **7**, 535.
3. Gabor, D. (1961). Light and information. (Presented as the Ritchie Lecture, University of Edinburgh, 1951), "Progress in Optics". **1**, 109. North-Holland, Amsterdam.
4. Brillouin, L. (1961). "Science and Information Theory". (Second Ed.). Academic Press, New York and London.
5. Tribus, M. and McIrvine, E. C. (1971). Energy and information. *Scient. Amer.*, **224**, No. 3, 179.
6. Maimon, S. R. (1971). The role of technology in forecasting and assessment in planning and resource allocation. Abstract of the 3rd SPSE Symposium on "Unconventional Photographic Systems". Page 107. Washington D.C.
7. Nakurama, K. and Sakata, T. (1971). Micro-imaging. *J. Soc. Phot. Sci. and Tech. Japan*, **34**, 30.
8. Pierce, J. R. (1972). Communication. *Scient. Amer.*, **227**, No. 3, 31.
9. Shannon, C. E. and Weaver, W. (1949). "The Mathematical Theory of Communication". University of Illinois Press, Urbana.
10. Elias, P. (1953). Optics and communication theory. *J. Opt. Soc. Amer.*, **43**, 229.
11. Fellgett, P. B. (1953). Concerning photographic grain, signal-to-noise ratio and information. *J. Opt. Soc. Amer.*, **43**, 271.
12. Fellgett, P. B. and Linfoot, E. H. (1955). On the assessment of optical images *Phil. Trans. Roy. Soc.*, **A247**, 369.
13. Black, G. and Linfoot, E. H. (1957). Spherical aberrations and the information capacity of optical images. *Phil. Trans. Roy. Soc.*, **A239**, 522.
14. Linfoot, E. H. (1959). Information theory and photographic images. *J. Photogr. Sci.* **7**, 148.
15. Woodward, P. M. (1953). "Probability and Information Theory, with Applications to Radar". Pergamon Press, Oxford.
16. Rosenfeld, A. (1968). "Picture Processing by Computer". Technical Report No. 68-71, Computer Science Center, University of Maryland.
17. Andrews, H. C. (1970). "Computer Techniques in Image Processing". Academic Press, New York and London.
18. Altman, J. H. and Zweig, H. J. (1963). Effect of spread function on the storage of information on photographic emulsions. *Photogr. Sci. Eng.*, **7**, 173.
19. Levi, L. (1958). On the effect of granularity on dynamic range and information content of photographic recordings. *J. Opt. Soc. Amer.*, **48**, 9.
20. Riesenfeld, J. (1967). Relationship between information capacity and resolving power. *Photogr. Sci. Eng.*, **11**, 415.
21. Nelson, C. N. (1972). Photographic system as a communication channel. *Appl. Opt.*, **11**, 87.
22. Lehmbeck, D. R. (1967). Experimental study of the information storage properties of extended range film. *Photogr. Sci. Eng.*, **11**, 270.
23. Bartletson, C. J. and Witzel, R. F. (1967). Source coding of information. *Photogr. Sci. Eng.*, **11**, 263.

24. Lamberts, R. L. and Higgins, G. C. (1966). Digital data recording on film by using superposed grating patterns. I. General theory and procedures. Lamberts, R. L. II. Analysis of the system. Blackmer, L. L., Van Kerkhove, A. P. and Baldwin, R. III. Recording and retrieval techniques. *Photogr. Sci. Eng.*, **10**, 209, 213, 263.
25. Bestreiner, F., Deml, R. and Greis, U. (1970). The capacity of carrier frequency (CF) photographic systems for alphanumeric information storage. *Photogr. Sci. Eng.*, **14**, 1.
26. Bestreiner, F., Greis, U. and Weierhausen, W. (1972). Alphanumeric storage capacity of a defocused Fourier hologram and a Fourier hologram matrix. *Photogr. Sci. Eng.*, **16**, 4.
27. Linfoot, E. H. (1961). Equivalent quantum efficiency and the information content of photographic images. *J. Photogr. Sci.*, **9**, 188.
28. Jones, R. C. (1961). Information capacity of photographic films. *J. Opt. Soc. Amer.*, **51**, 1159.
29. Shaw, R. (1961). An investigation of the informational properties of photographic emulsions. Thesis, Cambridge University.
30. Shaw, R. (1962). The application of Fourier techniques and information theory to the assessment of photographic image quality. *Photogr. Sci. Eng.*, **6**, 281.
31. Shaw, R. (1962). Information in the photographic image. *Perspective*, **4**, 69.
32. Shaw, R. (1963). Photon fluctuations, equivalent quantum efficiency and the information capacity of photographic images. *J. Photogr. Sci.*, **11**, 313.
33. Jones, R. C. Information capacity of radiation detectors. I. *J. Opt. Soc. Amer.*, **50**, 1166 (1960); II. *J. Opt. Soc. Amer.*, **52**, 1193 (1962).
34. Shaw, R. (1963). The equivalent quantum efficiency of the photographic process. *J. Photogr. Sci.*, **11**, 199.
35. Ścibor-Marchocki, R. (1961). Contrapositive. *J. Opt. Soc. Amer.*, **51**, 1029.
36. Shaw, R. (1968). Image quality criteria for photographic detectors. Proceedings of the S.I.R.A. Symposium on "Application of Optical Transfer Function", Page 17. London.
37. Biedermann, K. and Frieser, H. (1965). On the information capacity of photographic layers. *Optik*, **23**, 75.
38. Langner, G. (1966). The measurement of the information capacity of photographic layers. *Phot. Korres.*, **102**, 101.
39. Lebedev, D. S. (1965). The application of information theory to photographic systems. *Zh. Nauch. Prikl. Fotogr. Kin.*, **10**, 62.
40. Jones, R. C. (1962). Information capacity of a beam of light. *J. Opt. Soc. Amer.*, **52**, 493.
41. Jones, R. C. (1963). Information capacity of radiation detectors and of light. *Appl. Opt.*, **2**, 351.
42. Hisdal, E. (1969). Exact values of the information content of photon beams. *J. Opt. Soc. Amer.*, **59**, 921.
43. Takahasi, H. (1965). Information theory of quantum-mechanical channels. *In* "Advances in Communication Systems". **1**, 227. Academic Press, New York and London.
44. Linfoot, E. H. (1960). Informational considerations in the design of astronomical spectrographs. *Mon. Not. Roy. Astr. Soc.*, **121**, 115.
45. Levi, L. (1963). Photographic emulsions as computer storage media—especially with CRT readout. *Appl. Opt.*, **2**, 421.
46. McCamy, C. S. (1965). On the information in a microphotograph. *Appl. Opt.*, **4**, 405.

47. Arps, R. B. (1969). Entropy of printed matter at the threshold of legibility. Thesis, Stanford University.
48. Arps, R. B. (1971). The statistical dependence of run-lengths in printed matter. Proceedings of the Braunschweig Coding Conference.
49. Clark, G. L. and Zarem, A. M. (1967). Communication aspects of high-speed photography. *J. Soc. Motion Pict. Tel. Engrs.*, **76**, 1183.
50. Trinder, J. C. (1970). Entropy of aerial photography. *Photogramm. Eng.*, **36**, 1172.
51. Bershad, N. J. (1969). Resolution, optical-channel capacity and information theory. *J. Opt. Soc. Amer.*, **59**, 157.
52. Frieden, B. R. (1970). Information, and the restorability of images. *J. Opt. Soc. Amer.*, **60**, 575.
53. Dorion, G. H. and Wiebe, A. F. Chapter 5, entitled: Photochromic materials in information storage and optical signal processing. "Photochromism". Focal Press, London, 1970.
54. Beiser, L. (1966). Laser beam and electron beam extremely wideband information storage and retrieval. *Photogr. Sci. Eng.*, **10**, 222.
55. Znamensky, V. B., Kukarov, G. V. and Strukov, V. S., (1972). Investigation of the possibilities for the use of holographic methods to increase the recording density on microfilm. *Zh. Nauch. Prikl. Fotogr. Kin.*, **17**, 419.
56. Tubbs, M. R. (1973). Reversible holographic recording for optical information storage. *Opt. and Laser Tech.*, **5**, 155.
57. Beckman, J. E. (1966). Application of information theory to the evaluation of two image intensifier tubes. "Advances in Electronics and Electron Physics", **22A**, 369. Academic Press, London and New York.
58. Kahan, E. A. and Cohen, M. (1969). Comparison of the efficiency of image recording with a Spectracon and with Kodak IIa-0 emulsion. "Advances in Electronics and Electron Physics", **28B**, 725. Academic Press, London and New York.
59. Fanchenko, S. D. and Frolov, V. A. (1970). Actual sensitivity of multistage image converters. *Prib. Tekh. Eksper.*, 210.
60. Vendrovsky, K. V. (1973). Information capacity of photographic emulsions. *Zh. Nauch. Prikl. Fotogr. Kin.*, **18**, 331.

Exercises

(1) In scientific photography it may sometimes be required: (a) to store as much information as possible in the smallest image area; (b) to record incoming information at the highest possible rate. Give typical examples of practical situations where these criteria might be applicable, and discuss the choice of photographic material to be used.

Discuss with simple numerical examples how for a given recording system it is possible to interchange informational criteria of bits cm^{-2}, bits erg^{-1}, bits $photon^{-1}$ and bits s^{-1}.

(2) An information source has four states (A, B, C and D), with probabilities as shown, and three binary codes may be used to transmit these states:

State	Probability	Code 1	Code 2	Code 3
A	0·5	11	0	0
B	0·25	10	10	1
C	0·125	01	110	10
D	0·125	00	1110	01

If messages are transmitted as a long string of successive binary digits, which of these codes would be unsatisfactory from the viewpoint of decoding? Of the other two, which can be used to transmit messages with the minimum number of binary digits on average? Under what circumstances might the other be more useful for photographic storage?

(3) An information source has an alphabet of n possible states, the ith occurring with probability p_i. Shannon defined the uncertainty associated with such a source as

$$H(p_1, p_2, p_3, \ldots, p_n) = -k \sum p_i \log p_i,$$

where k is a constant.

Starting from this equation, show that the information capacity of an N-cell, M-level recording medium may be defined as $C = N \log_2 M$.

(4) What are the factors which influence the choice of cell-size and number of recording levels for the photographic storage of information? Distinguish between information capacity and information content, and explain why in principle alphanumeric recording cannot make use of the maximum information capacity of a storage medium.

It is estimated that the average daily world output of published information is around 10^{12} bits. What area of film would be necessary to store this information, assuming a cell-size of 4×4 μm^2 and 8 distinguishable density levels?

(5) Derive a general expression for the entropy of an mth-order Markov information source with a q-letter alphabet, and discuss the similarities with printed text.

Make a rough comparison of the information capacities of a 10^5-word book, a single television frame, and 1 cm^2 of microfilm, and discuss the coding implications to make use of these capacities.

(6) A sample of microfilm is found to have the following characteristics under normal conditions of use:

$$D_{fog} = 0.15, \quad D_{max} = 1.95;$$
$$G = 0.81 \ \mu m^2 \ D^2, \text{ at } D = 0.90 \text{ above fog.}$$

For the purposes of this problem it may be assumed that G is independent of density level, and that a $k = 10\sigma$ level-separation criterion is adopted. Plot a graph showing the relationship between the information capacity and cell-size, and indicate on this graph the cell-sizes corresponding to binary, 4-level, 8-level and 16-level recording.

If the film has a spread function which limits the cell area to a minimum of 37×37 μm^2, calculate the reduction in storage capacity if for convenience binary rather than multilevel recording is adopted. If alphanumeric recording is found to be possible with a cell-size of 150×150 μm^2, compare the information storage with the values calculated above.

(7) Make rough estimates of the bit-rates associated with the written and spoken word. Would you expect it more likely that the redundancy in a language optimizes speech-rate or reading-speed?

Describe an experiment to determine how the limiting redundancy of photographically-reproduced text is related to the system MTF, using a suitable statistical sample of "standard observers", and assuming the influence of image noise is negligible.

(8) Discuss the general relationship between NEQ, DQE, and information capacity and recording rate.

An optimally-coded low-amplitude exposure signal has a Wiener spectrum which may be assumed to be flat within the spatial frequency range of 90–100 cycles/mm at the image surface, and contains negligible information outside these frequencies. There is a choice between films A and B, with A having an MTF approximating to a triangle function with cut-off at 120 cycles/mm, and B having an MTF in the form of a gaussian distribution with 0·5 modulation at 60 cycles/mm. Use a graphical method to determine which of the two films will record most information per unit exposure time, assuming that both processes have identical low-frequency DQEs.

Would your conclusion still be the same if the low-frequency DQE of film A was 2%, while that of B was 1%? With these latter DQE values determine the informational efficiency of each film (i.e., as defined by equation (37)).

(9) Explain in general terms the principles of coding to transmit signals over noisy channels.

A noisy channel can transmit 16 different states, 8 of which can always be decoded uniquely, but due to noise the other 8 can never be distinguished from each other with certainty. If the channel can transmit one state per second, compare the bit rate with that for the equivalent noise-free case. If a source has a 16-symbol alphabet, explain the coding necessary to transmit the rate calculated in the noisy case.

(10) For a certain film an alphanumeric symbol must occupy an area of 1600 μm^2 in order to be legible, while the minimum area of a cell for multilevel recording is 100 μm^2. The minimum and maximum density levels are 0·04 and 2·64 above base, and the Selwyn granularity coefficient is 0·72 $\mu m\ D$, at a density of 1·34 above base.

Making reasonable assumptions, calculate the information capacities in bits cm^{-2} for alphanumeric and multilevel recording.

Appendix

This appendix contains two tables of 1024 Gaussian random variables with zero mean and unit variance. The numbers were generated using a standard computer subroutine (see reference 30 in Chapter 6). In the first table, Process A, the random variables are uncorrelated; Process B is a correlated set of random variables, with an autocorrelation function nominally equal to the triangle function, $\Lambda(\xi/10)$, the maximum correlation distance being 10 intervals. The numbers should be read off along each row.

Process A

```
 0.621 -0.578 -0.408  0.872  0.520  0.180  0.289 -0.691 -0.109  0.753
 0.544 -0.174  2.358  0.599 -1.596 -0.285  1.743 -0.256 -0.491 -0.995
 1.252  1.820 -0.667  1.922 -0.443  0.956  0.146  0.193 -1.107  0.519
-1.470  0.474 -1.474  0.549 -0.398  1.116  0.163  1.861 -1.153  0.941
-1.131  1.406 -2.068 -1.724 -0.693  0.359  0.487 -0.299 -0.752  1.542
-0.830 -0.212  0.587  0.322 -0.063  0.263  1.003  0.680  0.997 -0.725
 2.036 -0.620 -1.084 -2.232 -1.045 -0.495 -0.500 -0.452 -0.587  0.394
-0.135  0.024 -1.480 -0.269  0.901 -0.629 -0.853 -1.320 -0.782 -0.610
 2.041 -1.104  0.437 -1.597 -1.546  0.673 -0.155  0.566 -0.991 -0.775
 0.840  0.917  0.957 -1.466  0.877 -0.766 -1.071  1.024 -2.019 -0.037

 1.524  0.663 -1.446  1.307  0.754 -0.058 -1.039  0.084 -1.558 -0.188
-0.876  0.165  1.917 -0.645  0.237  0.074  1.328  0.151 -0.072 -0.394
 0.015  0.113  0.231  0.013  0.094  0.298 -1.144 -1.145  0.442  0.300
-0.489 -1.057  1.022 -0.393  0.553  0.855  1.371  0.918  0.778 -1.539
 0.451 -0.832  1.008 -0.046  1.001 -0.857 -0.516  0.008 -0.945 -0.440
 0.520  0.008  0.524 -0.252  1.798 -1.093 -2.422 -0.573  2.186  0.673
 2.818 -0.186 -0.457 -0.375  1.555  1.002 -0.000 -0.175 -0.258  0.796
 0.014 -1.792  0.346  0.149 -1.619  0.881  1.043 -1.032  0.535  0.220
-0.988 -0.338 -0.108 -0.456  0.865  1.184  0.561 -0.260 -0.996 -0.063
-1.119 -0.076  0.159  0.665 -0.697  0.993 -1.959  0.682  0.441  0.890

 1.365  0.536 -0.710  0.028  1.014 -1.662  1.427 -0.421  0.863  0.449
-0.465 -0.152  0.105  0.664 -1.168 -0.199 -0.409  0.358 -0.268 -0.141
-0.418  1.342  1.024 -1.157 -0.808  1.043  1.201 -0.597 -0.381  0.299
-0.980  1.616  1.032 -1.683 -0.521  1.104  0.313  1.384 -0.223  0.956
-1.096 -1.529 -0.334 -1.282  1.434 -0.328  2.185 -0.453 -1.938 -0.142
 0.013  1.512 -2.096  1.199 -0.619 -0.983  0.403 -1.226  0.163  0.320
 1.600 -1.342  0.248  0.064  1.700  0.242  0.219 -0.306 -1.282 -0.060
 1.777  0.346 -1.158 -0.970  2.023 -1.124 -0.252 -1.660  1.000  1.365
-0.060  0.249 -0.918 -0.907 -0.484  0.719 -0.394 -1.164  0.586 -1.029
 1.477 -1.580  2.296  0.388  0.197  1.425 -0.296  0.450 -0.604  0.735

-0.167 -1.236 -1.721  0.340 -0.104 -0.653 -2.140 -2.040 -0.869 -0.703
-1.464 -1.096 -0.545 -0.429 -0.757  1.073 -0.932 -1.861  1.335  0.121
-1.548  0.272 -0.112 -0.472  0.471  0.127  0.264  0.172 -0.462 -0.013
-0.219 -0.979  0.702  1.938  0.890  1.097  0.396  1.635  0.816  0.928
-0.728  1.560 -0.444  0.845 -0.740 -1.056  1.819 -0.086 -0.815 -0.571
-0.285  0.474 -1.801  1.455  0.504 -1.316 -2.111  0.379 -0.475 -0.056
 0.096  0.174  1.373  2.146 -0.543 -0.909 -0.356 -0.169 -1.704  0.354
 0.975 -0.853  2.098  0.025  1.657 -0.207  0.734  0.018 -0.502 -0.908
-0.246  0.447 -0.827  0.929  0.841 -0.078 -1.093  1.050  0.845  0.797
-0.675 -0.220 -0.744  0.059  0.897 -0.602 -0.091  0.591  0.229 -1.538

 0.862  0.052 -1.473 -1.554  0.042 -0.769 -1.159  0.328 -1.620 -0.636
-0.716 -0.668  0.476  0.702  0.667  0.341 -0.375 -0.653 -0.996 -0.013
-0.743 -1.302  0.727 -1.293  0.509 -0.017 -1.571  0.008 -0.183  1.340
 1.039  1.067  1.721  0.586  1.222  2.283 -1.007  0.619  0.377  0.135
 1.748  2.244 -0.448  0.138  0.357 -1.231  1.174 -1.043 -1.111 -0.039
 0.275 -0.001 -0.187  0.789 -0.079  0.333  1.233  1.077 -0.057  1.019
 0.578  0.654  0.825  1.132 -1.285  1.701 -0.620 -1.003 -1.595 -0.027
-0.296  1.730  0.249  2.723 -0.272  0.301 -0.765  0.959  1.345 -0.054
 0.301  1.156 -0.692  1.947 -0.038  1.243  0.976  0.093  0.340 -1.375
-0.531  3.097  0.601 -1.296  1.242 -0.604 -1.513  0.490 -0.960  0.056
```

```
-0.031   0.177   0.430   1.755   0.385  -0.526  -1.229   0.450   0.333  -0.615
-0.111  -0.293   1.329  -0.099  -0.520  -1.544   0.664   0.561  -0.990   2.083
-0.192   0.938  -0.001  -0.179  -0.109   0.075   0.220  -0.734   0.671   0.057
-0.991  -1.249   0.983  -0.518  -0.493   0.667   0.756   0.372   1.791   0.501
-1.473  -1.002  -0.815   0.622   0.844   0.457   0.257   0.947  -1.257  -1.051
 0.075   0.187   1.108   0.283   1.753  -0.386  -1.478  -1.013   1.657   1.205
-0.572  -1.191  -0.464   0.280   0.982  -0.847   1.001   0.185   1.753   0.563
-0.850   1.186   0.236  -0.271  -0.326   0.345  -1.382   1.626  -0.263   0.365
 0.650  -0.117  -0.763  -1.011  -2.441  -0.088   1.678  -0.272  -1.833   1.217
 0.625  -0.835   0.267   2.188   1.798  -0.176  -0.815   0.926   0.279  -0.277

-0.052  -1.726   0.622   0.495   0.476   1.089   0.368  -0.551  -0.091  -0.702
-0.757  -1.707   0.540   0.424  -0.564   2.038   0.331   1.315   0.068  -2.018
 0.568   0.132   2.241   0.471  -1.878   0.740  -0.769   1.412  -0.912  -0.647
-2.216  -1.061   0.841   0.422   0.213   1.300  -0.318  -1.408   0.378   0.922
-1.368   0.220   0.772  -1.724  -0.861  -1.137   0.781  -0.890   1.223  -1.395
-0.957  -0.614  -0.114  -1.382   1.780  -1.203   1.317   1.221  -1.560  -0.943
-1.227  -0.348   0.559  -1.754   2.034   1.707   1.142  -3.408   0.147   0.582
-0.572   0.470   0.310   1.087  -0.654  -1.823   2.213  -1.707  -0.980   0.361
-0.313   0.453  -1.029   0.332   0.550   0.412   1.500   0.173  -1.578   0.595
-0.689  -0.304  -0.421  -0.140  -0.696  -0.682   1.896   1.439  -0.984  -1.465

 0.405   1.215   0.401   1.312   2.185  -1.764   1.880   0.882   0.682   0.999
 0.328   0.132  -1.760   1.016   0.004   1.347   0.427  -0.609  -1.084   1.054
-0.443  -0.205  -1.422   1.273   0.492  -0.501  -0.522  -0.514   0.080  -0.033
-0.322   2.264   0.844   1.356  -1.371  -0.609  -0.156  -0.632   0.156  -0.485
 0.702  -1.228   0.399  -0.108   1.952   1.865  -0.473   1.479   1.535  -1.561
 0.569   0.054   0.184   0.160   1.554   0.379   1.712  -0.930  -0.625  -0.615
 0.065   1.233   0.341   0.669  -0.882  -2.121  -0.252  -0.822  -0.028  -1.435
-1.132   0.382  -1.313   0.949  -1.785  -0.432   0.604   1.279   0.824  -0.171
-0.670  -0.237   0.094  -0.528  -1.039   0.882   0.204   1.945   1.086  -0.459
 0.832   0.403  -0.526   0.020  -1.749   0.210  -0.168   0.488  -1.217  -0.094

-0.556   0.256   0.834  -0.992   0.166   1.274  -0.385   0.096   0.805  -1.167
 1.353   1.160   0.554   0.296   1.172  -0.317  -0.737  -0.484   0.169  -0.647
-2.321  -0.204  -0.459  -0.929   0.776  -1.220  -0.637   0.003  -1.263  -1.424
 2.078  -0.858   0.108  -0.669  -1.671   0.020   0.922   1.862  -0.552  -0.148
-0.006   1.005  -0.626   0.091   0.292   0.375   0.702  -1.369   0.523  -0.368
 1.901   0.446  -0.581   0.206  -0.168   1.856  -0.438   2.029  -0.532   0.493
 0.574  -1.552   0.271   0.432   1.144   0.006   0.110   0.321  -0.259  -0.571
-0.581   1.280  -1.838   1.292  -1.463  -1.059   0.835   0.164   0.525   0.494
 1.792  -0.920  -0.868  -0.525   1.174   0.082  -0.106   1.189  -2.705   0.531
 0.004   1.921   1.233   0.374   0.418   0.555  -0.433   1.035   0.309  -1.218

-1.082   0.084   1.462   1.204   2.446   0.096   0.234   1.102  -0.046   0.166
 0.801  -0.122   0.217   0.707   0.801  -0.966  -0.465   0.219  -1.604  -0.567
-0.228   0.372   0.407   0.759   0.494   0.121  -0.014  -1.808   0.940   0.303
 1.382  -0.194  -1.351   0.457  -0.929  -1.523   0.098   0.135  -0.988   1.349
-1.453  -1.325   0.513   0.684  -0.294  -0.398  -1.070  -0.654  -0.550   0.350
 0.099  -0.802   0.307  -1.010  -0.361  -0.606   2.541  -0.343   1.644   0.552
-0.517  -0.589  -0.679   0.583   0.601   1.472  -0.776  -0.410  -1.223  -0.188
-0.604  -2.429  -0.888  -0.575  -1.000  -0.717   1.238  -2.113  -2.314  -0.205
-0.565  -1.363   0.050  -0.235   1.649   0.982   0.777   1.541   1.174   0.528
-0.510   1.450   1.024   0.109   0.480  -0.243  -0.755  -1.345   0.461  -0.197

-0.486   0.867   1.104   0.457  -2.924   0.879  -2.512   1.337  -0.702   0.127
-0.381   0.453   2.281   0.336  -1.632   0.661   0.180   2.297  -0.711  -0.506
-1.325   0.282   0.983   1.190
```

Process B

```
 0.435  0.412  0.533  1.361  1.279  0.645  0.506  0.942  1.072  0.957
 0.434  0.646  1.243  0.337  0.733  1.079  1.450  0.972  1.106  0.922
 1.375  0.560  0.157 -0.084 -0.495 -0.482 -0.434 -0.428  0.071  0.057
 0.183  0.285  0.564  0.386 -0.294 -0.383 -0.609 -0.512 -1.159 -1.039
-0.859 -0.769 -1.253 -0.458  0.154  0.343  0.314  0.468  0.761  1.285
 0.606  1.464  1.342  0.842  0.077 -0.216 -0.443 -0.893 -1.232 -1.706
-1.371 -2.021 -1.828 -1.947 -1.359 -0.777 -0.817 -0.922 -1.182 -1.241
-1.541 -0.890 -1.228 -0.654 -1.051 -1.784 -1.394 -1.185 -0.621 -0.683
-0.733 -1.092 -0.487 -0.331 -0.292  0.433  0.002 -0.272 -0.135 -0.442
-0.222 -0.017 -0.093 -0.812  0.018 -0.019  0.193  0.203 -0.079  0.059

 0.014 -0.705 -0.854  0.153 -0.431 -0.586 -0.547  0.162  0.182  0.627
 0.565  0.832  0.816  0.312  0.509  0.466  0.533 -0.207 -0.595 -0.441
-0.233 -0.384 -0.734 -0.497 -0.619 -0.482 -0.315  0.438  1.055  1.156
 0.605  0.887  0.954  0.950  1.054  1.188  0.676  0.111 -0.162 -0.677
-0.348 -0.327 -0.076 -0.221 -0.283 -0.044 -0.115 -0.685 -0.859  0.078
 0.411  1.099  1.041  0.747  0.711  0.638  1.265  1.990  2.109  1.377
 1.414  0.575  0.094  0.334  0.491 -0.459 -0.495 -0.183 -0.439 -0.202
-0.374 -0.674 -0.239 -0.375 -0.556  0.187  0.278  0.134  0.365 -0.094
-0.178 -0.218 -0.139 -0.059  0.276 -0.191 -0.248 -1.003 -0.720 -0.290
-0.005  0.738  0.921  0.661  0.471  0.983  0.188  1.202  0.872  0.998

 0.866  0.318  0.112  0.356  0.546 -0.107  0.331 -0.219  0.014 -0.324
-0.501 -0.487 -0.039  0.236 -0.309 -0.201  0.170  0.652  0.366  0.333
 0.464  0.296  0.378  0.380  0.223  0.309  0.327  0.062  0.655  0.702
 0.898  0.864 -0.078 -0.636 -0.516  0.069 -0.360  0.201 -0.349 -0.863
-1.191 -0.859  0.051 -0.327  0.416 -0.199 -0.395 -0.929 -1.160 -0.531
-0.393  0.082 -0.772 -0.070 -0.410  0.284  0.651  0.596  0.871  0.439
 0.325  0.378  0.883  0.462  0.153  0.250 -0.159 -0.300 -0.705 -0.022
 0.404 -0.146 -0.175 -0.103 -0.084 -0.835 -0.283 -0.326 -0.177 -0.301
-1.018 -0.557 -1.105 -0.143  0.245  0.449  0.660  0.690  1.173  0.817
 1.345  0.852  0.955 -0.247 -0.261 -0.351 -0.973 -1.525 -2.271 -2.350

-2.780 -3.169 -3.127 -2.775 -3.005 -3.200 -2.684 -2.322 -2.269 -1.609
-1.362 -1.387 -0.978 -0.848 -0.861 -0.494 -0.777 -0.419  0.190 -0.348
-0.388  0.010 -0.365 -0.121  0.600  0.725  1.016  1.055  1.493  1.876
 2.157  2.005  2.765  2.422  2.095  1.607  0.963  1.388  0.873  0.385
-0.064  0.069 -0.256 -0.662 -0.480 -0.107 -0.185 -1.362 -1.222 -1.121
-0.967 -0.852 -0.942  0.008  0.215 -0.098  0.023  0.549  0.385  0.017
 0.139  0.402  0.095  0.312 -0.323  0.336  0.546  0.872  0.928  1.288
 0.910  0.545  0.934  0.058  0.329  0.084  0.123 -0.424 -0.115  0.288
 0.799  0.670  0.471  0.496  0.235  0.252  0.095  0.395  0.258  0.073
-0.626 -0.166 -0.084 -0.302 -0.785 -1.041 -1.091 -1.410 -1.489 -2.043

-1.773 -2.245 -2.460 -1.877 -1.202 -1.015 -0.683 -0.448 -0.743 -0.557
-0.370 -0.379 -0.568 -0.493 -1.090 -1.138 -1.245 -1.603 -1.403 -1.160
-0.755 -0.222  0.488  0.785  1.347  1.561  2.249  2.418  2.601  2.769
 2.408  2.620  2.973  2.323  2.189  1.931  0.879  1.532  1.035  0.589
 0.537  0.096 -0.576 -0.498 -0.303 -0.433  0.035  0.052  0.686  1.002
 1.319  1.409  1.605  1.908  2.011  1.650  2.059  1.505  0.882  0.422
 0.109 -0.153  0.169 -0.003  0.473  0.777  0.358  0.314  0.901  1.782
 1.773  1.952  1.780  1.499  1.266  1.336  1.618  2.139  1.880  1.579
 1.184  0.934  1.515  1.903  0.932  1.315  0.762  0.017  0.136 -0.254
 0.175  0.324 -0.549 -0.601  0.313  0.056  0.080  0.165  0.153  0.540
```

APPENDIX 385

```
 0.339  0.315  0.174  0.443 -0.112 -0.383 -0.687 -0.121 -0.087 -0.483
 0.324  0.300  0.668  0.270  0.246  0.369  0.854  0.721  0.333  0.831
 0.224 -0.015 -0.669 -0.375 -0.476 -0.591 -0.414 -0.254  0.077  0.413
 0.546  0.401  0.475 -0.063  0.278  0.678  0.615  0.466  0.638 -0.274
-0.739 -0.275  0.081  0.656  0.555  0.827  0.574  0.055 -0.531  0.341
 1.016  0.823  0.410 -0.060 -0.061 -0.292 -0.430  0.312  0.670  0.699
 0.507  0.424  1.136  1.345  1.181  0.789  1.146  0.433  0.864  0.261
 0.201  0.650  0.260 -0.039 -0.261 -0.894 -1.024 -0.108 -0.676 -1.146
-0.891 -0.898 -1.113 -0.805  0.153  1.421  1.395  0.649  1.008  1.640
 1.192  0.990  0.723  0.830  0.323 -0.073  0.306  0.660  0.218  0.107

-0.020 -0.231 -0.226 -0.250 -0.272 -0.583 -0.299 -0.310  0.249  0.296
-0.098  0.299  0.850  1.359  1.373  0.980  0.591  0.261  0.291 -0.003
 0.408 -0.426 -0.783 -1.202 -1.216 -0.590 -0.423 -0.288 -1.132 -0.746
-0.276 -0.022  0.361  0.340 -0.303 -0.624 -1.353 -1.025 -0.869 -0.617
-1.310 -1.187 -1.437 -1.702 -1.599 -0.809 -0.828 -0.668 -0.036 -0.869
-0.734 -0.814 -0.735 -0.533 -0.645 -0.569  0.302  0.250 -1.136 -0.625
-0.168  0.028  0.272  0.198  1.048  0.244 -0.813 -0.492  0.017 -0.320
-0.387 -0.309 -0.314 -0.715 -0.941 -0.581  0.088 -0.125  0.438  0.259
 0.329  0.216 -0.010  0.172  0.030 -0.342 -0.670 -0.551 -0.173  0.005
-0.611 -0.284  0.170  0.416  0.851  1.713  1.389  1.385  1.218  1.717

 2.454  2.431  2.107  1.461  1.372  0.719  1.650  1.215  0.769  0.241
 0.257  0.026 -0.075  0.026  0.103  0.249 -0.304 -0.588 -0.559 -0.211
-0.536 -0.500  0.239  0.917  0.942  0.385  0.352  0.462  0.426  0.449
 0.314  0.620 -0.425 -0.558 -0.996 -0.001  0.739  0.644  1.276  1.689
 1.367  1.327  1.711  1.646  1.726  1.607  1.162  1.816  1.096  0.449
 0.732  0.581  0.934  0.981  1.134  0.405 -0.344 -0.932 -0.900 -0.721
-0.966 -1.325 -1.579 -2.075 -1.991 -2.261 -1.755 -1.499 -0.870 -0.615
-0.237 -0.099 -0.284  0.137 -0.305 -0.082  0.311  0.191  0.391  0.469
 0.383  0.833  1.024  0.839  1.003  0.790  0.589  0.478  0.042 -0.648
-0.538 -0.954 -0.998 -0.590 -0.893 -0.320 -0.002 -0.067 -0.184  0.422

 0.100  0.672  0.943  0.859  1.244  1.546  1.069  0.964  0.791  0.600
 0.756 -0.344 -0.753 -1.056 -1.423 -1.541 -1.812 -1.782 -1.636 -2.065
-2.298 -0.981 -1.177 -1.007 -0.929 -1.661 -1.290 -0.823 -0.267 -0.054
 0.328 -0.296  0.262  0.042  0.270  0.857  0.964  0.898 -0.069  0.252
 0.186  0.757  0.590  0.603  0.638  0.500  0.943  0.602  1.619  1.303
 1.561  1.164  0.566  0.821  0.889  1.281  0.728  0.891  0.380  0.462
 0.144 -0.202  0.646  0.015  0.272 -0.509 -0.827 -0.610 -0.657 -0.423
-0.104  0.606 -0.052  0.238 -0.306  0.484  0.825  0.543  0.850 -0.117
-0.106 -0.641  0.209  0.838  1.108  0.881  1.023  0.925  0.879  1.781
 1.258  0.933  0.383  0.451  0.700  1.307  1.169  1.369  1.389  1.283

 1.697  2.260  2.199  1.826  1.678  1.185  0.867  0.658  0.394 -0.073
-0.292 -0.600 -0.452 -0.395 -0.379 -0.471 -0.146 -0.011 -0.618  0.144
 0.404  0.886  0.716  0.190  0.100 -0.327 -0.819 -0.785 -0.203 -0.780
-0.467 -1.316 -1.654 -1.097 -1.028 -0.838 -0.501 -0.851 -1.087 -0.956
-1.255 -0.790 -0.634 -0.696 -1.203 -1.223 -1.285 -0.204 -0.111  0.546
 0.606  0.421  0.485  0.190  0.667  0.955  1.577  0.584  0.564 -0.294
-0.516 -0.542 -1.093 -1.155 -1.502 -1.981 -2.636 -2.033 -2.543 -2.870
-2.875 -2.863 -2.544 -2.263 -2.161 -1.368 -0.860 -0.998  0.096  1.140
 1.359  1.376  2.218  2.509  2.612  2.263  1.896  1.437  0.574  0.360
 0.143  0.150 -0.025 -0.001  0.103 -0.916 -0.580 -1.106 -0.303 -0.651

-0.554 -0.522 -0.646 -0.294 -0.330  0.057 -0.009  0.797  1.084  1.082
 0.892  0.610  0.558  0.170  0.426  0.929  0.949  1.095  0.401  0.830
 0.976  1.225  0.829  0.619
```

Author Index

(*Numbers in italic refer to pages where references are listed.*)

A

Abrahamsson, S., 325, *340*
Abramowitz, M. and Stegun, I., 195, *228*, *258*, *272*
Adams, G. F., *see* A. A. Friesem, A. Kozma and —
Akcasu, A. Z., 292, *316*
Altman, J. H., 287, *315*, *340*, *341*
—, Grum, F. and Nelson, C. N., 49, *65*
— and Stultz, K. F., 320–322, *340*
— and Zweig, H. J., 357, 360–363, 372, *376*
see also: Schmitt, H. C. and —
Ames, A. E., 130, *149*, 151
see also: Bird, G. R., Jones, R. C. and —
Amoss, J. and Davidson, F. C., 171, *186*
Andrews, H. C., 200, 201, *228*, 352, *376*
Arm, M., *see* Stark, H., Bennet, W. R. and —
Arps, R. B., 374, *378*

B

Bacik, H., Coleman, C. I., Cullum, M. J., Morgan, B. L., Ring, J. and Stephens, C. L., 174, *187*
Bahler, W. W., *see* Barnes, J. C. — and Johnston, G. J.
Baker, E. A., 128, *148*
Baldwin, R., *see* Blackmer, L. L., Van Kerkhove, A. P. and —
Ballantyne, J. G., 265, *273*
Barnard, T. W., 171, *186*
Barnes, J. C., Johnston, G. J., and Moretti, W. J., 238, *270*
—, Bahler, W. W. and Johnston, G. J., 238, *270*
Barr, G. R., Thirtle, J. R. and Vittum, P. W., 238, *270*

Barrows, R. S. and Wolfe, R. N., 53, 54, *65*, 238, *270*
Bartletson, C. J. and Witzel, R. F., 364, *376*
Baudry, P., Desprez, R. and Preteseille, D., 223, *229*
Baum, W. A., 168, 169, *186*
see also: Wilcock, W. L. and —
Bayer, B. E., 96, *113*, 303, 304, *316*
— and Hamilton, J. F., 125, 127, 131, *148*
see also: Hamilton, J. F. and —
Becherer, R. J. and Geller, J. D., 335, *341*
Beckman, J. E., 375, *378*
Beiser, L., 375, *378*
Benarie, M. M., 135–137, *149*
Bendat, J. S. and Piersol, A. G., 292, *316*
Bennet, W. R., *see* Stark, H., — and Arm, M.
Benton, S. A. and Kronauer, R. E., 303, 305, *316*
Berg, W. F., 35, 51, 52, *64*, 135–137, *149*, 255, *272*, 333, *341*
Berry, C. R. and Skillman, D. C., 43, 45, *64*
Bershad, N. J., 374, *378*
Berwart, L., 306, 307, *316*
Bestreiner, F., Deml, R. and Greis, U., 364, *377*
—, Greis, U. and Weierhausen, W., 364, *377*
Biedermann, K., 302, *316*
— and Frieser, H., 373, *377*
— and Johansson, S., 249, 254, *272*, *273*, 340, *341*
see also Johansson, S. and —
Bingham, C., Godfrey, M. D. and Tukey, J. W., 292, *316*
Bird, G. R., 111, *113*
—, Jones, R. C. and Ames, A. E., 106, 108, *113*, 169, *186*
Black, G. and Linfoot, E. H., 346, *376*

Blackman, E. S., 255, *272*, 335, *341*
Blackman, R. B. and Tukey, J. W., 292, *315*
Blackmer, L. L., Van Kerkhove, A. P. and Baldwin, R., 364, *377*
Born, M. and Wolf, E., *229*, 325, 327, *340*
Bossung, J. W., *see* Galburt, D., Jones, R. A. and —
Bracewell, R., *229*, 246, *271*
Brillouin, L., 344, *376*
Brock, G. C., 264, 265, *273*
Brown, F. M., Hall, H. J. and Kosar, J., *65*
Brown, T. J., 145, *149*
Bumba, V., *see* Pospíšil and —
Burton, P. C., 51, *64*, 76, *112*
Burton, W. M., Hatter, A. T. and Ridgeley, A., 63, *65*
Buschmann, H. T., 249, *273*

C

Cain, D. G., 173, *187*
Castro, P. E., Kemperman, J. H. B. and Trabka, E. A., 303, 305, *316*
Catchpole, C. E., 173, *187*
Chambers, R. P. and Courtney-Pratt, J. S., 171, *187*
Champeney, D. C., *229*
Chanter, J. B., *see* Farnell, G. C. and —
Charman, W. N., 267, *273*, 333, *341*
Chibisov, K. B., 145, *149*
Clarke, G. L. and Zarem, A. M., 374, *378*
Cohen, M., *see* Kahan, E. A. and —
Coleman, C. I., *see* Bacik, H., *et al.*
Coltman, J. W., 243, *271*
Cooley, J. W. and Tukey, J. W., 200, *228*
Coughlin, J. F., *see* Jones, R. A. and —
Courtney-Pratt, J. S., 171, *186*
 see also: Chambers, R. P. and —
Cronin, D. J. and Reynolds, G. O., 329, 330, *342*
Cullum, M. J., *see* Bacik, H., *et al.*

D

Dainty, J. C., 241, *271*
Davenport, W. B. and Root, W. L., *229*
Davidson, F. C., *see* Amoss, J. and —
De Belder, M., 173, *187*
— and Kerf, J. De, 292, 293, *316*
—, De Kerf, J., Jespers, J. and Verbrugghe, R., 261, *273*

—, Jespers, J. and Verbrugghe, R., 241, *271*
 see also: Verbrugghe, R. and —
De Kerf, J., *see* De Belder, M. and —
 see also: De Belder, M., —, Jespers, J. and Verbrugghe, R.
Deml, R., *see* Bestreiner, F., — and Greis, U.
DePalma, J. J. and Gasper, J., 261–263, *273*
 see also: Wolfe, R. N., Marchard, E. W. and —
Derr, A. J., 46, *64*
 see also: Kleinsinger, I. J., — and Giuffre, G. F.
Desprez, R. and Pollet, J., 241, *271*
 see also: Baudry, P., — and Preteseille, D.
Dillon, P. L., *see* Marchant, J. C. and —
Djurle, E., *see* Ingelstam, E., — and Sjögren, B.
Doerner, E. C., 308, 309, *317*, 334, *341*
Dorion, G. H. and Wiebe, A. F., 375, *378*
Duffieux, P. M., 190, *228*

E

Edgar, R. F., Lawrenson, B. and Ring, J., 241, *271*
Edinger, J. R., *see* Lauroesch, T. J., *et al.*
Elias, P., 190, 204, *228*, 346, *376*
—, Grey, D. S. and Robinson, D. Z., 190, *228*
Emberson, D. L., *see* Wilcock, W. L., — and Weekley, B.
Enochson, L., *see* Otnes, R. K. and —
Ericson, R. H. and Marchant, J. C., 40, 59, *65*
Eyer, J. A., 239, *270*

F

Falconer, D. G., 171, *186*
Fallon, J. D., 329, 330, *342*
Fanchenko, S. D. and Frolov, V. A., 376, *378*
Farnell, G. C., 36, 37, 53, *64*
— and Chanter, J. B., 36, *64*, 117–123, 128, *148*
— and Solman, L. R., 43, *64*

—, Saunders, A. E. and Solman, L. R., 109, *113*
Feller, W., 127, *148*
Fellgett, P. B., 28, *31*, 62, *65*, 135–137, *149*, 169, 176, 177, 181, *186*, *187*, 191, *228*, 276, 288, *315*, 346, *376*
— and Linfoot, E. H., 346, 365, 366, *376*
Findeis, G., *see* Frieser, H. and —
Fogel, S. J., *see* Katz, J. and —
Fried, D. L., 181, *187*
Frieden, B. R., 375, *378*
Friesem, A. A., Kozma, A. and Adams, G. F., 249, *272*
Frieser, H., 141, *149*, 190, *227*, 236, 260, 261, *270*, *272*, 290, *315*
— and Findeis, G., 290, *315*
— and Klein, E., 76, *112*
— and Metz, H. J., 236, *270*
see also: Biedermann, K. and —
Frolov, V. A., *see* Fanchenko, S. D. and —
Fulmer, G. G., *see* Lauroesch, T. J., *et al.*

G

Gabor, D., 344, *376*
Galburt, D., Jones, R. A. and Bossung, J. W., 336–339, *341*
Garbe, W. F., *see* Lamberts, R. L., Straub, C. M. and —
Gaspar, J., *see* DePalma, J. J. and —
Geller, J. D., *see* Becherer, R. J. and —
see also: Kofsy, I. L., — and Miller, C. S.
Gendron, R. G., *see* Goddard, M. C. and —
Gilmore, H. F., 260, *272*
Giuffre, G. F., *see* Kleinsinger, I. J., Derr, A. J. and —
Goddard, M. C. and Gendron, R. G., 243, *271*
Godfrey, M. D., *see* Bingham, C., — and Tukey, J. W.
Goodman, J. W., 111, *113*, 229
—, Miles, R. B. and Kimball, R. B., 111, *113*
Gorokhovskii, Y. N. and Levenberg, T. M., 41, *64*
—, Grigor'yev, A. G., Ivanov, A. M. and Stepochkin, A. A., 321, *340*

Greis, U., *see* Bestreiner, F., Deml, R. and —
see also: Bestreiner, F., — and Weierhausen, W.
Grey, D. S., *see* Elias, P., — and Robinson, D. Z.
Griesmer, J. J., *see* Hill, E. R. and —
Grigor'yev, A. G., *see* Gorokhovskii, Y. N., *et al.*
Grimes, D. N., 329, 330, *341*
Grum, F., 243, *271*
see also: Altman, J. H., — and Nelson, C. N.
Gurney, R. W. and Mott, N. F., 34, *64*
Guttman, A., 169, 170, *186*

H

Haas, R. C., *see* Kinzly, R. E., — and Roetling, P. G.
Haase, G., 124, *148*
— and Müller, H., 260, *273*
Hacking, K., 276, 300, *315*
Hall, H. J., *see* Brown, F. M., — and Kosar, J.
Hamilton, J. F., *65*, 106, 108, *113*, 126, *148*
— and Bayer, B. E., 125, 127, 131, *148*
— and Marchant, J. C., 93, *112*, 174, *187*
see also: Bayer, B. E. and —
Hariharan, P., 249, *272*
Hartley, R. V. L., 344, 347, *376*
Hatter, A. T., *see* Burton, W. M., — and Ridgeley, A.
Haugh, E. F., 96, *113*
Hayen, L. and Verbrugghe, R., 171, *187*
Hendeberg, L. O., 239–243, 254, 255, *270*, *271*, *272*
Herrnfeld, F. P., 320, 321, *340*
Higgins, G. C., 240, *270*
— and Stultz, K. F., 59, *65*
see also: James, T. H. and —; Jones, L. A. — and Stultz, K. F.; Lamberts, R. L. and —; Zweig, H. J., — and MacAdam, D. L.
Hill, E. R. and Griesmer, J. J., 111, *113*
Hisdal, E., 373, *377*
Hoag, A. A. and Miller, W. C., 168, *186*
Hopkins, H. H., 333, *341*
Howell, H. K., *see* Kohler, R. J. and —

I

Ingelstam, E., 212, *228*
—, Djurle, E. and Sjögren, B., 232, *270*
Ivanov, A. M., *see* Gorokhovskii, Y. N., *et al.*

J

Jakeman, E., 294, *316*
James, T. H. and Higgins, G. C., *65*
 see also: Mees, C. E. K. and —
Jenkins, G. M. and Watts, D. G., 292, *316*
Jennison, R. C., *229*
Jespers, J., *see* De Belder, M., — and Verbrugghe, R.
 see also: De Belder, M., De Kerf, J., — and Verbrugghe, R.
Johansson, S. and Biedermann, K., 336, *341*
 see also: Biedermann, K. and —
Johnston, G. J., *see* Barnes, J. C. — and Moretti, W. J.
 see also: Barnes, J. C., Bahler, W. W. and —
Jones, L. A., Higgins, G. C. and Stultz, K. F., 282, *315*
Jones, R. A., 245, 246, *271*, 334, *341*
— and Coughlin, J. F., 336, *341*
— and Yeadon, E. C., 245, *271*
 see also: Galburt, D., — and Bossung, J. W.; Yeadon, E. C., — and Kelly, J. T.
Jones, R. C., 28, *31*, 62, *65*, 162–167, *185*, *186*, 191, 210, 214, *228*, 258, *272*, 276, 278, 280, 290, 306, *315*, 365, 368, 372, 373, *377*
 see also: Bird, G. R., — and Ames, A. E.

K

Kahan, E. A. and Cohen, M., 375, *378*
Kammerer, W., *see* Shepp, A. and —
Katz, J., and Fogel, S. J., *65*
Keene, G. T., *see* Lauroesch, T. J., *et al.*
Keitz, H. A. E., 46, *64*
Kelly, D. H., 232, 240, *270*
Kelly, J. T., *see* Yeadon, E. C., Jones, R. A. and —
Kemperman, J. H. B., *see* Castro, P. E., — and Trabka, E. A.

Kendall, M. G. and Stuart, A., 285, *315*
Kerwich, T. F., *see* Lauroesch, T. J., *et al.*
Kimball, R. B., *see* Goodman, J. W., Miles, R. B. and —
Kinzly, R. E., 328–330, *341*, *342*
—, Haas, R. C. and Roetling, P. G., 335, *341*
Kirillov, N. I., 135, 136, 142, 143, *149*
Klein, E., 76, *112*, 124, *148*, 252, *273*
— and Langner, G., 142, *149*, 280, 290, 300, *315*
 see also: Frieser, H. and —
Kleinsinger, I. J., Derr, A. J. and Giuffre, G. F., 333, *341*
Kofsy, I. L., Geller, J. D. and Miller, C. S., 340, *341*
Kohler, R. J. and Howell, H. K., 171, *187*
Kosar, J., *see* Brown, F. M., Hall, H. J. and —
Kozma, A., *see* Friesem, A. A., — and Adams, G. F.
Kowaliski, P., *65*
Kraus, W., 323, *340*
Krochman, J., 46, *64*
Kronauer, R. E., *see* Benton, S. A. and —
Kukarov, G. V., *see* Znamensky, V. B., — and Strukov, V. S.

L

Lamberts, R. L., 241, 267–269, *271*, *273*, 364, *377*
— and Higgins, G. C., 364, *377*
—, Straub, C. M. and Garbe, W. F., 243, *271*, 336–340, *341*
Langner, G., 96, 101, 102, *113*, 252, 272, 303, 306, *316*, 373, *377*
— and Muller, R., 241, *271*, 336, *341*
 see also: Klein, E. and —
Lauroesch, T. J., Fulmer, G. G., Edinger, J. R., Keene, G. T. and Kerwich, T. F., 265, *273*
Lawrenson, B., *see* Edgar, R. F., — and Ring, J.
Lawton, W. H., Trabka, E. A. and Wilder, D. R., 303, 304, *316*
Lebedev, D. S., 373, *377*
Lehmbeck, D. R., 364, *376*
Leistner, K., 340, *341*

AUTHOR INDEX

Levenberg, T. M., *see* Gorokhovskii, Y. N. and —
Levenson, G. I. P., 135, *149*
Levi, L., 180, *187*, 357, 374, *376*, *377*
Lighthill, M. J., *229*
Linfoot, E. H., *229*, 346, 365, 368, 370, 374, *376*, *377*
— *see also:* Fellgett, P. B. and —; Black, G. and —
Lohmann, A. W., 308, *317*
Lorber, H. W., 233, 247, *270*
Lubberts, G., 311, *317*
Luneberg, R. K., 190, *228*

M

MacAdam, D. L., 135–137, *149*
— *see also:* Zweig, H. J., Higgins, G. C. and —; Zweig, H. J., Stultz, K. F. and —
Maimon, S. R., 345, *376*
Mandel, L., 104, *113*
Marathay, A. S. and Skinner, T. J., 111, *113*
Marchand, E. W., 210, *228*
— *see also:* Wolfe, R. N., — and De-Palma, J. J.
Marchant, J. C., 155, *185*
— and Dillon, P. L., 96, *113*
— and Millikan, A. G., 155, *185*
— *see also:* Ericson, R. H. and —; Hamilton, J. F. and —
Marriage, A., 36, 37, *64*, 118–123, *148*
— and Pitts, E., 282, *315*
— *see also:* Pitts, E. and —
McCamy, C. S., 374, *377*
McGee, J. D. and Wheeler, B. E., 145, *149*
McIrvine, E. C., *see* Tribus, M. and —
Mees, C. E. K. and James, T. H., 34, 39, 41, 53, *64*, 219, *229*, 239, 252, 253, 265, *270*
Metz, C. E., Strubler, K. A. and Rossmann, K., 204, *228*
Metz, H. J., *see* Frieser, H. and —
Meyer, R., 135, *149*
Middleton, D., *see* Van Vleck, J. H. and —
Miles, R. B., *see* Goodman, J. W., — and Kimball, R. B.
Miller, C. S., *see* Kofsy, I. L., Geller, J. D. and —

Miller, W. C., 169, *186*
Millikan, A. G., *see* Marchant, J. C. and —
Mitchel, R. H., *see* Vander Lugt, A. and —
Mitchell, J. W., 34, *64*
Morawiski, T., *see* Romer, W. and —
Moretti, W. J., *see* Barnes, J. C., Johnston, G. J. and —
Morgan, B. L., *see* Bacik, H., *et al.*
Mott, N. F., *see* Gurney, R. W. and —
Müller, H., *see* Haase, G. and —
Müller, R., *see* Langner, G. and —
Murty, M. V. R. K., 243, *271*

N

Nakamura, K. and Sakata, T., 346, *376*
Nelson, C. N., 54, 55, *65*, 240, *270*, 364, *376*
— *see also:* Niederpruem C. J., — and Yule, J. A. C.; Altman, J. H., Grum, F. and —
Niederpruem, C. J., Nelson, C. N. and Yule, J. A. C., 48, *64*
Nitka, H. F., 173, *187*
Nutting, P. G., 41–46, *64*, 76

O

O'Neill, E. L., 306, *317*
Ooue, S., 139, 140, *149*, 241, 248, *270*
Otnes, R. K. and Enochson, L., 292, *316*

P

Papoulis, A., 219, *228*
Paris, D. P., 258, *272*
Perrin, F. H., 49, 56, 60, *65*, 232, *270*
Picinbono, M. B., 96, *112*, 303, *316*
Pierce, J. R., 346, *376*
Piersol, A. G., *see* Bendat, J. S. and —
Pitts, E. and Marriage, A., 282, *315*
— *see also:* Marriage, A. and —
Pollet, J., *see* Desprez, R. and —
Porteous, R. L. and Shaw, R., 169, 170, *186*
Pospíšil, J. and Bumba, V., 243, *271*
Powell, P. G., 239, 258, *270*
Preteseille, D., *see* Baudry, P., Desprez, R. and —
Pryor, P. L., 171, *186*
Ptashenchuk, V. M., *see* Vendrovsky, K. V., Veitzman, A. I. and —

R

Rabedeau, M. E., 249–251, *272*
Reynolds, G. O. and Smith, A. E., 327–329, *341*
  see also: Cronin, D. J. and —
Reynolds, G. T., 145, *149*
Richardson, W. W., *see* Smith, A. G., Schrader, H. W. and —
Ridgeley, A., *see* Burton, W. M., Hatter, A. T. and —
Riesenfeld, J., 363, *376*
Ring, J., *see* Bacik, H., *et al.*
  see also: Edgar, R. F., Lawrenson, B. and —
Riva, C., 288, 290, 292, 300, *315*
Robinson, D. Z., *see* Elias, P., Grey, D. S. and —
Roetling, P. G., 308, *317*
  see also: Kinzly, R. E., Haas, R. C. and —
Romer, W. and Morawiski, T., 43, *64*
Root, W. L., *see* Davenport, W. B. and —
Rose, A., 28, *31*, 135–139, *149*
Rosell, F. A., 140, *149*
Rosenau, M. D., *see* Scott, F. and —
Rosenblum, W., 174, 175, *187*
Rosenfeld, A., 352, *376*
Rossmann, K., 105, *113*, 173, *187*, 308–311, *317*
  see also: Metz, C. E., Strubler, K. A. and —

S

Sakara, T., *see* Nakurama, K. and —
Saunders, A. E., *see* Farnell, G. C., — and Solman, L. R.
Savelli, M., 96, *112*, 306, *317*
Schade, O. H., 96, *113*, 135–137, *149*, 181, 182, *187*, 190, *228*, 232, 239, *269*, *270*, 276, *315*
Schmidlin, F. W., 140, *149*
Schmitt, H. C. and Altman, J. H., 303, *316*, 330, *341*
Schrader, H. W., *see* Smith, A. G., — and Richardson, W. W.
Schwarzschild, K. and Villiger, W., 322, *340*
Ścibor-Marchocki, R., 365, *377*
Scott, F., 241, *271*
  — and Rosenau, M. D., 252, *272*
  —, Scott, R. M. and Shack, R. V., 245, *271*
Scott, R. M., *see* Scott, F., — and Shack, R. V.
Selwyn, E. W. H., 58, *65*, 96, *112*, 130, 135, 136, *149*, 190, 218, 220, *228*
Shannon, C. E., 344–357, 365, *376*
  — and Weaver, W., 197, *228*, 346, *376*
Shack, R. V., *see* Scott, F., Scott, R. M. and —
Shaw, R., 6, 9, 10, *30*, 61, 63, *65*, 71–111, *111*, *112*, *113*, 121–132, 142, 144, *148*, *149*, 156–160, 169, 171, 176–185, *185*, *186*, *187*, 241, 271, 280, 282, 306, 311–314, *315*, *316*, 365–372, *377*
  — and Shipman, A., 62, 63, *65*, 169, *186*, 302, *316*
  see also: Porteous, R. L. and —
Sheberstov, V. I., *see* Vendrovsky, K. V. and —
Shepp, A. and Kammerer, W., 171, *186*
Shin, M. C. H., *see* Swing, R. E. and —
Shipman, A., *see* Shaw, R. and —
Siedentopf, H., 58, *65*, 96–100, 103, *112*, 303, *316*
Silberstein, L., 76, *112*, 130, *149*
  — and Trivelli, A. P. H., 76, 93, *112*
Simonds, J. L., 49, *64*, 240, *270*
Sjögren, B., *see* Ingelstam, E., Djurle, E. and —
Skillman, D. C., *see* Berry, C. R. and —
Skinner, T. J., *see* Marathay, A. S. and —
Smith, A. E., *see* Reynolds, G. O. and —
Smith, A. G., Schrader, H. W. and Richardson, W. W., 169, *186*
Smith, H. M., 291, *315*
Solman, L. R., *see* Farnell, G. C. and —
  see also: Farnell, G. C., Saunders, A. E. and —
Soule, H. V., 141, *149*
Southwold, N. K. and Waters, W. G., 321, 322, *340*
Spencer, H. E., 73, *112*, 123, 124, 141, *148*
Stark, H., Bennet, W. R. and Arm, M., 291, 292, *315*
Steel, B. G., *see* Wall, F. J. B. and —
Stegun, I., *see* Abramowitz, M. and —
Stephens, C. L., *see* Bacik, H., *et al.*

AUTHOR INDEX

Stepochkin, A. A., *see* Gorokhovskii, Y. N., *et al.*
Straub, C. M., *see* Lamberts, R. L., — and Garbe, W. F.
Strübin, H., 255, *272*
Strubler, K. A., *see* Metz, C. E., — and Rossmann, K.
Strukov, V. S., *see* Znamensky, V. B., Kukarov, G. V. and —
Stuart, A., *see* Kendall, M. G. and —
Stultz, K. F. and Zweig, H. J., 307, *317*
  see also: Higgins, G. C. and —; Altman, J. A. and —; Jones, L. A., Higgins, G. C. and —; Zweig, H. J., — and MacAdam, D. L.
Swank, R. K., 311, *317*
Swing, R. E., 328, 329, *341*
— and Shin, M. C. H., 248, *272*

T

Takahasi, H., 374, *377*
Tatian, B., 204, *228*
Thirtle, J. R., *see* Barr, G. R., — and Vittum, P. W.
Thiry, H., 248, 249, *271, 272*, 291, 306, *315*
Toy, F. C., 76, *112*
Trabka, E. A., 96, 101, *113*, 280, 283, 284, 303, 304, *315, 316*
  see also: Lawton, W. H., — and Wilder, D. R.; Castro, P. E., Kemperman, J. H. B. and —
Tribus, M. and McIrvine, E. C., 345, *376*
Trinder, J. C., 374, *378*
Trivelli, A. P. H., 76, *112*
  see also: Silberstein, L. and —
Tubbs, M. R., 375, *378*
Tukey, J. W., *see* Cooley, J. W. and —; Blackman, R. B. and —; Bingham, C., Godfrey, M. D. and —

V

Valentine, R. C., 174, *187*
— and Wrigley, N. G., 174, *187*
Vander Lugt, A. and Mitchel, R. H., 249, 250, *272*, 274
Van Kerkhove, A. P., *see* Blackmer, L. L., — and Baldwin, R.

Van Vleck, J. H. and Middleton, D., 294, *316*
Veitzman, A. I., *see* Vendrovsky, K. V., — and Ptashenchuk, V. M.
Vendrovsky, K. V., 372, *378*
— and Sheberstov, V. I., 135–137, *149*
—, Veitzman, A. I. and Ptashenchuk, V. M., 63, *65*, 313, *317*
Verbrugghe, R. and De Belder, M., 300, *316*
  see also: Hayen, L. and —; De Belder, M., Jespers, J. and —; De Belder, M., De Kerf, J., Jespers, J. and —
Vilkomerson, D. H. R., 291, 292, *315*
Villiger, W., *see* Schwarzschild, K. and —
Vittum, P. W., *see* Barr, G. R., Thirtle, J. R. and —

W

Wall, F. J. B. and Steel, B. G., 292, 300, 301, *316*, 321, *340*
Waters, W. G., *see* Southwold, N. K. and —
Watts, D. G., *see* Jenkins, G. M. and —
Weaver, W., *see* Shannon, C. E. and —
Webb, J. H., 76, *112*, 116–118, 123–125, 128, 130, 141, *148, 149*
Weekley, B., *see* Wilcock, W. L., Emberson, D. L. and —
Weierhausen, W., *see* Bestreiner, F., Greis, U. and —
Weingartner, I., 329, 331, *341*
Weinstein, F. S., 171, *186*
Wheeler, B. E., *see* McGee, J. D. and —
Whittaker, E. T., 197, *228*
Wiebe, A. F., *see* Dorion, G. H. and —
Wiener, N., 190, *228*, 276, *314*
Wilcock, W. L. and Baum, W. A., 173, *187*
—, Emberson, D. L. and Weekley, B., 145, *149*
Wilder, D. R., *see* Lawton, W. H., Trabka, E. A. and —
Witzel, R. F., *see* Bartletson, C. J. and —
Wolf, E., *see* Born, M. and —
Wolfe, W. L., 155, *185*
Wolfe, R. N., Marchand, E. W. and DePalma, J. J., 261, 263, *273*

Woodward, P. M. 346 356, *376*
Wrigley, N. G., *see* Valentine, R. C. and —

X
Xuong, N., 325, *340*

Y
Yeadon, E. C., Jones, R. A. and Kelly, J. T., 255–258, *272*
  see also: Jones, R. A. and —
Yu, F. T. S., 171, *186*, 306, *316*
Yule, J. A. C., *see* Niederpruem, C. J., Nelson, C. N. and —

Z
Zarem, A. M., *see* Clark, G. L. and —
Znamensky, V. B., Kukarov, G. V. and Strukov, V. S., 375, *378*
Zweig, H. J., 28, *31*, 69, 71, 93, 106, 111, *111*, *112*, *113*, 162–164, 167, 168, 172, *186*, *187*, 191, *228*, 276, 283, *315*
—, Higgins, G. C., and MacAdam, D. L., 162, *186*
—, Stultz, K. F. and MacAdam, D. L., 314, *317*
  see also: Stultz, K. F. and —; Altman, J. H. and —

Subject Index

Abel transform, 210, 214, 223
Adjacency effects, 53, 237
  chemical spread function, 239
  effect on modulation transfer function, 239, 254
  negative feedback, 239
  three-stage model, 240
Airy diffraction pattern, 327
Agitation, effect on modulation transfer function, 254
Aliasing, 205, 296
Alignment, in microdensitometry, 338
Amplification
  c-type and z-type
    effect on characteristic curve, 92, 93
    effect on image noise, 103
    effect on detective quantum efficiency, 110, 171
    factors for several imaging processes, 36
  of photon noise, 178
Amplitude transmittance–log exposure curve, 51
Aperture stop, 320
Arrays of receptors
  regular, 1
  random, 17
Astronomy
  recording problems, 158
  special techniques, 168
Autocorrelation function, 220
  influence of scanning aperture, 221
  measurement, analogue, 288
    digital, 292
  relationship with Selwyn granularity, 283
  relationship with Wiener spectrum, 222
Autocorrelation theorem, 196
Automatic focussing in microdensitometry, 336
Average power limitation, 356

Binary digit, 347
Binary recording, 360
  experimental results, 362
Binary source, 349
Bit, 344, 347
Bose–Einstein distribution, 175

Callier coefficient, 39
  dependence on density, 41
  in microdensitometry, 303, 331
Capacity, information. *See* Information capacity
Capture cross section of receptors, 43
Cascading of modulation transfer functions, 267
Cell size
  dependence on spread function, 360
  influence on information capacity, 357
Central limit theorem 219
  communication, 352
  capacity, 352, 355
  capacity for Gaussian signals and noise, 356
Characteristic curve
  basic theory, 76
  effect of c-type amplification, 93
  effect of z-type amplification, 92
  experimental data, 47, 49
  influence of fog, 94
    exposure distribution in layer, 94
    grain shape, 92
    grain size distribution, 83, 88
    grain sensitivity distribution, 81, 129
    grain volume absorption, 89
    threshold of photon receptors, 80
Checkerboard model (uniform grain array)
  detective quantum efficiency, 69
  image noise, 101
Chemical development, 44
Chemical spread function, 239

van Cittert–Zernike theorem, 327
Close-packed model. *See* Checkerboard model
Coding, 363
Colour films, Wiener spectrum of, 230
Coherence
  degree of, 326
  in microdensitometry, 326
Coherent light
  effect on image noise, 302
  in modulation transfer function measurement, 247
Communication channel, 352
Communication theory. *See* Information theory
Comparative noise level, 6, 22, 28. *See also* Detective quantum efficiency
Computational methods
  Fourier transform, 198
  Monte Carlo techniques, 260
  Wiener spectrum, 288
Continuous entropy, 349
Contrast, 48
Contrast detectivity, 163
Contrast transfer function, 234
  variation with exposure, 234, 235
Convolution
  of spread functions, 207
  theorem, 196
Correlation. *See* Autocorrelation function
Covering power, 45
Cross section, of absorption, 45
Crowding, effect on granularity, 102, 303

D-log$E$ curve. *See* Characteristic curve
Decoding, 363
Defocus, effect on optical transfer function, 332
Degree of coherence, 326
Delta function, 192
Density, 13
  fluctuations. *See* Image noise
  of unexposed emulsion, 51
  specular/diffuse, 39
Density–exposure curve, 50
  X-ray exposures, 50
Density–log exposure curve. *See* Characteristic curve
Depth variation of exposure, 52
Detection techniques, 168

Detective quantum efficiency, 28
  cascade processes, 145
  effect of c-type amplification, 172
  effect of z-type amplification, 110
  electron exposures, 173
  experimental values, 29, 30, 62, 63, 160, 167, 170, 174
  influence of fog, 110
    layer parameters, 107
    grain sensitivity distribution, 72, 107, 132
    grain size distribution, 74, 108
    spatial randomness of receptors, 72
  and information capacity, 366
  multilevel grains, 6, 22
  random grain array, 70, 72
  spatial frequency domain, 3
  and signal-to noise, 152
  two stage processes, 145
  and ultimate sensitivity, 141
  uniform grain array, 69, 72
  of visual process, 29
Detectivity, 162
  energy, 163
  contrast, 163
Development
  effect on detective quantum efficiency, 62
  effect on MTF, 253
Development effects. *See* Adjacency effects
Diazo process
  modulation transfer function, 251
  relative speed, 49
  amplification factor, 36
Diffraction limited lens, optical transfer function, 331
Digital computation
  autocorrelation function, 288
  Fourier transform, 198
  Wiener spectrum, 288
Dirac delta function, 192
Direct development, 44
DQE. *See* Detective quantum efficiency
Drum scanners, 325

Eberhard effect, 64, 236
Edge effect, 54, 236, 254
Edge spread function, 211, 245
Effective exposure, 232, 236
Effectively incoherent illumination, 327

SUBJECT INDEX

Efficiency
　detective quantum. *See* Detective quantum efficiency
　quantum, 28
　responsive quantum, 28
Eigenvector analysis, 49
Electron beam exposure
　detective quantum efficiency, 173
Electronography, 174
Electrophotography
　amplification factor, 36
　relative speed, 49
Elementary layers, 51
Emulsions
　grain sensitivity distribution, 37, 124. *See also* Quantum sensitivity distribution
　grain size distribution, 38. *See also* Grain size distribution
　scattering models, 258, 260
Encoding, 363
English language as a Markov source, 351
Energy detectivity, 163
Energy levels in silver bromide crystal, 34
Entropy theory. *See* Information theory
Equivalent quantum efficiency. *See* Detective quantum efficiency
Ergodic random process, 216
　Wiener spectrum of, 222
Errors, in measurement of image noise, 286, 295
Error function, 287
Error rate, 353, 360
Exposure
　addition and multiplication, 144, 165
　latitude, 48
　quantum fluctuations of, 2
　sinusoidal, 210
　three dimensional distribution in emulsion, 52
　units, 46
Exposure scale, 46
Extrinsic quantum efficiency, 134

Fast Fourier transform, 200
Focus
　effect on optical transfer function, 332
　automatic, in microdensitometers, 336
Fog, 48

Fog grains
　influence on characteristic curve, 94
　detective quantum efficiency, 110
Fourier transform, 190
　aliasing, 205, 296
　computational methods, 198
　in coherent optical systems, 247
　examples, 193
　fast, 200
　effects of truncation, 203
　properties, 196
　sampling theorem, 197
　two dimensional, 194
Flare light in microdensitometers, 322
Fractional grain counts
　Farnell and Chanter method, 117
　relation to grain sensitivity distribution, 118
Fraunhofer diffraction plane, 247
Frequency
　Nyquist, 197, 204
　spatial, 191, 214
　response. *See* Modulation transfer function

Gamma, 48
　variation with wavelength, 53
Gaussian distribution, 218, 220
　of density fluctuations, 218, 219
Generalized Fourier transform, 192
　examples, 194
Grain sensitivity distribution. *See* Quantum sensitivity distribution
Grain shape
　influence on characteristic curve, 92
Grain size
　and speed, 37
Grain size amplification
　influence on characteristic curve, 92
　detective quantum efficiency, 110
　image noise, 103
Grain size distribution
　experimental data, 38
　influence on characteristic curve, 83, 88
　　detective quantum efficiency, 74, 108
　　image noise, 102
Grain volume absorption
　influence on characteristic curve, 89

Granularity. *See also* Image noise and Wiener spectrum
  isotropic, 223
  measurement, 284
  influence of scanning aperture, 16, 59, 277
  scale value, G, 17, 22, 58, 284
  Selwyn, 58
  standard error in σ_A, 286
  syzygetic, 282
Granularity–density curve, 60, 61, 101
Gurney–Mott theory, 34

Hankel transform, 195, 214, 223
H and *D* curve. *See* Characteristic curve
Holographic films
  amplitude transmittance–exposure curve, 51
  modulation transfer function, 249
  noise characteristics, 302

Ideal image, 1
Ideal detector, 1
Image,
  ideal, 1
  latent, 33
Image characteristics. *See* Characteristic curve, Image noise and Detective quantum efficiency
Image intensifiers, 145
Image noise. *See also* Granularity and Wiener spectrum
  basic theory, 96
  influence of grain sensitivity distribution, 101
    grain size amplification (z-type), 103
    grain size distribution, 102
    layer parameters, 99, 102
  Siedentopf formula, 98
  upper and lower bounds, 102, 305
Imaging processes
  amplification factors, 36
  relative speeds, 49
Information capacity
  and detective quantum efficiency, 366
  of various recording processes, 345
Information content, 363
  and energy requirements, 345
Information theory
  applied to photographic systems, 357, 364

Information theory (*cont.*)—
  and Fourier analysis, 364
  Markov sources, 350
  power limited signals, 356
  practical applications, 374
  zero memory sources, 348
Input–output relationships, 68
  image resolution, 232
  signal-to-noise, 152
  See also Characteristic curve, Modulation transfer function, and Detective quantum efficiency
Intermittancy effect, 35
Intrinsic quantum efficiency, 134
Isotropic granularity, 223

Kohler illumination, 320, 325

Laser light. *See* Coherent light
Latent image
  formation of, 33
Latitude. *See* Exposure latitude
Least squares restoration filter, 335
Lens, optical transfer function of diffraction limited, 331
Light scattering
  effect on characteristic curve, 94
  modulation transfer function, 233
Limit of sensitivity, 135
Line spread function, 209
Linear microdensitometer, 329
Linearity, 232
  in microdensitometry, 328
Low contrast aprpoximation, 234

Magnification, role in optimum detection, 159
Markov information source, 350
Mean square density fluctuation. *See* Granularity, Image noise
Measurement
  grain sensitivity distribution, 116
  mean square density fluctuation, 284
  modulation transfer function, 241
  Wiener spectrum, 284
Microdensitometers, 320
  automatic focussing, 336
  correction for degradation, 243, 334
  correction functions, 335
  digitised, 324
  double beam, 323
  linear, 329

SUBJECT INDEX

Microdensitometers (cont.)—
 single beam, 320
 transfer function, 330
Microdensitometry
 effect of partial coherence, 325
 slit misalignment, 334
 stray light, 322
 reduction of granularity, 339
Misalignment, scanning slit, 334
Models
 of imaging detectors, 1, 68
 modulation transfer function, 233, 256
 one stage, 233
 three stage (Kelly), 240
 two stage, 236
 Wiener spectrum, 303
Modulation
 applied, 264
 exposure, 211
Modulation transfer function
 cascading, 267
 coherent light, 247
 experimental data, 250, 251, 254
 influence of development, 253
 image noise, 255
 measurement, 241
 Monte Carlo techniques, 260
 relation to spread functions, 212, 244, 258
 theory, 211, 233, 258
Monte Carlo techniques
 modulation transfer function, 260
 sensitivity distribution, 125
Moire fringes, as test objects, 241
Morse code, 353
Mottle
 print, 307
 X-ray, 300
MTF. *See* Modulation transfer function
Multidensity recording for information storage, 360
Multiexposure techniques, 171
Multilevel grains
 characteristic curve, 12
 detective quantum efficiency, 16
 image noise, 17
Mutual intensity, 326

Negative binomial distribution, 127
Noise, image. *See* Image noise, Granularity

Noise
 photon, 2, 176
 in communication systesm, 353
Noise-equivalent contrast, 163
Noise-equivalent energy, 163
Noise-equivalent number of quanta (NEQ), 156
 experimental data, 156
 and detective quantum efficiency, 156, 160, 161
 and photon noise, 171, 176
 signal-to-noise optimisation, 156
 X-ray exposure, 171
Noise-equivalent quantum efficiency. *See* Detective quantum efficiency
Noiseless channel, 352
Nutting formula, 43
Nutting model of density, 39
Nyquist frequency, 197, 204

One dimensional scans of granular patterns, 277
One stage model of image formation, 233
Opacity, 13
Optical transfer function, 212. *See also* Modulation transfer function
 defocussed lens, 332
 diffraction limited lens, 331
 microdensitometer, 330
 slit misalignment, 334
Optimisation of signal-to-noise, 158
Optimum restoration filter, 335
OTF. *See* Optical transfer function
Overlapping of grains
 effect on granularity, 304

Partially coherent light
 exposure, 302
 in microdensitometry, 325
Peak power limitation, 357, 365
Phase transfer function, 213, 256
Photometric units, 46
Photon receptors, 1
Photon statistics, 2
 effect on image noise, 176
Photopolymerisation processes
 amplification, 36
 relative speed, 49
Physical development, 44
Point spread function, 56, 207
Poisson distribution, 2

Post-exposure techniques, 167, 169
Power spectrum. *See* Wiener spectrum
Practical resolution criteria, 263
Pre-exposure techniques, 165
  experimental results, 170
Primary grain, 307
Print mottle, 307
Printing
  transfer of granularity, 307
  transfer of detective quantum efficiency, 146
Probability density function, 217, 220
Probability distribution function, 217

Quantum efficiency
  detective. *See* Detective quantum efficiency
  equivalent. *See* Detective quantum efficiency
  noise-equivalent. *See* Detective quantum efficiency
  responsive, 28, 156
Quantum fluctuation, 2
Quantum limited pictures, 138
Quantum mottle, 300
Quantum sensitivity, 35
  interpretation of, 117, 125
Quantum sensitivity distribution, 36
  experimental data, 37, 124
  extrinsic, 134
  influence on characteristic curve, 81, 129
    detective quantum efficiency, 72, 107, 132
    image noise, 101
  intrinsic, 134
  measurement, 116
  Monte Carlo techniques, 125
  two parameter model, 126
Quantum yield, 35

Radiation detectors
  amplification factors, 36
  relative speeds, 49
Radiometric units, 46
Random grain array
  detective quantum efficiency, 70, 72
Random grain defect, 182
Random walk calculation. *See* Monte Carlo techniques
Reciprocity law failure, 35

Rectangle function, 192
Redundancy, 351
Resolution
  effect of receptor area, 24
  and speed, 27, 137
Resolution criteria, 263
Resolution test charts, 263
Resolving power, 263
Responsive quantum efficiency, 28, 156
Root mean square density fluctuation.
  See Granularity and Image noise
Rotational symmetry
  point spread function, 210
  autocorrelation function, 223

Sampling theorem, 197
Saturation defect, 182
Sayce chart, 244
Scale
  density, 48
  exposure, 48
Scale value of granularity, G. *See* Granularity
Scanning aperture
  modulation transfer function of, 332
  effect on granularity, 16, 59
Scattering in emulsion layers, 45, 260
Schwarzchild–Villiger effect, 322
Selwyn granularity, 58
  variation with density, 60
  scanning aperture, 59, 220
Sensitometric properties, 47
Sensitization
  spectral, 53
  chemical, 124, 130
Sensitivity
  variation with wavelength, 53
  ultimate, 135
Sensitivity distribution. *See* Quantum sensitivity distribution
Shannon's theory for a noiseless channel, 352
Shift theorem, 196
Siedentopf formula, 58, 98, 303
Sifting property of delta function, 192
Signal-to-noise, 153
  and detective quantum efficiency, 152
  optimisation of ,158
Silver, structure of, 44
Silver bromide, energy levels, 34
Sinc function, 192

SUBJECT INDEX

Sine-wave response. *See* Modulation transfer function
Single bar contrast, 265
Single level grains, 34, 68
Sinusoidal exposures, 211
  methods of producing, 241
Slit misalignment, 334
Solution-physical development, 44
Source, entropy of, 348, 351
Spatial frequency, 191
  two dimensional, 214
Speed, 48
  and grain size, 37
  of several imaging processes, 49
  and resolution, 27, 137, 140
  maximum possible, 135
Spectral sensitivity, 34, 53
Spread function
  empirical expressions for, 258
  experimental data, 56
  influence on information capacity, 357
  line, 209
  point, 56, 207
  relation to modulation transfer function, 212, 214, 244
Stationary systems, 207
Statistical isotropy, 223
Statistical stationarity, 216
Stirling's formula, 80
Stray light in microdensitometers, 322
Structure of developed silver, 44
Symbol, 346
System, 205
  linear, 206
  stationary, 207
Syzygetic granularity, 282

Test objects
  modulation transfer function, 242, 244
  resolving power, 263
Three stage model of image formation, 240
Threshold defect, 182
  effect on detective quantum efficiency, 183
Threshold sensitivity of receptors
  effect on characteristic curve, 80
  effect on image noise, 101
  effect on detective quantum efficiency, 72, 107, 132

Threshold modulation curve, 264
Transfer function. *See* Modulation transfer function and Optical transfer function
Transmittance
  amplitude, 51
Transmittance–exposure curve, 51
Transmittance fluctuations, relation to density fluctuations, 283
Two stage model of image formation, 236
Two stage processes
  detective quantum efficiency, 145
  image noise, 307

Ultimate sensitivity
  detective quantum efficiency, 141
  estimates of, 135
  symposium, 135
Unconventional processes
  amplification factors, 36
  models, 111
  relative speeds, 49
  speed and resolution of, 140
Undetermined multipliers, method of, 356
Unexposed emulsion density, 51
Uniform grain array, 69

Variable-area test charts, 242
Variance, 218
  of Poisson distribution, 7
Vesicular processes, 35, 140

White noise, 249
Wiener spectrum, 222
  aliasing, 296
  amplitude transmittance, 291, 302
  and autocorrelation function, 222
  for coherent light exposures, 302
  colour films, 230
  and detective quantum efficiency, 311
  errors of measurement, 286, 295
  influence of scanning aperture, 223, 277
  measurement, analogue, 288
    digital, 292
  models, 303
  practical results, 299
  relationship to scale value G, 222, 280

Wiener spectrum (*cont.*)—
  relationship to Selwyn granularity, 222, 280
  transfer of, 307
  variation with mean density, 301
  X-ray exposures, 300, 308
Wiener–Khintchin theorem, 222

X-ray exposures, density–exposure curve, 50
X-ray mottle, 300, 308
X-ray quanta, interaction with receptors, 174

Zero memory information source, 348